P9-DTI-707

Cadmium in the Environment

Second Edition

Cadmium in the Environment

Second Edition

Lars Friberg
Magnus Piscator
Gunnar F. Nordberg
Tord Kjellström
Karolinska Institute
Stockholm, Sweden

with technical and editorial assistance from
Pamela Boston
Karolinska Institute

Published by

CRC PRESS, Inc.
18901 Cranwood Parkway • Cleveland, Ohio 44128

Library of Congress Cataloging in Publication Data
Main entry under title:

Cadmium in the environment.

Earlier ed., by L. Friberg, M. Piscator, and G. Nordberg, published in 1971.
Bibliography
1. Cadmium – Toxicology. 2. Environmental health. I. Friberg, Lars. Cadmium in the environment. [DNLM: 1. Cadmium – Analysis. 2. Cadmium – Toxicity. QV290 F897c]
RA1231.C3C33 1974 615.9'25'662 73-88622
ISBN 0-87819-018-X

International Standard Book Number 0-87819-018-X

Library of Congress Card Number 73-88622
Printed in the United States

THE AUTHORS

Lars T. Friberg, M.D., is professor and chairman of the Department of Environmental Hygiene of the Karolinska Institute and the National Environment Protection Board (up to 1972, the National Institute of Public Health), Stockholm, Sweden.

Dr. Friberg's degrees are from the Karolinska Institute of Stockholm (M.D., 1945, and Doctor of Medical Sciences, 1950).

Prior to his present appointment, Dr. Friberg held different positions in internal medicine and industrial toxicology. He serves in various consulting capacities to Swedish governmental agencies, e.g., the National Board of Health and Welfare and the National Environment Protection Board. He is a member of the World Health Organization Advisory Panel on Occupational Health and Chairman of the Subcommitte on the Toxicology of Metals under the Permanent Commission and International Association of Occupational Health. During 1967 he was visiting professor at the Department of Environmental Health of the University of Cincinnati in Ohio, U.S.

Dr. Friberg has published over 100 papers on epidemiology and toxicology. Most papers on toxicology concern metals, particularly mercury and cadmium. He was chairman of a Swedish expert committee for the evaluation of risks from methyl mercury in fish. He is one of the authors of the review *Mercury in the Environment,* published by CRC Press, Cleveland, Ohio, U.S., in 1972.

Magnus Piscator, M.D., is associate professor at the Department of Environmental Hygiene at the Karolinska Institute, a position he has held since 1969.

Dr. Piscator's degrees are from the Karolinska Institute of Stockholm (M.D., 1958, and Doctor of Medical Sciences, 1966).

Prior to his present appointment, Dr. Piscator held different positions in physiological chemistry, internal medicine, and environmental hygiene. He serves as consultant to the Department of Environmental Hygiene of the National Environment Protection Board and is a member of the World Health Organization Expert Advisory Panel on Food Additives. During 1973 he was visiting scientist at the Human Studies Laboratory, National Environmental Research Center, Environmental Protection Agency, Research Triangle Park, N.C., U.S.

Dr. Piscator has published about 40 papers on metal toxicology and proteinuria, most concerning cadmium.

Gunnar F. Nordberg, M. D., is associate (teaching) professor at the Department of Environmental Hygiene at the Karolinska Institute, Stockholm.

Dr. Nordberg earned his medical degrees at the Karolinska Institute of Stockholm (M.B., 1962; Doctor of Medical Sciences, 1972).

Dr. Nordberg has held various positions within environmental hygiene and internal medicine. He has been engaged as a temporary advisor to the W.H.O. and NATO Scientific Committee, and is a consultant to the National Environment Protection Board.

During 1974 Dr. Nordberg served as a visiting research scientist at the Department of Pathology of the University of North Carolina at Chapel Hill, the National Environmental Research Center, and the National Institute of Environmental Health Sciences, U.S. He has been involved in the work of the Subcommittee on the Toxicology of Metals under the Permanent Commission and International Association of Occupational Health.

Dr. Nordberg is a coauthor of *Mercury in the Environment*; in connection with the preparation of this publication, he visited institutions in several countries, including the U.S.S.R., for collection of pertinent data.

Dr. Nordberg's research, oriented toward the toxicity of metals, especially cadmium and mercury, has resulted in about 40 publications.

Tord Kjellström is research assistant at the Department of Environmental Hygiene at the Karolinska Institute, Stockholm.

Mr. Kjellström holds a Bachelor of Medicine degree from the Karolinska Institute, 1966, and a Master of Mechanical Engineering degree from the Royal Institute of Technology, Stockholm, 1967.

During 1968 Mr. Kjellström was a research student at Tokyo University, and during 1969 and 1970 research assistant at the Department of Social Medicine, Uppsala University. Since 1971 he has been working at the Department of Environmental Hygiene at the Karolinska Institute, mainly in cadmium research.

TABLE OF CONTENTS

Chapter 1

INTRODUCTION

The second edition of *Cadmium in the Environment* updates the earlier review on cadmium carried out under a contract (No. CPA 70-30) between the U.S. Environmental Protection Agency and the Department of Environmental Hygiene of the Karolinska Institute, Sweden. The earlier report to the EPA, with modifications, formed the basis for the first edition of *Cadmium in the Environment*, which was published by CRC Press in 1971. The collaboration between the two institutions continues (contracts 68-02-0342 and 68-02-1210) and has resulted in a second report to the EPA in 1972. The project officer from the U.S. Environmental Protection Agency has been Robert J. M. Horton, M.D.

The second edition of *Cadmium in the Environment* presents and in some cases reevaluates all information contained in the first edition, in the second EPA report, and subsequent findings, published as well as unpublished, up to and including part of 1973. Like the first edition, the present work focuses upon information essential to the understanding of the toxic action of cadmium and the relationship between exposure and effects on human beings and animals.

Through repeated personal contact with several Japanese researchers, including a 5-week visit to Japan by one of the authors (Dr. Tord Kjellström), it has been possible to obtain and evaluate much data from Japan which would not have been accessible otherwise. Papers published in Japanese could be taken into account as well since Dr. Kjellström speaks and reads Japanese.

We express our gratitude to the Environmental Agency of Japan, particularly to Dr. Yoshimasa Yamamoto, Chief of the Section of Environmental Health and Public Hazards, as well as to the Prefectural Institutes of Hygiene and Departments of Public Hazards in Fukushima, Gumma, Hyogo, Nagasaki, Miyagi, and Toyama as well as the Japanese Association of Public Health. Valuable information and assistance have been received from several independent researchers. Kenzaburo Tsuchiya, M.D., Professor of Preventive Medicine and Public Health, School of Medicine, Keio University, assisted us in planning and executing valuable visits in Japan.

Chapter 2

PROBLEMS OF ANALYSIS

Biological effects of cadmium cannot be discussed without taking into account the analytical methods used. When concentrations of cadmium in air, water, food, body fluids, and organs are compared to each other or related to effects, one must be sure that a proper method is being used for the determination of cadmium; otherwise erroneous conclusions may be drawn.

The following properties of an analytical method are of interest:

1. **Specificity** – The method used must measure only cadmium. If other substances interfere, they must be either eliminated or a correction must be made allowing for their presence.
2. **Sensitivity** – The minimum amount or concentration that the method will detect.
3. **Precision** – The error of the method, often expressed as the coefficient of variation, is evaluated by a number of analyses on the same sample over a certain period of time. Precision is often dependent on the concentration.
4. **Repeatability** – By running duplicates of the samples and by having the deviation of the two values from their mean expressed as a percentage, repeatability is obtained.
5. **Accuracy** – This means the systematic deviation from the true values.

Many methods have been used for the determination of cadmium, but during the last two decades four methods have dominated: (1) colorimetric determination after wet digestion and extraction with dithizone (dithizone methods), (2) emission spectroscopy, (3) neutron activation, and (4) atomic absorption spectrophotometry. The last years have brought such new developments as (5) electrochemical methods (anodic stripping polarography, pulse polarography, anodic stripping voltammetry), and (6) X-ray analysis after proton irradiation. Information about analytical methods can also be found in an article by Yamagata and Shigematsu, 1970, and in a recent monograph, *Cadmium, the Dissipated Element*, by Fulkerson et al., 1973.

2.1 DITHIZONE METHODS

The basic principle is that cadmium forms a stable complex with dithizone at an alkaline pH. The many other metals that form complexes with dithizone must first be eliminated. Examples of dithizone methods that have been used for the determination of cadmium are the ones by Church, 1947, Saltzman, 1953, and Smith, Kench, and Lane, 1955. With regard to precision etc., there is scarce information in the literature. If proper steps are taken to remove other metals, the method must be regarded as being specific, but, on the other hand, there is always a risk of contamination with cadmium from reagents during all the steps. If blanks and standards are treated exactly in the same way as the samples, this error will be eliminated.

Dithizone methods have been used for the determination of cadmium in urine, where normally only a few micrograms per liter are found, and for the determination of cadmium in foodstuffs and organs. The results obtained by this method seem to compare favorably with results obtained by other methods (see Table 2:1). In the following chapters, the dithizone method will be regarded as a useful method for determination of cadmium in biological material.

2.2 EMISSION SPECTROSCOPY (SPECTROGRAPHY)

The basic principle is that by exciting metal atoms by high energy sparks or electric arcs, the atoms will emit light of a characteristic spectral distribution. The spectral lines may be recorded on a film, which makes it possible to get a qualitative or quantitative estimate of the metals. Tipton et al., 1963, studied the precision for determinations in organs and found that the coefficient of variation was 19% at concentrations of 220 to 3,400 μg Cd/g in the ash. The repeatability varied between 5 and 7% for liver and kidney. The sensitivity was 50 μg Cd/g in ash.

Geldmacher-v. Mallinckrodt and Pooth, 1969,

TABLE 2:1

Cadmium Concentrations in 13 Urines (μg/l) Determined in 6 Laboratories

	1	2		3			4	5	6
		a	b	a	b	c			
1	16.6	41.0	12.7	19.4	14.6	9.2	12	12.3	13.8
2	16.0	47.0	12.4	23.0	18.6	16.8	15	12.0	16.4
3	24.0	0	10.4	12.3	9.3	8.4	7	7.3	7.2
4	12.0	6.8	8.9	12.1	9.3	8.4	7	6.0	7.7
5	16.0	12.0	14.6	28.3	24.8	23.2	18	16.5	21.4
6	190.4	153	150.7	178.0	154.8	147.4	182	152.9	194.8
7	53.4	100	48.1	64.8	51.6	44.2	47	49.9	45.8
8	173.6	125	138.3	143.6	128.3	111.6	148	139.2	126.1
9	68.4	56	55.5	63.1	53.2	43.2	48	36.8	24.6
10	39.4	63	38.9	47.1	35.5	31.6	41	35.6	28.3
11	5.4	50	2.4	7.8	4.5	4.2	6	4.0	4.1
12	128	165	125.8	141.9	108.4	105.3	93	136.7	119.2
13	22	26	18.1	17.6	12.9	11.8	20	13.2	13.6

The following methods were used:
Laboratory 1 : Dithizone method
Laboratory 2a: Dithizone – atomic absorption. Perkin-Elmer apparatus
Laboratory 2b: Dithizone – atomic absorption. Hitachi apparatus
Laboratory 3a: APDC-MIBK extraction (without ashing) – atomic absorption
Laboratory 3b: Dithizone – atomic absorption
Laboratory 3c: Dithizone method
Laboratory 4 : Dithizone – atomic absorption
Laboratory 5 : Dithizone – atomic absorption
Laboratory 6 : APDC-MIBK – atomic absorption

From Japanese Association of Public Health, 1970c.

have also used a spectroscopic method for the determination of metals, including cadmium, in biological material. After wet ashing, the metals were extracted with diethylammonium-diethyldithiocarbamate in chloroform and after removal of the solvent the analysis was performed. They stated that they could detect 1 μg of cadmium, but made no further studies on the method. Some of the values reported by this research team (e.g., Section 3.4.1 concerning cadmium in food) have not been included by the present authors. This exclusion has been made partly because the values have been higher than the top of the range found by other investigators and partly because the analytical methods used have not been sufficiently well described and validated. Imbus et al., 1963, determined cadmium in urine and blood after extraction with dithizone. The lower limit of sensitivity was 0.1 μg/sample and by using 25 ml of blood or 250 ml of urine, about 0.4 μg could be detected in 100 ml and 1,000 ml, respectively.

The sensitivity has been improved during recent years and at present 0.02 μg can be detected in samples with modern equipment. By using 10-ml samples, concentrations of 0.2 μg/100 ml blood can thus be determined.

It seems that some earlier spectrographic methods have had a relatively low sensitivity and that accurate quantitative determinations of cadmium could only have been obtained in organs with relatively large concentrations, such as liver and kidneys.

2.3 NEUTRON ACTIVATION

The basic principle is that by irradiation of the sample with the elementary particles, chiefly neutrons, radioactive nuclides will be obtained. Elements can be identified with the aid of their energies and half-lives. It is often necessary to introduce separation steps by chemical methods. This is a method that should be quite specific for cadmium, but until the late 1960's neutron activation had only been used in a few investigations. The last years have witnessed a growing number of reports on this subject (Westermark and Sjostrand, 1960, Axelsson and Piscator, 1966a, Piscator and Axelsson, 1970, Plantin, 1964, Lieberman and Kramer, 1970, Wester, 1965, 1971, Henke, Sachs, and Bohn, 1970, Lucas, Edgington, and Colby, 1970, Ljunggren et al., 1971, Livingston, 1972, Linnman et al., 1973, Kjellström et al., 1974).

2.4 ATOMIC ABSORPTION SPECTROPHOTOMETRY

The basic principle is that the sample in solution is passed into a high temperature burner, so that atoms in the ground state will pass through a path of light emitted from a lamp with a cathode of a specific metal. If the sample contains the actual metal, the atoms from that metal will absorb the light and the absorbance can be read on a meter. Usually an air acetylene flame is used, but it has been claimed that air propane can be used as well (Lener and Bibr, 1971a).

At present, atomic absorption is the most commonly used method for the determination of cadmium. The sensitivity is high and concentrations as low as 0.005 μg/g can be determined in pure water solutions. There has been much optimism about atomic absorption, but the difficulties have often been overlooked. Pulido, Fuwa, and Vallee, 1966, have thoroughly discussed the determination of cadmium in biological material. They studied interference by other substances and found that phosphate in concentrations above 0.1 *M* could decrease the absorbance and that sodium chloride in concentrations above 0.01 *M* could increase absorbance. By using a hydrogen lamp, they could allow for the presence of NaCl in serum and urine and obtain more true values.

The importance of sodium chloride as a source of error in determining cadmium has been documented also by Piscator, 1971. Figure 2:1 shows false concentrations of cadmium (in a solution containing negligible amounts of cadmium) in relation to concentrations of sodium chloride in the solution.

The specificity of cadmium determinations with atomic absorption is thus dependent on the amount of NaCl and other compounds in the sample, but still cadmium in urine and serum has been determined in several investigations without

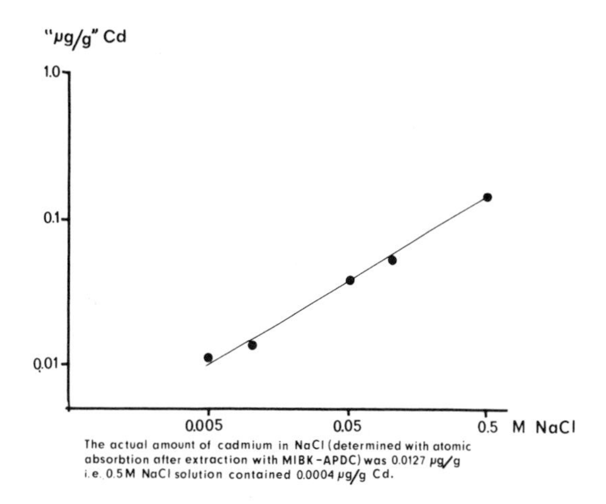
"μg/g" Cd
1.0
0.1
0.01
0.005
0.05
0.5
M NaCl
The actual amount of cadmium in NaCl (determined with atomic absorbtion after extraction with MIBK-APDC) was 0.0127 μg/g
i.e. 0.5M NaCl solution contained 0.0004 μg/g Cd.

FIGURE 2:1. False concentrations of cadmium in sodium chloride solutions of different strengths. (From Piscator, 1971.)

taking the presence of interfering substances into account. For example, this has resulted in high normal urine values being reported by Schroeder et al., 1967, and Tipton and Stewart, 1970, about 40 and 100 μg/day, respectively. A normal value of about 1 μg/day has been given in other reports (see Table 4:6).

If interfering substances are not present, the accuracy is very high. Pulido, Fuwa, and Vallee, 1966, reported on two samples with a concentration of 0.165 and 0.051 μg/g of cadmium (National Bureau of Standards' sample) where they found an accuracy of 99.4 and 104.5%. Repeatability was 2% when nine duplicate analyses of cadmium in metallothionein were done. When atomic absorption was compared with emission spectrography, values were obtained from 95 to 113% of those obtained with the latter method.

The difficulties encountered when investigating cadmium in biological materials at a low concentration of cadmium are not found to the same degree when analyzing human tissues with a normally high cadmium content such as kidney and liver. Repeatability was 3.2% for determination of cadmium in liver and kidneys as reported by Morgan, 1969, and similar values were found by Piscator, 1971, for cadmium in renal cortex.

Attempts have been made to improve the flame method by using cups or boats, which are inserted into the flame after a preliminary drying and ashing of the sample. Such methods, which involve small samples, have been developed especially for blood (Delves, 1970, Hauser, Hinners, and Kent, 1972, and Ediger and Coleman, 1973). Since the true values in blood are not known, it is still difficult to evaluate the accuracy of these methods.

During recent years, it has been recognized that it might sometimes be necessary to extract cadmium into an organic solvent before analysis. This will give a higher concentration, and interfering salts will be eliminated. One method is to digest the sample with acid, adjust the pH to around 3, and extract it with a chelating agent such as ammoniumpyrrolidine dithiocarbamate (APDC) in methylisobutylketone (MIBK). Such a method for the determination of cadmium in urine and serum has been described by Lehnert, Schaller, and Haas, 1968.

Another way to eliminate interfering substances is to use ion-exchange. For determination of cadmium in urine and blood, Vens and Lauwerys, 1972, passed urine acidified with 1.5 *N* HCl or hemolyzed blood in 1.5 *N* HCl through a basic resin, which binds cadmium. Elution was performed with 2 *N* perchloric acid and cadmium was then determined with conventional atomic absorption spectrophotometry. The coefficient of variation was 10% at a cadmium concentration of 1.5 μg/l. The accuracy is not known. The results are shown in Tables 4:5 and 4:6. It seems that their values for urinary cadmium are of the same order as in other reports from Europe, whereas the method seems to have given relatively high values for cadmium in blood.

In a Japanese report (Japanese Association of Public Health, 1970c) it is seen that most laboratories used first an extraction of the digested sample with dithizone and then once again extracted cadmium with acid before determination with atomic absorption was done. Table 2:1 shows the results from analyses in six different laboratories of urines from cadmium-exposed workers and patients with Itai-itai disease. There is good agreement among most of the laboratories.

Similar interlaboratory studies have been performed in Japan on rice samples (Yamagata et al., 1971). Ten laboratories participated, using a standard method of analysis for determination of cadmium in rice. Some of the steps were wet ashing with nitric acid and sulfuric acid, extraction with APDC-MIBK, and measurement by atomic absorption. The means for two rice samples were 0.176 and 2.33 μg/g and the ranges 0.14 to 0.20 and 2.20 to 2.75 μg/g, respectively, giving a coefficient of variation of about 10% for both samples. Since it is difficult to obtain homogenous rice samples, these results indicate a high degree of precision in these Japanese laboratories at this relatively high concentration level. In one of the laboratories, a comparison was made between a method involving low temperature ashing followed by extraction into acid and the above mentioned standard method (chloroform used instead of MIBK). Eight determinations on the same rice sample gave 0.188 (S.D. 0.005) and 0.189 (S.D. 0.014) μg/g, respectively.

It is impossible to evaluate specificity and accuracy from the comparisons among the Japanese laboratories because all of them used atomic absorption after extraction procedures. No methodological studies have been published in which specificity and accuracy of the methods have been investigated.

Recently, a flameless atomic absorption method has been developed (Kjellström et al., 1974) for determination of cadmium in grains. Radioactive ^{109}Cd was added to 2-g and 4-g samples of wheat. Radioactive analysis was compared with flameless atomic absorption and conventional atomic absorption (with and without extraction in MIBK/APDC system). The deuterium lamp technique was used for background correction. The flameless atomic absorption was performed in the following way. The sample was dry ashed twice and the ash dissolved in 1-*M* HNO_3; 20 μl of this solution was injected into a Perkin-Elmer HGA-70 heated graphite atomizer. This method showed a higher repeatability than conventional atomic absorption and the recovery (average 95%) was about 10% higher than for the method using extraction. Figure 2:2 shows the result of flameless atomic absorption analysis of cadmium. The accuracy of the new method for cadmium analysis was studied by Linnman et al., 1973, who compared it with neutron activation (method of Ljunggren et al., 1971). The results are shown in Figure 2:3. A good correlation between the methods is shown at cadmium concentrations in wheat between 20 and 200 ng/g.

2.5 ELECTROCHEMICAL METHODS

Basically related to the classic polarographic methods, these methods have been improved to increase their sensitivity. Anodic stripping voltammetry (ASV), which utilizes either mercury electrodes or solid electrodes, has been used in

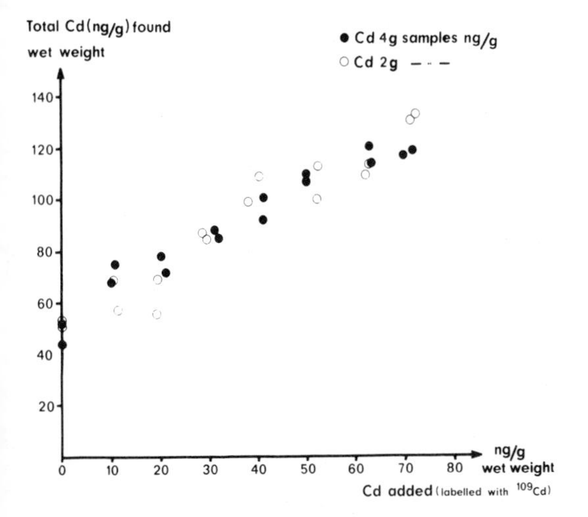

FIGURE 2:2. Result of flameless atomic absorption cadmium analysis of two different sizes of wheat samples to which cadmium (labeled with ^{109}Cd) had been added. (From Kjellström et al., 1974.)

several investigations for the determination of cadmium and other metals in water (Whitnack and Sasselli, 1969, Allen, Matson, and Mancy, 1970), in food (Hundley and Warren, 1970), and in blood serum (Sinko and Gomiscek, 1972).

The ASV method was compared with another method only in the report involving the food samples (Hundley and Warren, 1970). Here it was compared with atomic absorption spectrophotometry, but it was not stated how this latter method was performed. There seems to have been a poor agreement between the two methods and these data do not help to assess the value of the ASV method.

Pulse polarography was used by Abdullah and Royle, 1972, to determine cadmium in sea water after preconcentration of metals on resins. Only one value is given for cadmium in sea water, 1.38 μg/l, and this method cannot be evaluated at present.

2.6 X-RAY ANALYSIS

Johansson, Akselsson, and Johansson, 1970, showed that it is possible to detect small amounts of elements by observation of characteristic X-rays after proton bombardment. Since this method is nondestructive, repeated analyses can be made on the same sample. As yet, the problem of quantitative determination of small amounts of cadmium in biological fluids has not been solved, but in the future this method may be of great value for cadmium analysis.

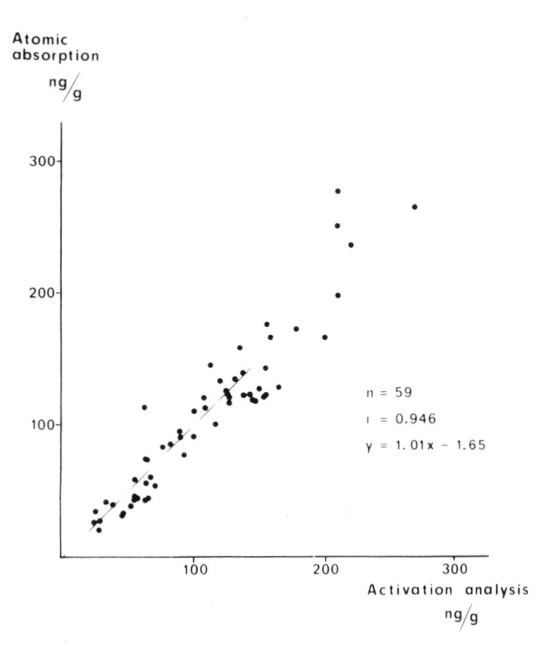

FIGURE 2:3. Result of analysis of cadmium in wheat using the flameless atomic absorption method and the neutron activation method. (From Linnman, L., Andersson, A. Nilsson, K. O., Lind, B., Kjellström, T., and Friberg, L., *Arch. Environ. Health,* 27, 45, 1973. With permission.)

2.7 CONCLUSIONS

There are several methods that can be used for the determination of cadmium in biological material such as blood, urine, organs, and food. At present the most commonly used method is atomic absorption spectrophotometry, but interference from salts such as NaCl may give erroneous results, especially when low concentrations are determined. Such interference may be diminished by using extraction techniques or background correctors.

To date, few strict methodological studies on accuracy and other important parameters have been carried out. This is a severe limitation, as witnessed by some recent reports on occurrence and metabolism of cadmium in which serious analytical errors must have been involved. It is therefore necessary to evaluate the analytical methods used before accepting data.

Except for Japanese investigations, no collaborative study has been performed by laboratories engaged in research on biological effects of cadmium. Such studies are necessary, as these would facilitate comparisons in the future of parameters such as cadmium in food, urine, or blood.

Chapter 3

OCCURRENCE, POSSIBLE ROUTES OF EXPOSURE, AND DAILY INTAKE

3.1 OCCURRENCE AND POSSIBLE ROUTES OF EXPOSURE

Cadmium is closely related to zinc and will be found wherever zinc is found in nature. The cadmium to zinc ratios will vary. In most minerals and soils, ratios of 1:100 to 1:1,000 have been found (Fulkerson et al., 1973). Zinc is an essential metal for most life forms (Underwood, 1962, Bowen, 1966, Schroeder et al., 1967, and Yamagata and Shigematsu, 1970). Thus, it is probable that no naturally occurring material will be *completely* free from cadmium.

Cadmium is obtained as a by-product in the refining of zinc and other metals. However, as it is difficult to separate zinc and cadmium, the latter will often be found in small amounts in commercially available zinc compounds, as pointed out by Schroeder et al., 1967.

Though cadmium has been recognized for only a relatively short period of time, copper, lead, zinc, and some other metals have been used for several thousand years. Thus, as soon as man started to produce metals, he also started to pollute the environment with cadmium. In this century cadmium and cadmium compounds have been used increasingly by industries, causing a sharp increase in environmental contamination. Cadmium will be emitted to air and water by mines, by metal smelteries, especially lead, copper and zinc smelteries, and by industries using cadmium in alkaline accumulators, alloys, paints, and plastics. The burning of oil and waste and scrap metal treatment will also contribute. The use in agriculture of fertilizers, either as chemicals or as sludge from sewage plants, and the use of cadmium-containing pesticides might also contribute to the contamination.

Some of the cadmium emitted to the air will be inhaled by people and animals, but most of it will be deposited in soil or water. The cadmium deposited in water may then increase the concentrations of cadmium in edible water organisms. In the event of flooding or irrigation, cadmium in water might also increase the concentrations in soil, in turn causing an increase of cadmium in agricultural products, such as rice and wheat.

The behavior of cadmium in nature has not been extensively studied; it is only recently that systematic investigations have begun on the transfer of cadmium between different compartments of the ecosystem (Fulkerson et al., 1973). It has been suspected that organic cadmium compounds could exist similar to the ones described for mercury, but it has not been possible to prove the existence of such compounds (Westermark, 1969). For a more extensive treatise on the occurrence of cadmium, reference is again made to the work of Fulkerson et al., 1973.

3.1.1 Cadmium in Air

3.1.1.1 In Air Generally

During the last decades, yearly determinations of cadmium in air have been performed in several areas in the United States by the National Air Sampling Network (NASN). In 1969 annual averages ranged from 0.006 $\mu g/m^3$ (San Francisco, Calif.) to 0.036 $\mu g/m^3$ (St. Louis, Mo.) in the 20 largest cities (NASN unpublished data). The highest annual average ever recorded in any city was 0.12 $\mu g/m^3$ in 1964. The city was El Paso, Tex., site of a big lead and zinc smeltery. Kneip et al., 1970, made daily determinations for 1 year at several sites in New York City and surrounding areas. They found that lower Manhattan and the Bronx had yearly mean levels of 0.023 $\mu g/m^3$ and 0.014 $\mu g/m^3$, respectively, whereas a nonurban site had 0.003 $\mu g/m^3$. These data agree fairly well with the data by NASN for New York in 1969, when the annual average was 0.014 $\mu g/m^3$. At 29 nonurban stations the annual averages in 1969 were all below 0.003 $\mu g/m^3$, which was the minimum detectable concentration with the method used (NASN, unpublished data).

In the Chicago area Harrison and Winchester, 1971, during 6 24-hr sampling periods from May to August at 50 sampling stations recorded cadmium concentrations in air from less than 0.005 $\mu g/m^3$ to 0.08 $\mu g/m^3$. Severs and Chambers, 1972, found 24-hr values during 1 day of up to 0.33 $\mu g/m^3$ in Houston, Tex. Creason et al., 1972, determined cadmium in air during 3 months in two urban and two suburban areas in Cincinnati, Oh. In each area sampling was made 25 and 100 ft, respectively, from a roadway. The geometric means in the two urban areas were 0.0032 and 0.0034 $\mu g/m^3$, respectively, and in the two suburban areas 0.0021 and 0.0017 $\mu g/m^3$, respectively. No difference was found with regard to distance from the roadway. Just and Kelus, 1971, found in Poland that annual averages in ten cities varied from 0.002 to 0.05 $\mu g/m^3$. Nagata et al., 1972, report on cadmium concentrations in ambient air in Tokyo during 1969 to 1971. Mean values over several months in different areas varied from 0.010 to 0.053 with a maximum 24-hr value of 0.53 $\mu g/m^3$. In the center of Stockholm, Sweden, weekly means of 0.005 $\mu g/m^3$ were found, whereas in a rural area far from cadmium-emitting factories, a monthly mean of 0.0009 $\mu g/m^3$ was found (Piscator, unpublished data). In Erlangen, West Germany, Essing et al., 1969, found a level of 0.0015 $\mu g/m^3$ (sampling time not stated).

It is important to know the state in which cadmium exists in air, but at present, no data concerning this question are available. Even if it can be presumed that cadmium oxides will constitute a large part of the airborne cadmium, the possibility that other compounds are formed cannot be excluded. Particle size is also essential for calculating deposition and absorption in the lungs. In Cincinnati, cadmium concentrations in the air and mass median diameters of the particles were measured downtown and in a suburb (Lee, Patterson, and Wagman, 1968). Concentrations averaged 0.08 and 0.02 $\mu g/m^3$, respectively, and MMD (mass median diameter) averaged 3.1 and 10 μm, respectively. In both areas about 40% of the particles were below 2 μm. In St. Louis, it was found that the average MMD was 1.54 μm when the cadmium concentration in air was 0.01 $\mu g/m^3$. Of the particles, 65% were below 2 μm (Lee et al., 1972).

3.1.1.2 In Air Around Point Sources

Investigations have shown that high cadmium concentrations may be found in the vicinity of known emission sources. In Sweden, weekly means of 0.3 $\mu g/m^3$ were recorded on several occasions 500 m from a factory using copper cadmium alloys. At a distance of 100 m from the source a monthly mean value of 0.6 $\mu g/m^3$ was recorded. The highest concentration found in a 24-hr sampling was 5.4 $\mu g/m^3$. In Japan, a 3-day mean of 0.56 $\mu g/m^3$ has been reported at a distance of about 500 m from a zinc smeltery (Ministry of Health and Welfare, 1971c). The maximum 24-hr average was 3 $\mu g/m^3$. Near another smeltery, 8-hr values of 0.16 to 0.32 $\mu g/m^3$ were obtained at a distance of 500 m from the source (Kitamura, personal communication). Further data from Japan will be given in Chapter 8. In the city of East Helena, Mont., the average concentration during a 3-month sampling period was 0.06 and 0.29 $\mu g/m^3$ at about 1,300 and 800 m, respectively, from a smeltery (Huey, 1972). The maximum 24-hr value was 0.7 $\mu g/m^3$. Thus, it seems that in areas around cadmium-emitting factories, cadmium concentrations in air several hundred times greater than those in non-contaminated areas will be found.

3.1.1.3 Dustfall

Background data on deposition of cadmium have been provided by Laamanen, 1972. Measuring cadmium in dustfall in a rural subarctic area of Finland, he found that the deposition varied from less than 0.001 mg/m^2/month (winter) to 0.006 mg/m^2/month (summer).

In the United States, Hunt et al., 1971, determined cadmium in dustfall in residential, commercial, and industrial areas in 77 cities and found monthly means of 0.040, 0.063, and 0.075 mg/m^2 in the respective areas.

The relationship between dustfall of cadmium and concentrations of cadmium in air has been studied by Pinkerton et al., 1972a. They found that dustfall values of 0.055 and 0.091 mg/m^2/month corresponded to concentrations of cadmium in the air of 0.0032 $\mu g/m^3$ and 0.0048 $\mu g/m^3$, respectively. It was also found that in an area with a monthly dustfall of cadmium from 0 to 0.12 mg/m^2 month, household dust contained between

4.25 and 12.5 μg Cd/g dust. Corresponding values in an area where the dustfall was from 0.04 to 0.22 mg/m^2/month were 8.75 to 14 μg/g dust.

Monthly deposition of cadmium has also been measured around some factories. The results from the years 1968 to 1970 at locations near a Swedish factory are shown in Table 3:1 (Olofsson, 1970). It should be noted that a considerable deposition was sometimes even found at 10 km from the source. Deposition measurements near a Japanese factory during a period of 6 months showed a mean value of 6.2 mg/m^2/month at a distance of 500 m from the source and a mean value of 1.8 mg/m^2/month at a distance of 1,400 m, whereas in two control areas in two cities, values of 0.1 and 0.4 mg/m^2/month, respectively, were found (Kitamura, personal communication). These data may be compared with similar studies in the city of East Helena, Mont., where values of 1 to 4 mg/m^2/month were found up to about 1,000 m from a cadmium-emitting smeltery (Huey, 1972).

3.1.1.4 Indirect Methods for Measuring Cadmium in Air

Deposited cadmium has been measured by determining cadmium in mosses (Rühling, 1969, Rühling and Tyler, 1970, 1972, and Olofsson, 1970). As seen in Figure 3:1, there is an obvious difference between the southern and northern parts of Sweden, Norway, and Finland (Rühling and Tyler, 1972). Figure 3:2 shows how the spreading of cadmium around a factory producing copper cadmium alloys can be determined by this technique. Moss analysis has also been used in Great Britain by Goodman and Roberts, 1971, and Burkitt, Lester, and Nickless, 1972, to study the distribution of cadmium pollution. The latter authors found about 50 μg Cd/g in moss at distances of 6 mi from a smeltery. This confirms that considerable air pollution by cadmium may spread to relatively long distances from emission sources. The spread of cadmium from a cadmium-emitting factory can also be studied by determining cadmium in leaves from Morus shrubs (mulberry), as shown in Figure 3:3 (Kobayashi, 1972).

Secular trends in cadmium deposition have been studied by comparing cadmium concentration measured in moss saved from several decades ago with that in moss from present times. The old

TABLE 3:1

Deposition of Cadmium Around a Cadmium-emitting Factory (mg/m^2/month). Emission of Cadmium: 460 kg/month

	Distance and direction from the source						
Period	0.1 S	0.3 NE	0.3 N	0.5 NNW	0.7 NE	1.0 SW	10.4 km ENE
5.7–7.8.68	(12.4)*	(1.1)			(0.4)	(0.8)	
9.8–18.9.68	28	1.8			0.4	2.1	
26.9–25.10.68	18	0.7			4.0	3.5	
4.7–31.7.69	(1.3)			(1.4)	(1.3)	(1.3)	
1.8–2.9.69	19			1.8	0.8	1.2	
5.9–9.10.69	8			3.0	2.0	1.1	
10.10–20.11.69	4			2.6	1.7	1.0	
29.4–8.6.70	8.8		1.2	0.5	0.6		0.2
9.6–2.7.70	40		1.9	0.7	0.7		<0.03
3.7–27.7.70	(4.6)		(0.9)	(0.1)	(<0.1)		(<0.1)
28.7–2.9.70	19		3.7	2.3	1.1		<2
2.9–31.10.70	7.7		4.7	2.5	1.3		0.7
2.10–2.11.70	14.8		5.3	2.4	0.8		0.3

*Values in parentheses refer to periods when the factory was closed completely or to a considerable degree.

From Olofsson, 1970.

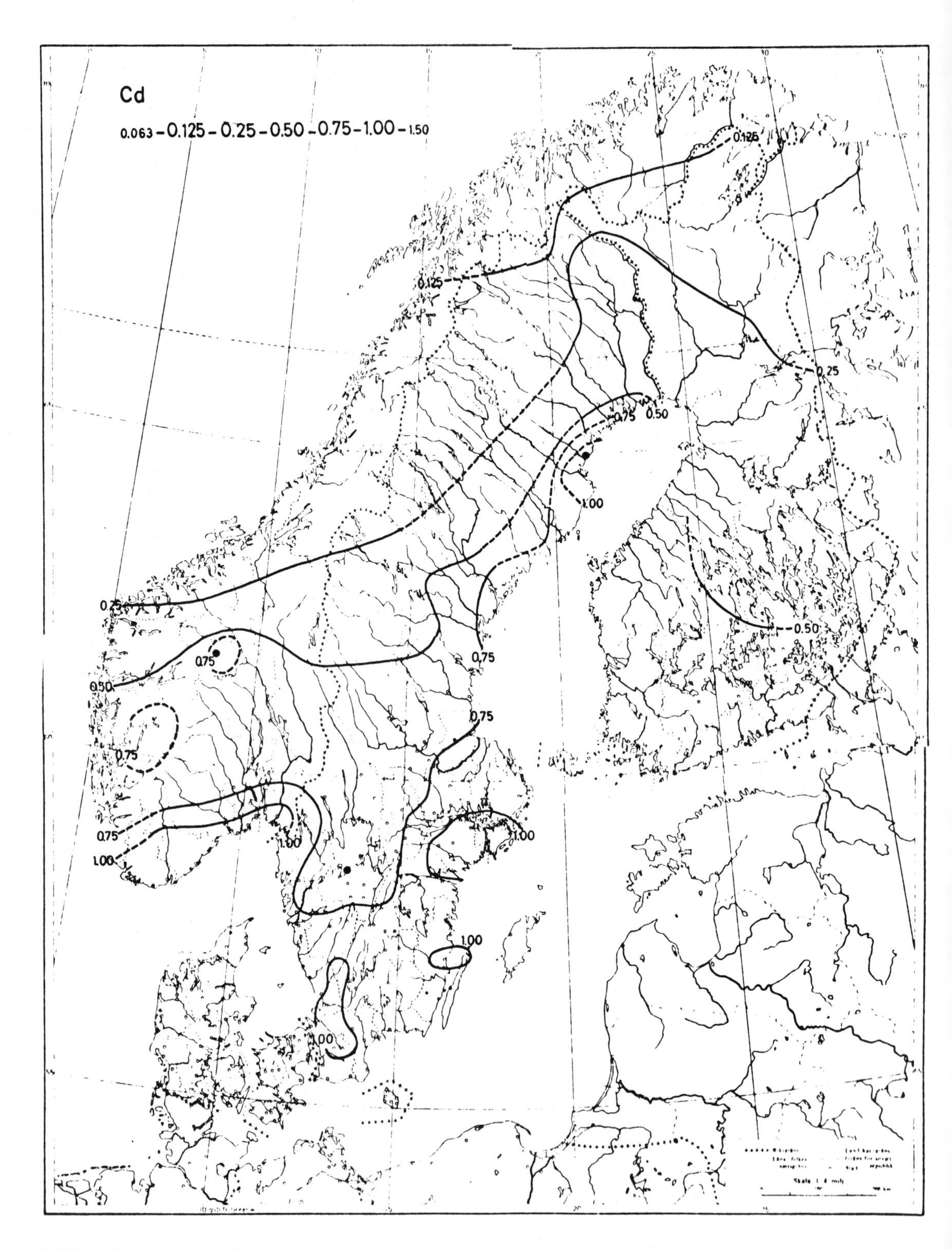

FIGURE 3:1. Isocurves for cadmium (μg/g dry weight) in moss in different parts of Scandinavia. Broken lines: values uncertain. Scale: 1 cm = ~90 km. (From Rühling, Å. and Tyler, G., Deposition of Heavy Metals Over Scandinavia, Dept. of Plant Biology, University of Lund, 1972. With permission.)

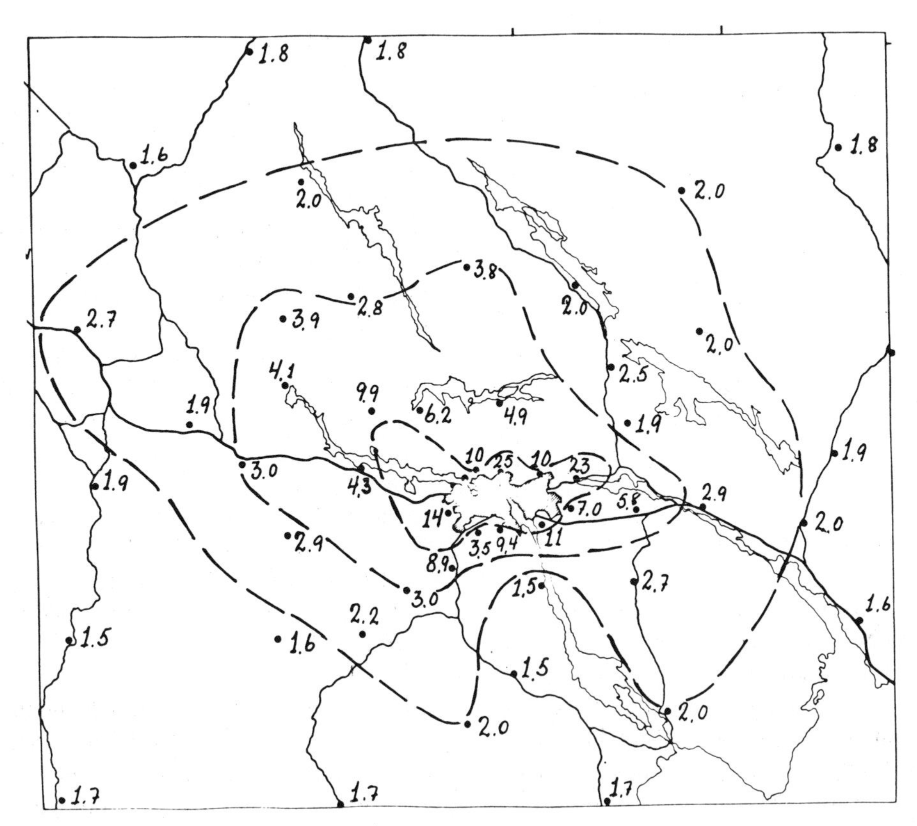

FIGURE 3:2. Concentration of cadmium in moss (μg/g dry weight) in an area surrounding a cadmium-emitting factory. The isocurves are at the 10, 3, and 2 μg/g levels. Scale: 1 cm = ~2 km. (From Olofsson, 1970.)

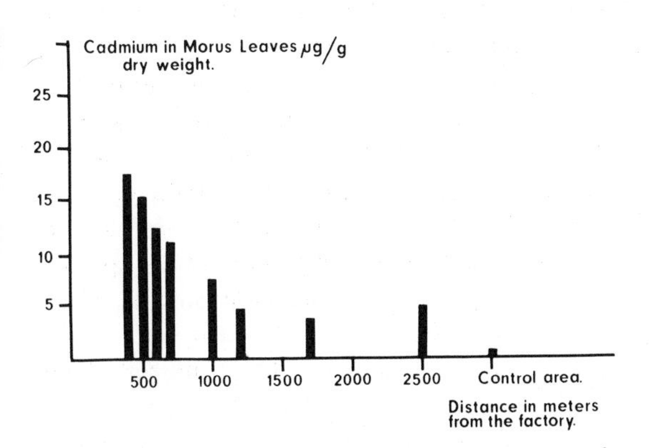

FIGURE 3:3. Concentration of cadmium in leaves of a Morus shrub in relation to distance from a cadmium-emitting factory in Annaka City, Japan. (From data by Kobayashi, 1972.)

samples consisted of moss from the years 1927–1942 (herbaria samples) from ten sites in the rural areas around the city of Uppsala, Sweden, and moss from three sampling sites in the city of Uppsala, collected in 1916–1917. It was found that in the rural areas around Uppsala there had been a mean increase from 0.2 to 0.3 μg/g in moss during the intervening 20 to 30 years. In the city a mean increase from 0.2 to 0.4 μg/g was observed during a 50-year period (Gelting and Pontén, 1971). The increase was consistent in several comparisons.

The influence of cadmium emissions upon cadmium concentrations in soil has been studied by Watanabe (personal communication). Test pots containing soil were placed 500 m from the emission source and in a control area. The results are shown in Table 3:2. Within a few years, a tenfold increase in the cadmium concentration occurred in the soil placed near the emission source. At that same location, deposited cadmium was measured as 6.2 mg/m^2/month during a 6-month period. Buchauer, 1973, found increased cadmium levels in soil up to a distance of at least 20 km from a zinc smeltery.

3.1.2 Cadmium in Water

In areas not known to be polluted by cadmium,

TABLE 3:2

Deposit of Cadmium in Soils Contained in Test Pots

Location	Time	Cd (μg/g)
0.5 km from emission source	0	0.7
	2 months	0.8
	1 year	6.2
	2 years	4.2
	3 years	9.4
Control area	0	0.7
	2 months	0.7
	1 year	0.7
	2 years	0.7
	3 years	0.6

Watanabe, personal communication.

values of less than 1 ng/g have been reported in water. Values exceeding 10 ng/g have been recorded both in natural waters and in water for consumption. An analysis of drinking water from 969 communities in the U.S. in 1969 showed that 0.2% of the samples exceeded 10 ng/g (McCabe et al., 1970). Increased amounts of cadmium can be due to the contamination of the water either by industrial discharges or by the metal or plastic pipes used in distribution (Schroeder et al., 1967). In drinking water, a W.H.O. working group (World Health Organization, 1973) recommended that cadmium concentrations should not exceed 5 ng/g.

In an investigation of 720 water samples from rivers and lakes throughout the U.S. in 1970, it was found that 42% were between 1 and 10 ng/g and 4% in excess of 10 ng/g (Durum, Hem, and Heidel, 1971). As described in two recent papers from Great Britain, cadmium has been determined in sea water, both on and off coast. Preston et al., 1972, found a geometric mean of 0.04 ng/g in 21 samples of filtrated sea water collected in the Irish Sea, whereas near the coast, in Liverpool Bay, a geometric mean of 0.41 ng/g water was obtained in 9 samples. This latter value is in good agreement with data by Abdullah, Royle, and Morris, 1972, who found a mean value of 0.27 ng/g in Liverpool Bay. According to Preston et al., 1972, the highest concentrations in the coastal waters around Britain were found in the North Sea, where the mean was 0.41 ng/g. In Oslofjorden, Norway, which is part of the North Sea, the average concentration in five samples of sea water was 0.19 ng/g (Rojahn, 1972).

Yamagata and Shigematsu, 1970, have pointed out that in rivers polluted by cadmium, the metal will often be undetectable in the water phase while large concentrations will be found in suspended particles and in the bottom sediments. This is especially true at neutral or alkaline pH. A similar finding was obtained in Sweden, where 500 m downstream from a cadmium-emitting factory 4 ng/g of cadmium were found in water, while 80 μg/g (dry weight) were found in the mud (Piscator, 1971). To avoid errors when determining the degree of contamination in water, cadmium in the suspended particles or the sediments must be measured. The contamination of rice fields surrounding the Jintsu River, the area in Japan where the Itai-itai disease has occurred, is probably due to the transport of cadmium-containing suspended particles to the paddy soil by irrigation with river water (Yamagata and Shigematsu, 1970).

3.1.3 Cadmium in Soil and Uptake by Plants

It has already been mentioned that both airborne and waterborne cadmium can cause increased concentrations of cadmium in soil. In areas not known to be polluted, the cadmium concentrations in soil have been reported to be less than 1 μg/g, with great variations (Schroeder et al., 1967, Yamagata and Shigematsu, 1970, and Fulkerson et al., 1973). Klein, 1972, studied an area in Michigan. He found that 70 soil samples from residential areas contained an average of 0.41 μg/g, 91 samples from agricultural areas an average of 0.57 μg/g, and 86 samples from an industrialized area an average of 0.66 μg/g. In Japan, levels of 1 to 69 μg/g have been found in rice fields in areas under observation for suspected contamination from cadmium (Yamamoto, 1972).

There are other ways by which soil can be contaminated with cadmium. Odén, Berggren, and Engvall, 1970, determined cadmium in sewage sludge from 56 plants in the southern and middle parts of Sweden. The median concentration of cadmium was 12 μg/g dry weight (range: 2 to 61 μg/g). Berrow and Webber, 1972, have reported on metals in sewage sludge from 42 sewage works, scattered throughout England and Wales. The samples were collected in 1964. Of these 42 samples, 7 contained more than 100 μg/g dry weight. The use of this sludge as fertilizer could increase cadmium concentrations in soil. Superphosphate fertilizers also contain relatively large amounts of cadmium, as pointed out by Schroeder and Balassa, 1963, and Schroeder et al., 1967.

Two important foodstuffs, rice and wheat, are able to take up considerable quantities of cadmium from soil. Kobayashi et al., 1970, added cadmium oxide to soil in pots where rice and wheat were growing. The results are shown in Table 3:3. It is noteworthy that wheat grains accumulated more cadmium than did rice. It can also be noted that concentrations above 10 μg/g in soil brought about less yield of both rice and wheat.

The cadmium uptake in radish and lettuce was studied by John, Van Laerhoven, and Chuah, 1972. Cadmium chloride was added to 30 different soil samples in pots (original cadmium concentrations not given) to give a concentration of 100 μg/g, and corresponding samples without added cadmium were used as controls. After a growing period of 3 weeks, a reduced yield of both radish and lettuce was found. Cadmium was determined by atomic absorption with correction for background absorption and light scattering. Cadmium concentrations on a dry weight basis in radish roots were on an average 387 and 7.4 μg/g, respectively, and in lettuce tops 138 and 2.3 μg/g, respectively.

In another experiment, John, 1972, studied the uptake of cadmium in radish roots and tops in unlimed and limed soil with an original cadmium concentration of 0.67 μg/g. Cadmium was added at concentrations ranging from 0 to 100 μg/g soil. Added cadmium in concentrations of 0.5 and 1 μg/g did not result in higher concentrations in tops or roots compared to controls. An increase was noted when added cadmium was 5 μg/g. Addition of calcium carbonate made it impossible to detect any cadmium uptake at all. In plants grown on soil without added cadmium, the cadmium concentration in roots was 5.4 and 5.1 μg/g dry weight in limed and unlimed soil, respectively.

The concentrations reported in the control plants are high. There is no doubt, however, that cadmium concentrations in these two foodstuffs greatly increased when grown on cadmium-contaminated soil. It was also shown that the uptake was related to pH, i.e., increased soil acidity resulted in higher cadmium concentrations in plants.

Williams and David, 1973, studied the uptake of cadmium into different plant species from superphosphate fertilizer and from cadmium chloride added to the soil in pot cultures. Uptake was 0.4 to 7% of the cadmium available in the soil. The addition of cadmium to soil, in either way, always increased the cadmium content in grain of cereals or in the edible portion of vegetables.

Linnman et al., 1973, studied the uptake of cadmium in wheat grown in test pots at different sewage sludge additions (corresponding to 0 up to 175 tons/ha) and different pH levels. All samples were analyzed for cadmium both by means of

TABLE 3:3

Uptake of Cadmium by Rice Plant and Wheat as well as Yield

Addition of Cd to soil (% of CdO)	Rice Cd (μg/g)			Wheat Cd (μg/g)	
	Yield (%)	Polished (10%)	Bran	Yield (%)	Whole grain
0	100	0.16	0.59	100	0.44
0.001	100	0.28	0.79	106	8.27
0.003	92	0.40	0.84	72	15.5
0.01	92	0.78	1.60	16	29.9
0.03	93	1.37	2.68	13	41.4
0.1	69	1.62	2.94	3	60.7
0.3	32	1.94	3.19	3	48.6
0.6	19	1.73	3.94	2	90.8
1.0	1	4.98*		1	139.0

*Unpolished

From Kobayashi et al., 1970.

flameless atomic absorption with background correction and by neutron activation (for a comparison of the methods see Chapter 2, Figure 2:3).

The results of the studies are shown in Table 3:4, where cadmium uptake in wheat in relation to added sewage sludge, concentration of added cadmium in soil, and pH of soil is given. It can be seen that a considerable cadmium uptake occurs in wheat when sewage sludge is used as a plant nutrient source. The lower the pH, the greater was the cadmium concentration in wheat. These results are from laboratory studies, and field studies are now needed. It should be mentioned that the two lower levels of added sludge, 6.5 and 19 tons (dry weight)/ha, were in agreement with Swedish recommendations in 1972 for annual applications (Linnman et al., 1973). In 1973 the recommendations were changed to lower levels because of the risk for cadmium uptake in crops. A maximum average addition of 1 ton (dry weight)/ha of sludge (containing 5 to 15 μg Cd/g sludge, dry weight) is now recommended (Swedish National Board of Health and Welfare, 1973). Sludge containing more cadmium than 25 μg/g (dry weight) should not be used at all for food crops.

3.1.4 Cadmium in Food

During later years, several investigations concerning the cadmium content of common foodstuffs have been performed (United States: Schroeder and Balassa, 1961; Schroeder et al., 1967; Corneliussen, 1970; Duggan and Corneliussen, 1972; West Germany: Kropf and Geldmacher-v. Mallinckrodt, 1968; Essing et al., 1969; Czechoslovakia: Lener and Bibr, 1970; Rumania: Rautu and Sporn, 1970; Japan: Ishizaki, Fukushima, and Sakamoto, 1970a; M. Fukushima, 1972; Yamagata and Iwashima, 1973 [Section 8.3.2.1]; United Kingdom: Thomas, Roughan, and Watters, 1972; Canada: Zook, Greene, and Morris, 1970). In Table 3:5, some basic foodstuffs are compared. It will be seen that these reports indicate that "normal" levels of cadmium in food are generally below 0.05 μg/g. In Table 3:6 data are compiled from the reports of Corneliussen, covering studies made on samples collected from 30 markets in 24 cities in the U.S. The data by Schroeder et al., 1967, have been excluded because in the investigation reported in 1967, atomic absorption spectrophotometry was used without any preliminary extraction. For example, cadmium levels in milk were reported to be 0.1 to 0.14 μg/g. Interference from sodium chloride might well have been the cause. The data by Kropf and Geldmacher-v. Mallinckrodt, 1968, have not been included because the values given by the spectrographic method they used seemed in error (see Section 2.4).

There have been several recent investigations of the cadmium content in fish. By analysis of 19 whole fishes from the Great Lakes, Lucas, Edgington, and Colby, 1970, found an average cadmium concentration of 0.094 μg/g wet weight (neutron activation analysis). In an extensive study from New York State, Lovett et al., 1972, examined 406 fishes from 49 freshwaters by the atomic absorption technique. The majority of samples of fish meat contained 0.02 μg/g or less,

TABLE 3:4

Cadmium Uptake in Wheat in Relation to Added Sewage Sludge, Cadmium Concentration in Soil, and pH of Soil. Means of Four Replicates. Atomic Absorption, μg/g dry weight.

CaO added (%)		0		0.1		0.2	
Sludge (ton/ha)	Added cadmium in soil (μg/g)	pH	Cadmium uptake (μg/g)	pH	Cadmium uptake (μg/g)	pH	Cadmium uptake (μg/g)
0	0	4.8	0.067	6.1	0.045	7.4	0.029
6.5	0.031	4.8	0.119	6.1	0.086	7.2	0.033
19	0.094	5.1	0.170	6.2	0.123	7.2	0.050
58	0.28	5.3	0.257	6.2	0.134	6.8	0.086
175	0.84	5.8	0.124	5.9	0.147	6.5	0.122

Data from tables in Linnman et al., 1973.

TABLE 3:5

Cadmium in Selected Food (μg/g wet weight) in Various Countries

Country	Potato	Tomato	Wheat flour	Milk	References
U.S.A.	0.001	0	0.07	0.0015–0.004	Schroeder and Balassa, 1961
Western Germany	0.039	0.015	0.047	0.009	Essing et al., 1969
Czechoslovakia	0.09		0.02	0.01	Lener and Bibr, 1970
Roumania	0.017	0.013			Rautu and Sporn, 1970
Japan (nonpolluted areas)	0.038	0.032	0.025	0.003	Ishizaki, Fukushima, and Sakamoto, 1970a
U.K.	0.08	0.06			Thomas, Roughan, and Watters, 1972
Canada			0.05–0.10		Zook, Greene, and Morris, 1970

TABLE 3:6

Cadmium Content (μg/g wet weight) in Different Food Categories in the U.S.A. Total Number of Samples: 30.

	Cadmium (μg/g wet weight†)			
	1968–1969		1969–1970	
Type of food	No. ≥0.01	Maximum	No. ≥0.01	Maximum
Dairy products	10	0.09	9	0.01
Meat, fish, and poultry	21	0.06	22	0.03
Grain and cereal products	27	0.08	27	0.06
Leafy vegetables	27	0.08	28	0.14
Legume vegetables	16	0.03	10	0.04
Root vegetables	24	0.08	27	0.08
Garden fruits	25	0.07	27	0.07
Fruits	15	0.38	10	0.07
Oils, fats, and shortening	27	0.13	28	0.04
Sugar and adjuncts	18	0.07	27	0.04
Beverages	8	0.04	9	0.04
Potatoes	–	–	29	0.08

†Cadmium was analyzed by atomic absorption and/or polarography at a sensitivity of 0.01 μg/g.

Data taken from Corneliussen, 1970; Duggan and Corneliussen, 1972.

and only in a few cases were concentrations above 0.1 μg/g found. In the trout, for instance, there was no relationship between age and cadmium concentrations. Havre, Underdal, and Christiansen, 1972, determined cadmium in meat of fish from a Norwegian fjord thought to be polluted by cadmium. Analyses were performed by atomic absorption spectrophotometry after extraction in an organic solvent. Cadmium concentrations in 21 samples of cod ranged from 0.001 to 0.041 μg/g, in 5 samples of pollack from 0.002 to 0.008 μg/g, and in 14 samples of flounder from 0.005 to 0.024 μg/g wet weight.

Taylor, 1971, reported on cadmium concentrations in samples of fish and fish products from Great Britain. Although the results are difficult to interpret because the analytical method is not mentioned, the following data are given. Cadmium concentrations in 48 samples of tuna fish were all below 0.2 μg/g, but some salmon meat was reported to contain more than 3 μg/g. The highest concentrations, up to 5.3 μg/g, were found in fish paste.

In a polluted area, Ishizaki, Fukushima, and Sakamoto, 1970a, reported values of 10 to 110 and 92 to 420 μg/g, wet weight, in livers of cuttlefish and shellfish. Pringle et al., 1968, reported that cadmium concentrations varied from 0.1 to 7.8 μg/g, wet weight, in oysters from the east coast of the U.S. and from 0.2 to 2.1 μg/g in oysters from the west coast. Reynolds and Reynolds, 1971, examined the cadmium content of crabs (dithizone method). Cadmium concentrations in the white meat were relatively low, 0.02 to 1.1 μg/g, while dark body meat contained from 2.5 to 8.6 μg/g, wet weight. Rice and wheat from most parts of the world contain less than 0.1 μg/g, but rice and wheat from contaminated areas in Japan have been shown to contain considerably higher concentrations, around 1 μg/g (see Sections 8.3.2.1 and 8.3.3). It is not always clear whether Japanese data on rice and wheat refer to wet or dry weight. However, this distinction is of minor importance, as the difference will be only about 10%. Liver and kidney, such as from calves or swine, often contain more than 0.05 μg/g, whereas meat usually contains less cadmium than 0.05 μg/g. An exception seems to be whale meat, as Ishizaki, Fukushima, and Sakamoto, 1970a, reported levels of 0.25 and 0.4 μg/g in two samples. Fish liver has been reported to contain 0.06 to 1.4 μg/g (Lucas, Edgington, and Colby, 1970) and 0.109 to 2.225 μg/g (Havre, Underdal, and Christiansen, 1972).

The cadmium content of German wine was investigated by Eschnauer, 1965 (polarographic method), who found a mean of 3.1 ng/g (1.3 to 4.1) in 13 wines. Bergner, Lang, and Ackermann, 1972 (atomic absorption after extraction in organic solvent), found a mean cadmium concentration of 2.9 ng/g (0.5 to 8) in 22 wines and 71 and 40 ng/g, respectively, in 2 wines from 1965. Essing et al., 1969 (atomic absorption after extraction in organic solvent), found means of 17 and 8 ng/g in 4 red wines and 4 white wines, respectively.

3.1.5 Cadmium in Cigarettes

That smoking is a source of exposure to cadmium has been recognized the last years. Szadkowski et al., 1969, found a mean content of 1.4 μg per cigarette when they determined the cadmium content of eight cigarette brands in West Germany. Cigarettes were smoked in a smoking apparatus with the following experimental parameters: two puffs a minute, each of 2 sec duration, suction volume, 30 ml with a pressure of 80 to 120 mm H_2O. The mainstream smoke particles were collected on a Cambridge filter and the gases in nitric acid. They found a mean value of 0.15 μg cadmium per cigarette in the particle phase of the eight brands and 0.03 μg in the gaseous phase. From their data it can also be seen that about 0.5 μg went into the sidestream smoke, i.e., about 35%.

Nandi et al., 1969, investigated six different brands of cigarettes. The mean total content was about 1.2 μg per cigarette. They also used a smoking machine but the experimental conditions were not described in detail. Intermittent suction was used 5 sec every minute to make the cigarette burn. The smoke was not collected on filters. Cadmium was only determined in the ashes and the filters of the cigarettes. With the method used they found that about 30% of cadmium was recovered in ash and filter, and they calculated that about 70% passed with the smoke. As they did not separate mainstream and sidestream smoke, it cannot be known how much cadmium was in the mainstream.

Linnman and Lind (unpublished data) studied the influence of different numbers of puffs per cigarette and different numbers of puffs per minute in one brand of cigarettes. They used a duration of 2 sec per puff, a suction volume of 35 ml. They found that one puff per minute and ten puffs altogether resulted in a collection of 0.14 μg cadmium per cigarette on a Cambridge filter. Two puffs per minute and 15 puffs altogether resulted in 0.19 μg per cigarette on the filter. The same result was obtained with 3 puffs per minute and 15 puffs altogether. The difference between the first value and the second and third ones was statistically significant. In American cigarettes Menden et al., 1972, found amounts of 1.56 to 1.96 μg per cigarette (atomic absorption spectrophotometry). By use of a smoking apparatus (35-ml puff, 2 sec/min) it was found that the particulate phase of the mainstream contained

0.10 to 0.12 μg per cigarette. All data agree well, showing that 0.1 to 0.2 μg of cadmium might be inhaled by smoking one cigarette. Smoking habits will to some extent affect the amount of cadmium in the mainstreams, as shown by the influence of puff intervals on the cadmium amounts.

Diverging results have been reported by Tomita, 1972, who studied 12 different Japanese brands with a smoking machine. Instead of measuring the amount of cadmium in the mainstream, he mentions that he collected and analyzed cadmium in the sidestream in three bottles. He also measured the amount of cadmium in the whole cigarette, butt, and ash. Through subtraction he determined the amount in the mainstream. Tomita states that between 30 to 50% of the cadmium in the cigarette may be inhaled by the smoker, which is a figure far above other reported data. The possibilities for methodological errors are obvious with the method used. The data on the amount of cadmium in whole cigarettes are between 1.35 and 2.5 μg, in accord with data from other countries.

Piscator and Rylander, 1971, exposed guinea pigs to free cigarette smoke for 1 month (one cigarette per day for 1 week, four cigarettes per day for 3 weeks). The method used was the one by Rylander, 1969. By means of a suction device, the animals were exposed for two sec every minute to the smoke from a puff of 2 sec duration. About ten puffs from each cigarette were given. In Table 3:7 the concentrations of cadmium and zinc in the renal cortex of exposed animals compared with controls are shown. The cadmium concentrations are significantly higher in the exposed animals than in the controls. There is also a difference in zinc concentrations.

Cadmium content in snuff has been determined (Baumslag, Keen, and Petering, 1971). American snuff contained 0.7 to 0.9 μg/g, while snuff used in South Africa contained 1.1 to 1.5 μg/g. This latter type of snuff is a mixture of powdered tobacco with ash of incinerated plants and herbs. The method used was atomic absorption.

TABLE 3:7

Concentrations of Cadmium and Zinc in Renal Cortex from Guinea Pigs Exposed to Cigarette Smoke for 1 Month. Means and Standard Deviations

		μg/g Dry weight	
	n	Cadmium	Zinc
Controls	15	8.4 ± 1.2	133 ± 9
		$p<0.025$	$p<0.0025$
Smokers	22	10.8 ± 2.8	151 ± 19

From Piscator and Rylander, 1971.

3.2 DAILY INTAKE OF CADMIUM

Estimates of the daily intake of cadmium based mainly on the data on cadmium concentrations in food have been made in several countries, as shown in Table 3:8. There is a marked discrepancy when the results of Schroeder et al., 1967, and Tipton and Stewart, 1970, are compared with those of the other reports.

They estimated the daily intake of cadmium to be 200 to 500 μg and about 170 μg, respectively. Also, Murthy, Rhea, and Peeler, 1971, have reported relatively high values, with a mean of about 90 μg/day, from different places in the U.S. It should be noted that atomic absorption without extraction of the metal was used in these investigations. As has been pointed out in Chapter 2, there is a risk of obtaining values that are too high with this particular method because of interference, e.g., from sodium chloride. It should also be noted that in nonpolluted areas in Japan, the intake of cadmium does not seem to differ markedly from the intake in the European countries included (Section 8.3.2.1).

Available data thus suggest that on an average, 50 μg of cadmium may be ingested in most countries, with probable variations from 25 to 75 μg/day. This value may be compared with a tolerable weekly intake of 400 to 500 μg, i.e., about 70 μg/day, recently proposed by a joint F.A.O./W.H.O. expert committee on food additives (W.H.O., 1972).

Cadmium in water will contribute very little to the daily intake at concentrations below 5 ng/g. A concentration in drinking water around 10 ng/g would increase the daily intake with 10 to 20 μg at a consumption of 1 to 2 liters per day, considered a normal intake of water for an adult human being. The contribution by drinking water to the total daily intake of cadmium will vary with age. The fraction of the daily intake that is derived from water may thus be greater for children in certain age groups.

In addition to conventional methods for estimating daily intake there is another approach that can be taken. The amount of cadmium excreted in

TABLE 3:8

Daily Intake of Cadmium Via Food in Different Countries

Country	Cd μg/day	Method	References
U.S.A.	4–60	Dithizone	Schroeder and Balassa, 1961
Western Germany	48	Atomic absorption after extraction	Essing et al., 1969
Roumania	38–64	Dithizone	Rautu and Sporn, 1970
Czechoslovakia	60	Dithizone or isotope dilution or atomic absorption	Lener and Bibr, 1970
Japan (nonpolluted area)	59	Dithizone or atomic absorption after extraction	Yamagata and Shigematsu, 1970
U.S.A.	92	Atomic absorption	Murthy, Rhea, and Peeler, 1971
U.S.A. (1969)	50	Dithizone or atomic absorption	Duggan and Corneliussen, 1972
U.S.A. (1970)	38	Dithizone or atomic absorption	Duggan and Corneliussen, 1972
Japan (nonpolluted area)	47	Method not stated	M. Fukushima, 1972

feces would consist of unabsorbed ingested cadmium to 90 to 95%, meaning that the fecal excretion would correspond approximately to the daily intake of cadmium. In a study in West Germany covering 23 persons, Essing et al., 1969, estimated the daily fecal excretion to be 31 μg. Another study in the same country by Szadkowski, 1972, gave the same result (31 μg Cd/day, n = 16). Tipton and Stewart, 1970, in a long-term balance study of three American men found a mean fecal amount of 42 μg Cd/day. Tsuchiya, 1969, referred to Japanese data from a noncontaminated area showing an average daily amount of 57 μg in feces from four men. These data for the most part confirm the figure of 50 μg given above for the daily intake. There is a tendency for a somewhat higher value in Japan and a somewhat lower one in Germany.

Unfortunately, no methodological studies on cadmium analysis in feces have been reported. There is reason to believe that feces analyses are rather reliable even without extraction methods since interfering salts are not present to the same extent as in urine.

Cadmium in air will also contribute very little in nonpolluted areas. If it is assumed that normally around 20 m^3 of air are inhaled every day, concentrations of 0.001 to 0.01 $\mu g/m^3$ would not give more than 0.02 to 0.2 μg per day via the respiratory route. The amount deposited in the lung would be less than that. It has been mentioned that concentrations of 0.1 to 0.5 $\mu g/m^3$ have been found around cadmium-emitting factories. These concentrations would cause an inhalation of 2 to 10 μg per day. This is still less than the oral intake, but, as will be discussed in the next chapter, the absorption rates for the gastrointestinal route and the respiratory route are quite different. Thus, exposure via air can be an important factor.

An additional source is smoking. Twenty cigarettes per day will probably cause the inhalation of 2 to 4 μg.

It should also be pointed out that there is a possibility of respiratory exposure to cadmium in areas contaminated by cadmium when in dry seasons deposited dust is again transferred to the air by whirlwinds, etc. Sporting areas and playgrounds for children could be sources of this type of exposure.

Finally, it is important to remember that contamination of food or beverages may still occur due to leakage in acid media from cadmium-glazed pottery and enameled steel (Guthenberg and Beckman, 1970). High leakage has been found especially in goods with red, orange, or yellow glazing (Guthenberg, Beckman, and Dich, unpublished data). A standard method for testing cadmium-glazed goods and maximum allowable leakage has been set in Sweden (Swedish National Food Administration, 1972).

Another source of cadmium could be cooling devices for soft drink vending machines. In one recent case (Nordberg, Slorach, and Stenstrom, 1973) the contamination in the water (16 mg Cd/l in solution and 40 mg Cd/l in precipitate) was such that acute gastrointestinal symptoms were induced. Later investigations on two other similar vending machines revealed cadmium concentrations of 0.004 and 0.5 mg Cd/l.

3.3 CONCLUSIONS

Human beings will be exposed to cadmium via food, water, and air. Exposure via food is the most important. In uncontaminated areas most foodstuffs will contain less than 0.05 μg Cd/g wet weight, and the average daily intake probably will be about 50 μg with regional and individual variations. Liver and kidney from animals and shellfish are among foodstuffs that may have concentrations larger than 0.05 μg/g, even under normal circumstances. When certain foodstuffs are contaminated by cadmium in soil and water, the cadmium concentrations may increase considerably. Among these are rice and wheat, in which concentrations around 1 μg/g have been reported. This may result in increases in daily intake of cadmium. The uptake from soil is pH dependent.

In water the normal concentration of cadmium is less than 1 ng/g. If the cadmium concentration in drinking water exceeds 5 ng/g, it might contribute significantly to the daily intake of cadmium.

Whereas the "normal" concentrations of cadmium in air, about 0.001 $\mu g/m^3$, will not contribute significantly to the daily intake of cadmium, in areas where cadmium-emitting factories are situated concentrations of 0.1 to 0.5 $\mu g/m^3$ (weekly or monthly means) have been recorded. This may result in the inhalation of 2 to 10 μg cadmium per day. It should be remembered that a considerably larger percentage of inhaled cadmium compared with ingested cadmium will be expected to be absorbed.

Smoking will also contribute to the daily intake. It can be expected that the smoking of 20 cigarettes per day will cause the inhalation of 2 to 4 μg of cadmium.

Chapter 4

METABOLISM

At a recent international symposium on accumulation of toxic metals (Task Group on Metal Accumulation, 1973), general aspects of absorption, transportation, distribution, excretion, and accumulation of metals were discussed. The term "absorption" was defined as "entry into the body by passage of the metal compound across a membrane." According to this definition (for example), a metal taken into the cells of the mucous lining of the gastrointestinal tract should be considered as absorbed metal, even if the metal passes again out into the gastrointestinal lumen and out with feces, without ever reaching the circulation. The term "systemic absorption" was deemed unnecessary. The term "absorption" in the following text will mean the amount or fraction of metal entering the circulation (i.e., systemic absorption). The term "retention" is sometimes used. If not otherwise specified, the present authors mean the amount within the body excluding that at the site of absorption (i.e., the amount absorbed less the amount excreted).

The amount of metal found in feces derives from the fraction of ingested material that passes unabsorbed through the gastrointestinal tract as well as from net gastrointestinal excretion. Net gastrointestinal excretion is defined as total gastrointestinal excretion minus the amount of metal reabsorbed. Gastrointestinal excretion in this context also includes possible excretion by glands in the upper parts of the alimentary tract.

Data on metabolism to be reviewed in the following chapter have been derived from experimental studies on animals and human beings, studies on populations exposed to cadmium to various extents, or studies on autopsy materials. All of the data have been obtained either by chemical analysis of cadmium or by means of radioactivity counting techniques applied to various radioisotopes of cadmium. The difficulties encountered in chemical analysis of cadmium have been discussed in detail in Chapter 2. In radioactive counting techniques, errors of a similar nature are not to be expected. When proper standardization is upheld, accurate values may be expected from the use of radioactive isotope techniques, even though equipment differs considerably. Unlike in the early 1950's, the now existing terminology distinguishes between ^{115}Cd and ^{115m}Cd. Although it is evident that in some experiments written up prior to the making of such a distinction the isotope now named ^{115m}Cd was mainly used, the present authors retain the term ^{115}Cd when discussing such experiments.

4.1 UPTAKE AND ABSORPTION

There is abundant evidence from human data that cadmium is found in different organs in concentrations increasing with age (see Section 4.3.2). Data from workers exposed to cadmium as well as from animal experiments show also that cadmium is absorbed after exposure. Even though certain skin penetration of soluble cadmium compounds can take place when they are applied in a solution onto the skin (Skog and Wahlberg, 1964), and even if the skin can then be a route of uptake in such cases, the respiratory and gastrointestinal tracts are still the two main routes of absorption in man.

4.1.1 Respiratory Deposition, Clearance, and Absorption

Some conclusions can be drawn from general knowledge about deposition and clearance of particulate matters in human beings (for example, see Task Group on Metal Accumulation, 1973). Apart from that, virtually all evidence of deposition, clearance, and absorption of cadmium compounds is based on animal experiments.

4.1.1.1 Respiratory Deposition

There are no data elucidating the immediate respiratory deposition after inhalation of cadmium or cadmium compounds. As cadmium exposure via inhalation will be in the form of an aerosol, there is every reason to believe that the deposition will follow general physical laws governing deposition of particulate matters (see, for example, Task Group on Lung Dynamics, 1966, *Air Quality Criteria for Particulate Matter,* 1969). Thus, particle size will have a decisive importance and in humans breathing at a moderate work rate (20 l/min), the deposition in the pulmonary compartment will vary from about 10 to about 50% for particles with a mass median diameter of from about 5 to 0.01 μm.

4.1.1.2 Respiratory Clearance

Although deposition patterns for particles of different sizes are fairly well known, clearance rates are not. However, certain general tendencies can be stated. Particles deposited on the bronchial mucosa, for example, are usually cleared by means of ciliar activity in less than 8 hr (Albert, Lippmann, and Peterson, 1971). Thus, particles with high probability of deposition in the tracheobronchial compartment (relatively large particles) should be cleared relatively fast. As regards particles deposited in the pulmonary compartment, the most important factors are their solubility and other physicochemical properties. Half-times in the lungs are reported from a few days to about a year for different substances (Task Group on Lung Dynamics, 1966). Recently, the lack of precise knowledge about clearance of particles has been pointed out (Task Group on Metal Accumulation, 1973).

Concerning cadmium specifically, no precise information seems to be available on the several clearance mechanisms and their rate constants. However, some data have been reported by Harrison et al., 1947. They exposed dogs via an atomizer to a 25% solution of cadmium chloride over a period of 30 min. The exposure varied between 280 to 360 mg Cd/m^3 with an LD_{90} of 320 mg Cd/m^3 or 9,600 min · mg/m^3. They measured the concentration of cadmium in lungs, kidneys, and liver at autopsies at different times after the exposure. The lung data are shown in Figure 4:1.

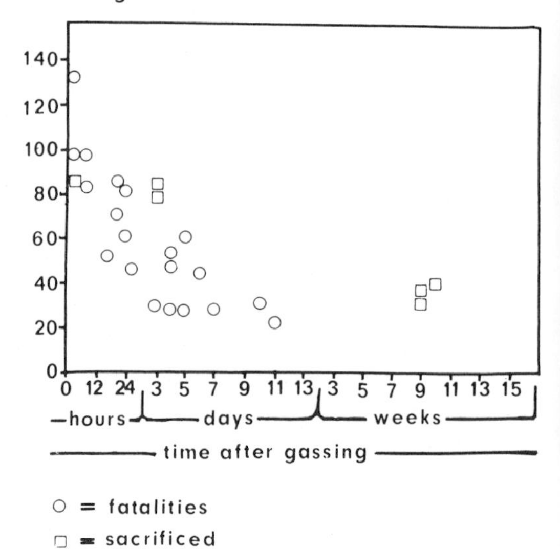

FIGURE 4:1. Concentrations of cadmium in lungs of dogs at intervals following exposure to cadmium chloride aerosols. (Modified from a figure by Harrison et al., 1947.)

As can be seen, clearance was rapid during the first 2 weeks with a half-time of approximately 5 days. With all probability, the initial clearance was still more rapid, even if this is not shown in the figures (no data from the very early postexposure period are reported). The long-term clearance seems to have been slow, as no further decrease was evident at the 10 weeks' observations. No information was given concerning particle size. The very high exposure, as a rule leading to the deaths of the animals, makes it impossible to extrapolate clearance rates for lower concentrations. An initial rapid clearance has also been reported by Gerard, 1944 (as quoted by Harrison et al., 1947). Gerard exposed mice to atomized radioactive cadmium chloride and found that about one fourth of the cadmium initially retained in the lungs remained there for 48 hr.

A report by Barrett, Irwin, and Semmons, 1947, should also be mentioned. They exposed groups of several animal species to cadmium oxide fumes in order to determine the toxicity of cadmium (see Section 5.1). They reported retention values for cadmium in the lungs at different times after the exposure. The percentage retention (based upon cadmium concentrations in the lungs at postmortem and exposure data) varied between about 5 and 20% with an average of about 11%. No systematic differences were found between values from the early postexposure period and from the period beginning 2 weeks later. The exposure conditions varied considerably, however, and few conclusions can be stated concerning clearance rates.

4.1.1.3 Respiratory Absorption

Abundant evidence verifies that cadmium can be found in different organs after industrial exposure in concentrations well above those found in the organs of "unexposed" subjects (see Section 4.3.2). Thus, absorption via inhalation takes place. The information about exposure and lung deposition, however, is so scarce that quantitative evaluations of absorption rates from such data cannot be made at this time.

In the above mentioned report by Harrison et al., 1947, the uptake in the kidneys and liver in dogs after an *acute* exposure was studied. The results are shown in Figures 4:2 and 4:3. An increase in the concentration in the kidneys occurred simultaneously with a decrease in the lungs. Thus, the concentration in the kidneys reached about the same value (per gram dry weight) as had been present in the lungs during the first stages following exposure. Since no information was given concerning total lung and kidney weights or the initial lung deposition of cadmium, a precise quantitative analysis is not possible. The data, however, do fit with a high absorption, even taking into consideration that the initial lung deposition certainly was considerably higher than is evident from Figure 4:1.

A conclusion that the absorption was high would also be reached if a calculation starts out from data concerning inhaled cadmium and recovered cadmium in liver and kidneys. The dogs were exposed to about 300 mg cadmium chloride/m^3 for 30 min and at least about 60 μg Cd/g dry weight (corresponding to about 20 μg Cd/g wet weight) was found in the liver (Figure 4:3) after a couple of days. The concentration in the kidneys was at least about 70 μg Cd/g dry weight (Figure 4:2), which would correspond to about 15 μg Cd/g wet weight. If it is assumed that about three fourths of the cadmium in the body (not counting the gut content) is in the liver and kidneys (see Section 4.3.4) and that the weights of the dogs were 20 kg, with a lung ventilation of 300 l per hour (Altman and Dittmer, 1964) and a liver weight of about 4% and kidney weights of about 0.4% of the body weight, the following tentative estimations could be made. About 50 mg cadmium would have been inhaled and about 20 mg of cadmium would have been retained in the body a few days after exposure. This would mean a retention of about 40% of the inhaled cadmium and an absorption amounting to a similar but slightly higher value.

In regard to the mode of absorption, it seems highly probable that a direct absorption was definitely the most important. The absorption was very rapid and studies of several animal species show that the absorption of cadmium via the gastrointestinal route is low (see Section 4.1.2.1).

Cd µg/g
dried weight
140
120
100
80
60
40
20
0
0 12 24 3 5 7 9 11 13 3 5 7 9 11 13 15
—hours— —days— —weeks—
time after gassing
○ = fatalities
□ = sacrificed

FIGURE 4:2. Concentrations of cadmium in kidneys of dogs at intervals following exposure to cadmium chloride aerosols. (Modified from a figure by Harrison et al., 1947.)

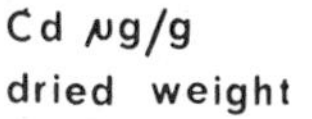

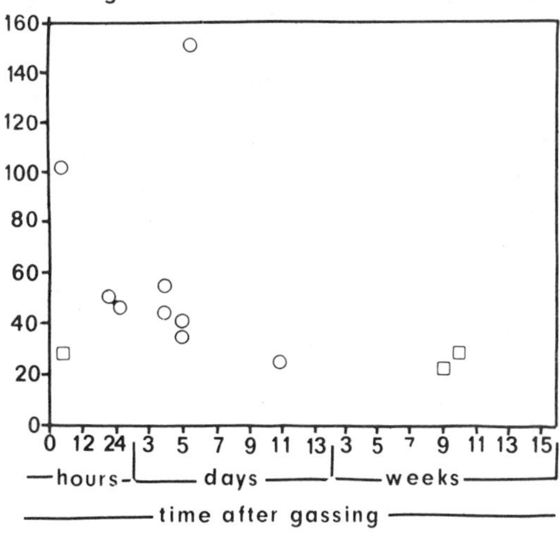

FIGURE 4:3. Concentrations of cadmium in liver of dogs at intervals following exposure to cadmium chloride aerosols. (Modified from a figure by Harrison et al., 1947.)

Potts et al., 1950, reported about cadmium absorption and retention in mice a few hours after a single inhalation exposure to a radioactive cadmium chloride aerosol (particle size less than 2 μm). The concentration of cadmium in the exposure chamber was on an average 100 mg/m^3 and the exposure time about 30 min. Because of several inconsistencies in the report, there are difficulties in interpreting the results as they are presented. It seems, however, that the data tend to show a retention of inhaled cadmium in internal organs (not counting the gut content) of about 10 to 20%.

Two *chronic exposure* studies, lasting for several months, both of which are discussed in detail in Section 5.2.2, should be mentioned. Friberg, 1950, in experiments with rabbits inhaling cadmium iron dust, found a high absorption as judged by concentrations of cadmium found in kidneys and liver. It can be calculated that the rabbits had inhaled, on an average, a total of about 190 mg of cadmium iron oxide dust, corresponding to about 120 mg of cadmium (data by Guyton, 1947, indicating a lung ventilation of about 50 l per hour for rabbits of about 2 kg, have been used in the calculations). The concentrations of cadmium in lungs, kidneys, and liver at autopsy were reported as generally varying from about 50 to 150, 300 to 700, and 100 to 400 μg/g wet weight, respectively (spectrographic determinations). With weights of approximately 25, 15, and 65 g for lungs, kidneys, and liver, respectively, the mean values for the organ amounts of cadmium can be estimated to have been as follows: the lungs contained about 2.5, the kidneys about 7.5, and the liver about 16 mg of cadmium, a total in these organs of about 25 mg retained cadmium. It can be expected that these organs contained about three fourths of the total body burden (see Section 4.3.4). It seems thus that the retention ($\frac{25 \cdot 10^4}{75 \cdot 120}$%) of the *inhaled* cadmium was high, about 30%. This, of course, means a considerably higher pulmonary absorption of cadmium in these chronically exposed rabbits.

Princi and Geever, 1950, made similar studies with dogs inhaling cadmium oxide dust and cadmium sulfide dust, respectively. Available data do not lend themselves to a detailed calculation like the one referred to above. The absorption must have been lower than in Friberg's studies, however, or the excretion of cadmium must have been very large, as the concentration of cadmium in the liver and particularly in the kidneys (dithizone method according to Church, 1947) was usually low. The kidney concentration varied from 21 to 133 μg/g – probably wet weight, but no information was given in the paper – in animals exposed to cadmium oxide, and from 0.2 to 43 μg/g in animals exposed to cadmium sulfide. Corresponding values in the liver were 12 to 47 and less than 0.2 to 10 μg/g. The reported kidney values are of the same magnitude as has been reported for human beings without occupational exposure (see Section 4.3.2.1). The liver values for dogs exposed to cadmium oxide dust are considerably higher than those found in nonoccupationally exposed human beings.

The difference in absorption found by Friberg as compared to that reported by Princi and Geever is not easily explained. Of some relevance could be the point that the U.S. cadmium dust differed from the Swedish dust in regard to solubility in water, the Swedish dust being more soluble (Ahlmark, Friberg, and Hardy, 1956).

Lewis et al., 1972a, studied the body burden of cadmium in smokers and nonsmokers. It is obvious from their findings that body burden increases with amount smoked (see Section 4.3.4).

If total body burden is assumed to be twice the cadmium amount accumulated in kidney, liver, and lung, 30 pack-years (packs per day × number of years of smoking) would mean an increase in body burden of about 12 mg cadmium. The following equation would then give the fraction retained:

$$\text{Retention (\%)} = \frac{100 \times 12{,}000\ \mu g}{30\ \text{packs/day} \times \text{years} \times 365\ \text{days/yr} \times 2\ \mu g/\text{pack}} = 54\%$$

In this example, the amount of cadmium inhaled per cigarette pack is assumed to be 2 μg (Section 3.1.5). If the amount of cadmium is assumed to be as high as 4 μg per pack, the retention will be as high as 27%. The actual absorption, of course, would be higher.

4.1.1.4 Conclusions

Cadmium is absorbed and retained to a considerable degree in the body after inhalation. The absorption is primarily directly from the lungs.

Observations on human beings are scarce but

data on smokers indicate that absorption of cadmium fumes may well be 25 to 50%. Animal experiments speak in favor of an absorption of between 10 to 40% of inhaled cadmium. A considerable difference might well exist for different cadmium compounds.

There is a great need for more detailed studies on the absorption and retention of different cadmium compounds after inhalation. Studies should be carried out on different species and in connection with acute as well as chronic exposures.

4.1.2 Gastrointestinal Absorption

4.1.2.1 In Animals

4.1.2.1.1 Single Exposure

Several reports have described the fate of a single oral dose of radioactive cadmium. Decker, Byerrum, and Hoppert, 1957, gave about 6.6 mg/kg of ^{115}Cd as the nitrate to rats in a stomach tube and after 8 and 24 hr found 98 to 99% in the stomach, the gut, and the feces. These results would indicate an absorption of 1 to 2%. Cotzias, Borg, and Selleck, 1961, gave 40 μCi ^{109}Cd as the chloride to mice by intubation and found that after 4 hr 0.5 to 8% had been absorbed (mean not given). Richmond, Findlay, and London, 1966, gave ^{109}Cd (dose and compound not stated) to mice by intubation. They found a mean retention after a few days of about 1% (range: 0.5 to 3%), which means that on an average at least 1% had been absorbed. Miller, Blackmon, and Martin, 1968, and Miller et al., 1969, gave goats ^{109}Cd-chloride (0.04 to 0.06 mg Cd per goat) in gelatin capsules and found that more than 90% of the dose had been excreted in the feces after 14 days. The absorption was not calculated, but as Miller et al., 1969, estimated, total body retention was 0.3 to 0.4% of the given dose and urinary excretion was insignificant. Thus, not more than a few percent of the dose can have been absorbed.

Suzuki, Taguchi, and Yokohashi, 1969, gave mice on a normal diet 0.08 mg of ^{109}Cd as the chloride by stomach needle. They followed the whole body retention (digestive tract with contents removed) between 0.5 and 164 hr in groups of three mice. Maximum retention was found after 11 hr, when 7% of the dose was in the body (range: 4.5 to 12%). After 164 hr, 1.6% (range: 1 to 2.3%) was present. Nordberg, 1971b, gave mice $^{109}CdCl_2$ by the oral route and found about 1.6% retention 14 to 16 days after dosing (for further details see Section 4.2.7). Ogawa et al., 1972a, gave mice a single oral dose of 1 μCi of $^{115m}CdCl_2$. The whole body retention after 24 and 48 hr was 7.3 and 2.7%, respectively. As gut contents are included in these figures, it can only be concluded that the retention was less than 2.7% after 48 hr. Kitamura (personal communication) found that about 2% was retained after 24 hr when mice were given a single oral dose of 250 μg cadmium nitrate.

Nordberg, Friberg, and Piscator, 1971, studied the retention of cadmium in monkeys (*Saimiri sciureus*) by means of whole body measurement after ingestion of $^{115}CdCl_2$ through stomach tube. Two monkeys were given 1.7 mg Cd/kg body weight and two other monkeys, 0.17 mg Cd/kg body weight. The monkeys were kept for 24 hr without solid foods prior to the exposure. For 2 hr preceding and for 4 hr following the dose, the monkeys were not allowed to have any food or water at all. The monkeys had retained 2.5 to 3.2% (average: 2.9%) of the ingested amount 10 days after ingestion. Four additional monkeys exposed to a single dose of only 1 μg/kg body weight have now been studied (Nordberg, unpublished data). The average whole body retention after 10 days was 1% (range: 0.5 to 1.5%) of the dose administered.

The retention found is considerably lower than the one reported by Suzuki et al., 1971, only 5 days after a similar exposure in a monkey. They found about 20% of the dose in the gastrointestinal tract at that time. These results may indicate that a considerable amount of unabsorbed cadmium will pass out via feces as late as between the 5th and the 10th day.

Moore, Stara, and Crocker, 1973, did not find any differences in liver or renal levels of cadmium among groups of rats given single doses of ^{115m}Cd as acetate, chloride, or sulfate.

4.1.2.1.2 Repeated Exposure

Decker et al., 1958, gave rats cadmium in drinking water in concentrations ranging from 0.1 to 10 μg/g during 6- and 12-month periods. (This experiment will also be discussed in Section 4.2.2.2.) They did not measure total body retention, but did determine the percentage of the administered dose retained in kidney and liver after 1 year. As these organs will accumulate most of the cadmium, this estimate may indicate the whole body retention. Less than 1% of the total ingested amount had been retained in these organs (0.3 to 0.5% at all exposure levels). If it is assumed that 50 to 75% of the total body burden is in liver and kidney (see Section 4.3.4), total retention would be less than 1%. As urinary excretion and excretion into the intestines are very low before any toxic effects are seen (see Section 4.2.4), this would mean that absorption would be somewhat higher than the retention.

Miller et al., 1967, administered about 5 mg Cd/kg to cows daily for 2 weeks. During the second week, cadmium excretion in feces was determined. Finding that 82% of the daily dose was excreted, they concluded that 18% was retained. Until more data have been obtained with regard to passage time of cadmium in the intestinal tract of cows, this figure must be regarded with caution, especially as the cow does not seem to accumulate cadmium to any large extent. Investigations on cows in cadmium-contaminated areas in Sweden have shown renal levels of cadmium around 1 to 2 μg/g wet weight, indicating a low absorption (Piscator, 1971).

Differences in retention after oral exposure to cadmium sulfate and cadmium stearate were found by Schmidt and Gohlke, 1971. They gave rats by stomach tube 15 mg Cd/kg of body weight of the compounds twice a week for 7 weeks. Cadmium levels in liver after 3, 7, and 16 weeks in rats that had received cadmium stearate were only 44, 32, and 51%, respectively, of the levels found in animals given cadmium as sulfate.

4.1.2.1.3 Influence of Dietary Factors upon Absorption of Cadmium

4.1.2.1.3.1 Calcium

It has been indicated in several reports concerned with the relationships among hardness of water, metal uptake, and toxicity that calcium may play an important role in influencing the absorption of metals from the intestines. Schroeder, Nason, and Balassa, 1967, gave rats 5 μg/g cadmium in "hard" and "soft" drinking water during 1 to 2 years. Since the diet contained adequate amounts of calcium, however, the difference in total intake of calcium was only about 5%. The authors found no difference between the groups in the accumulation of cadmium in liver and kidney and concluded that the hardness of the water was not in itself an important factor. Larsson and Piscator, 1971, gave female rats on low and high calcium diets 25 μg/g Cd as the chloride in drinking water for 1 and 2 months. They found that the rats on the low calcium diet had accumulated about 50% more cadmium in liver and kidney than the rats on the high calcium diet (Table 4:1). This indicates a higher absorption in the former group, as there is no reason to believe that there would be differences in excretion.

Kobayashi, Nakahara, and Hasegawa, 1971, gave rice containing 0.1 to 0.6 μg/g of cadmium to mice on low and normal calcium intake. After 70 weeks of exposure, animals on a low calcium diet had liver and kidney concentrations of cadmium 50 (0.1 μg/g Cd in rice) to 300 to 400% (0.6 μg/g) higher than mice on a normal calcium diet. Piscator and Larsson, 1972, gave rats on low and normal calcium diets cadmium in drinking water in concentrations ranging from 0 to 10 μg/g for 1 year. Calcium deficient animals retained about twice as much cadmium in liver and kidney as those on a normal calcium diet.

4.1.2.1.3.2 Vitamin D

Worker and Migicovsky, 1961, found that the uptake of cadmium in the tibia after an *oral* dose of ^{115}Cd was greater in rachitic chickens under treatment with vitamin D than in untreated chickens. They concluded that this difference was

TABLE 4:1

Concentrations of (1) Cadmium in Renal Cortex and Liver and (2) Inorganic Matter in Percent of Wet Weight of the Right Tibia*

		Cd (μg/g dry weight) Mean and standard deviation		Inorganic matter (percent of wet weight) Mean and standard deviation
Group	n	Renal cortex	Liver	Right tibia
Cd + Ca	6	136 ± 23	29 ± 5	47.67 ± 0.83
Cd – Ca	6	205 ± 23	49 ± 6	44.40 ± 0.52
– Ca	6	<5	<1.5	47.12 ± 0.58
Controls	6	<5	<1.5	48.01 ± 1.20

*Rats were treated for 2 months with cadmium and normal diet (Cd + Ca), cadmium and calcium deficient diet (Cd – Ca), calcium deficient diet (– Ca), and normal diet (controls). Cadmium was given at 25 μg/g in the drinking water to the cadmium-exposed groups.

From Larsson and Piscator, 1971.

due to the effect of vitamin D on the intestinal absorption of cadmium, as there was no difference between similar groups in uptake in bone when ^{115}Cd was *injected.*

4.1.2.1.3.3 Protein

Fitzhugh and Meiller, 1941, mentioned briefly that the toxicity of cadmium was increased by a low protein diet. Suzuki, Taguchi, and Yokohashi, 1969, gave mice low and high protein diets for 24 hr before and after an oral dose of ^{115}Cd-chloride. The low protein diet gave considerably higher levels of cadmium in kidney, liver and whole body, irrespective of calcium content in the diet, indicating an increase in the absorption of cadmium. Whole body retention was about 9% (range: 5 to 14) and 4.5% (range: 3 to 10) in mice fed low and high protein diets, respectively.

4.1.2.2 In Human Beings

Absorption can be studied through long-term balance studies on intake and excretion of cadmium or on intake and body burdens of cadmium found at autopsies. Long-term balance studies have been performed on three subjects (Tipton, Stewart, and Dickson, 1969, and Tipton and Stewart, 1970). Cadmium was determined daily by atomic absorption (without extraction) in the diet, in the urine, and in the feces during 140 to 347 days. The mean daily intake of cadmium was given as 170 μg, and the mean daily fecal content was about 40 μg. This would indicate an absorption of about 75%, an impossible figure (see below). The mean daily excretion via the urine was reported to be 100 μg. In Chapter 2 it is mentioned that sodium chloride is an interfering substance causing erroneously high values when cadmium is determined by atomic absorption. It is highly probable that the amount of sodium chloride in the diet and urine did interfere with the cadmium determinations. This is supported by the fact that much lower values for daily intake of cadmium and daily urinary excretion of cadmium (see Tables 3:8 and 4:6) have been obtained when the metals have been extracted before analysis. The values for cadmium in feces are within the range reported by other authors (see Section 4.3.3.2). Schroeder and Balassa, 1961, by using the dithizone method, estimated that the daily intake of cadmium in the United States could vary between 4 and 60 μg, depending upon foods chosen. Boström and Wester, 1968, found that in one healthy subject, studied for two periods of 5 days, the intake of cadmium was 12 μg per day in both periods, and the mean fecal excretion of cadmium was 5.3 and 4.5 μg, respectively. The determinations were done with neutron activation. These data do not lend themselves to an evaluation of absorption due to the short observation period.

Rahola, Aaran, and Miettinen, 1971, studied the fate of ^{115m}Cd given orally to five human male volunteers, age 19 to 50 years. They received single doses of 4.8 to 6.1 μCi of ^{115m}Cd, mixed with a calf kidney suspension. The total ingestion of cadmium was about 100 μg. During the first 3 to 5 days after administration, about 70% of the activity was eliminated, primarily in the feces. A rapid elimination continued until about 6% (4.7 to 7.0%) of the dose remained in the body. This indicates an average absorption of at least 6%.

Kitamura, 1972, reported on two balance studies performed on a 55-year-old man. In one study 5 mg of Cd $(NO_3)_2$ was administered in drinking water (10 μg/g Cd) during 1 day. The total fecal amount was collected during 15 days and cadmium excretion thus calculated. The absorption of cadmium was calculated as the amount given minus the accumulative excretion in feces. In order to establish a background fecal excretion of cadmium, the subject ate rice containing a total of 20 μg cadmium per day for 7 days, including the day before the experiment. Unfortunately, the cadmium concentration in feces before cadmium exposure was not measured. Apart from this rice, the subject ate an unidentified amount of other food with low cadmium content (Kitamura, personal communication). Kitamura estimated the background excretion to be 20 μg and subtracted this from the measured values. He calculated that the long-term absorption of this single dose in drinking water would be 5.34%. All data from Japan favor the conclusion that foodstuffs other than rice also contribute significantly to the total intake (Section 8.3.1.5). If the background had been 40 μg, which is in accordance with reported data on daily cadmium intake in Japan (Section 8.3.2.1), it can be calculated that long-term absorption would have been about 8%.

In another experiment, Kitamura, 1972, administered about 5 mg of $Cd(NO_3)_2$ in rice to the same person. The experiment was performed in a similar way but no background measurements were made and no standardized rice diet was provided (Kitamura, personal communication). Using the same background (20 μg/day), Kitamura calculated a long-term absorption of 1.35%. It can be calculated that, had the fecal background excretion instead been 40 μg/day, absorption would have been about 10%.

Alexander, Delves, and Clayton, 1972, made balance studies on children. There must have been serious errors in the analytical procedures because they found a mean urinary excretion far above what is seen even in exposed workers. The data cannot be used for an evaluation of absorption of cadmium.

4.1.2.3 Conclusions

Most animal data, albeit with large individual variations, indicate an absorption of about 2% of ingested cadmium. The absorption is increased considerably, with a factor of two or more, by a low calcium or a low protein intake.

An average absorption of about 6% (range: 4.7 to 7.0) has been seen in five humans given single doses of radioactive cadmium.

A retention rate of 10% or even more of ingested cadmium must be considered possible in the individual case under certain circumstances such as calcium or protein deficiency.

4.1.3 Placental Transfer

4.1.3.1 In Animals

Berlin and Ullberg, 1963, gave pregnant mice single intravenous ^{109}Cd-chloride injections of 10 μCi, carrier free ^{109}Cd on the 18th day after conception. They could not detect any cadmium in the fetus but could note an accumulation of cadmium in the placenta. Tanaka et al., 1972, gave pregnant mice single intravenous injections of 50 μCi of ^{115m}Cd as chloride (about 15 μg Cd per mouse) 24 to 36 hr before delivery. The mean uptake in newborns was 0.09% of the dose. After giving a single intravenous dose of 0.5 to 0.85 mg Cd/kg to pregnant hamsters on the 8th day of gestation, Ferm, Hanlon, and Urban, 1969, found relatively high concentrations of cadmium both in the placenta and in the fetus. The dose was considerably larger than the one given by Berlin and Ullberg but similar to the one given by Tanaka et al. In addition, Ferm, Hanlon, and Urban administered the metal at that period during which the major organ systems are being established in the hamster embryo and the embryologic activity is high, whereas Berlin and Ullberg injected the cadmium during the last few days of pregnancy.

By exposing rats from the day of conception to cadmium oxide dust in concentrations of about 3 mg/m^3, Cvetkova, 1970, found that the livers of the embryos after 22 days contained more than twice the amount of cadmium contained in controls.

Large doses of cadmium will destroy the placenta, especially the fetal part, as shown by Parízek, 1964, who gave rats subcutaneous injections of 4.5 mg Cd/kg on the 17th to 21st days of gestation. Parizek, 1965, showed that 2.5 mg/kg by the subcutaneous route had the same effect. He also noted that the pregnant rat was more susceptible to the action of cadmium than the nonpregnant rat. Parizek et al., 1968, found that the simultaneous administration of sodium selenite protected the fetus from the destructive action of cadmium, but increased the uptake of cadmium in the placenta. Holmberg and Ferm, 1969, found that selenite protected the hamster embryo after intravenous injection of 2 mg Cd/kg. Of additional interest is the fact that chelating agents such as EDTA increase the transplacental passage of cadmium (Eybl, Sýkora, and Mertl, 1966b).

4.1.3.2 In Human Beings

Henke, Sachs, and Bohn, 1970, analyzed liver and kidney samples from newborns in West Germany by neutron activation and found between 4 and 20 ng/g wet weight in four kidneys. The concentration was less than 2 ng/g in the liver. This means that the total content of cadmium in the newborn will be less than 1 μg. It also means that the fetal liver and kidney concentrations are about 1,000 times lower than what could be expected in the mother (see Section 4.3.2.1).

Analysis by atomic absorption of liver, kidney, and brain from fetuses (Kyoto, Japan) of 85 to 185 days of gestational age showed cadmium concentrations of from <0.02 to 0.22, <0.02 to 0.062, and <0.02 to 0.144 μg/g wet weight, respectively (Chaube, Nishimura, and Swinyard, 1973). These values are higher than those reported from West Germany, which may be due to the higher level of exposure in Japan. Since atomic absorption was used without elimination of interfering substances, analytical errors may be present.

Concentrations in the placentas of 44 Swedish women were found to be less than 0.010 μg/g wet weight by a spectrographic method (Piscator,

1971). Even though the methods would not merit an exact evaluation, the indication is that not more than about 5 μg will accumulate in the placenta, which weighs about 500 g. In comparison, a human kidney will usually accumulate about 100 μg during a 9-month period (see Section 4.3.2.1).

4.1.3.3 Conclusions

Animal experiments indicate that the placenta constitutes a barrier against transfer of cadmium when small doses are given. However, when large doses are given, cadmium may destroy the placental barrier and enter the fetus. In human newborns, the total body content of cadmium is small, less than 1 μg. In areas where the intake of cadmium is elevated, higher body burdens may be found.

4.2 TRANSPORT, DISTRIBUTION, AND EXCRETION OF CADMIUM IN ANIMALS

Cadmium will be found in blood, internal organs, and excreta after absorption following exposure via air, oral intake, or injection. In the following sections, the metabolism after both single and repeated exposure will be considered. The distribution of cadmium after respiratory exposure has been discussed in Section 4.1.1.3 and will not be further treated in this section.

4.2.1 Uptake to and Clearance from Blood

In Section 4.1.2 it was mentioned that after oral exposure only a few percent of the dose was absorbed. Attempts to determine levels of radioactive cadmium in blood after a single dose have not been successful, so only cadmium in blood after injection will be considered in this section.

4.2.1.1 Fate of Cadmium in Blood After Single Injection

4.2.1.1.1 First Hours After Injection

Intravenous route – During the first hours after injection, cadmium is found mainly in the plasma, as seen in dogs (Walsh and Burch, 1959), in rabbits (Kench, Wells, and Smith, 1962), and in rats (Perry et al., 1970). The last mentioned authors found 98 to 99% of the blood cadmium in the plasma 0.5 and 8.5 min after injection, and Kench, Wells, and Smith found 90 to 80% 9 min and 1 hr, respectively, after injection. Perry et al. found that about two thirds of the plasma cadmium was dialyzable during the first 8.5 min.

The clearance of cadmium from the plasma is characterized by an initial rapid phase followed by a slower phase, as shown in Figure 4:4 from Walsh and Burch, who gave dogs about 0.15 mg ^{115}Cd/kg as cadmium nitrate. Perry et al. also found a rapid clearance in rats during the first 8.5 min after an injection of 0.2 mg ^{115}Cd/kg. Kench, Wells, and Smith gave a large dose, about 4 mg ^{115}Cd/kg, as the sulfate to two rabbits and found that after 9 min about 45% of the dose remained in the plasma, and after 1 hr 6.5%. As they gave a lethal dose and one rabbit died within 6 hr, their data cannot be regarded as representative for the rabbit.

Perry et al. found a small uptake of about 2% of the blood cadmium in the erythrocytes after 0.5 min. Though Walsh and Burch mentioned that a small but constant amount of cadmium was found in the cells, they did not report the values.

Intraperitoneal route – After an intraperitoneal injection of soluble cadmium compounds, the maximum blood levels are reached within very short periods. In studies by Johnson and Miller, 1970, who gave rats ^{109}Cd (0.01 to 0.02 mg/kg) as the chloride, the maximum concentration was reached after 5 min. In similar studies (Perry and Erlanger, 1971) maximum blood levels were found 8 min after injection (rats, ^{115}Cd as chloride, 0.06 to 0.96 mg Cd/kg). In the last mentioned experiment, it was calculated that 3% of the injected dose was circulating in the blood. In the study by Johnson and Miller, the blood level 80 min after the injection was only 12% of the maximum value, indicating a rapid initial clearance from the blood. Perry and Erlanger reported that the concentrations 120 min after the injections had decreased to 44 and 16% of the maximum concentrations after injections of 0.24 and 0.96 mg Cd/kg, respectively. They further stated that only a few percent of the cadmium in the blood was in the red blood cells and that part of the plasma cadmium was dialyzable.

Subcutaneous route – After a subcutaneous injection of soluble cadmium compounds, cadmium will soon appear in the blood, and maximum levels have been reported after 10 min by Johnson and Miller, 1970, and after 60 min by Eybl, Sýkora, and Mertl, 1966a. Johnson and Miller gave rats 0.01 to 0.02 mg ^{109}Cd/kg as the chloride. They found that there was a rapid initial

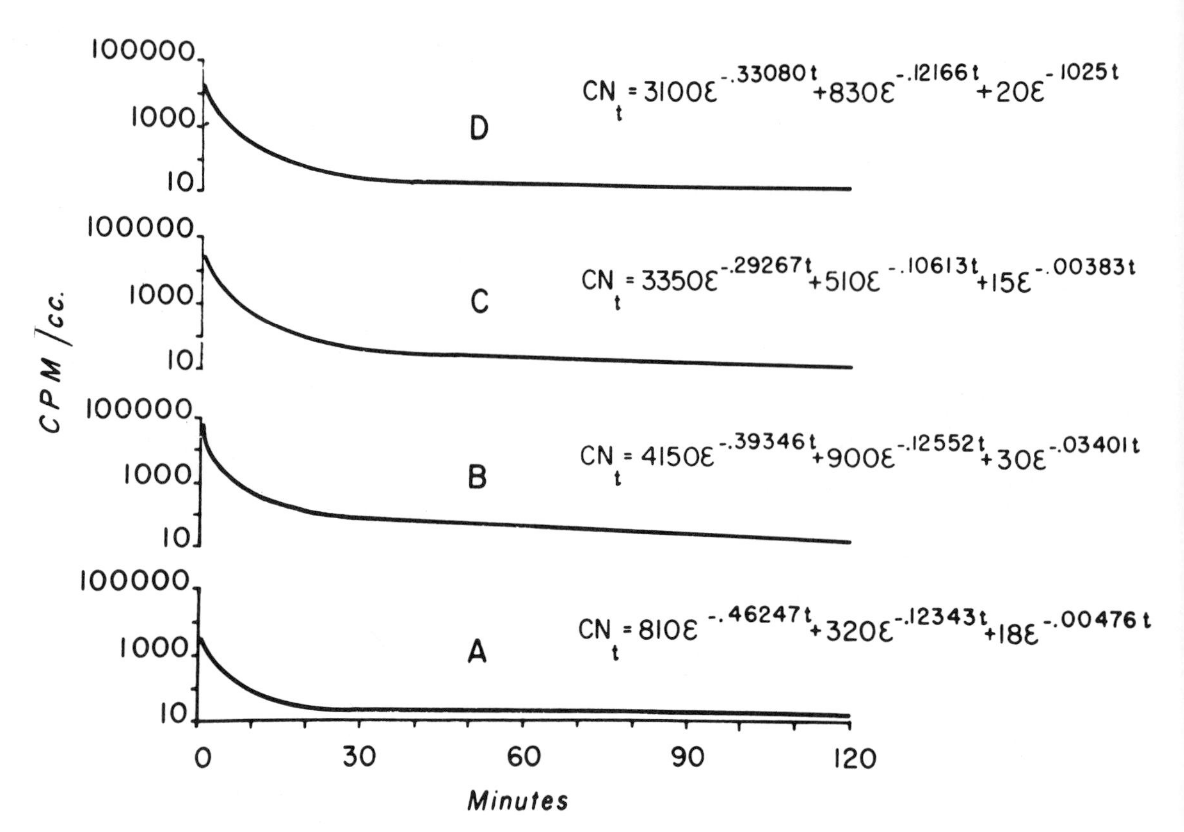

FIGURE 4:4. Disappearance of cadmium from plasma after an intravenous injection of ^{115}Cd nitrate. (From Walsh and Burch, 1959.)

clearance, so that 60 min after injection the concentration in whole blood was only 16% of the 10-min value. Eybl, Sýkora, and Mertl gave rats 1.25 mg ^{115}Cd/kg as the chloride and found that the concentration in blood after 1 hr was higher than after 20 min. Hence, these results are different, probably due simply to the considerably larger dose given by these latter authors, i.e., the larger the injection, the greater the deposition at the injection site.

During the first 12 hr following injection, blood levels decrease, as shown by Eybl, Sýkora, and Mertl, 1966a, by Gunn, Gould, and Anderson, 1968a, who gave mice 1.4 mg ^{109}Cd/kg as the chloride, and by Lucis, Lynk, and Lucis, 1969, who gave rats a tracer dose of ^{109}Cd as the chloride. This decrease is mainly due to a decrease in plasma levels of cadmium, as shown in Figure 4:5, for which data are taken from experiments by Eybl, Sýkora, and Mertl, 1966a, and Lucis, Lynk, and Lucis, 1969. In Figure 4:5, an initial uptake in the blood cells can be seen. The figure also elucidates that the clearance from the cells during the first 12 hr is smaller than the clearance from the plasma, so that after 12 hr, there is almost the same concentration in cells as in plasma.

The decrease in the plasma levels and increase in red cell levels have also been observed by Shaikh and Lucis, 1972a, and Nordberg, 1972a. Nordberg gave mice single subcutaneous injections of ^{109}Cd as the chloride, corresponding to 1 mg Cd/kg body weight. Cadmium concentrations in cells and plasma were studied from 20 min to 96 hr after injection. The results, which are in agreement with the data by other authors, given in the preceding paragraph and displayed in Figure 4:6, show an initial maximum in red cells after about 4 hr followed by a decrease and a subsequent increase up to 96 hr. In plasma there was a rapid decrease from 20 min to 48 hr. After that there was an increase in plasma levels. Nordberg also studied the distribution of cadmium in the red cells by separating hemolysates on G-75 Sephadex® columns. After 20 min most of the cadmium was in high molecular weight fractions whereas only a minor part was in the fraction corresponding to

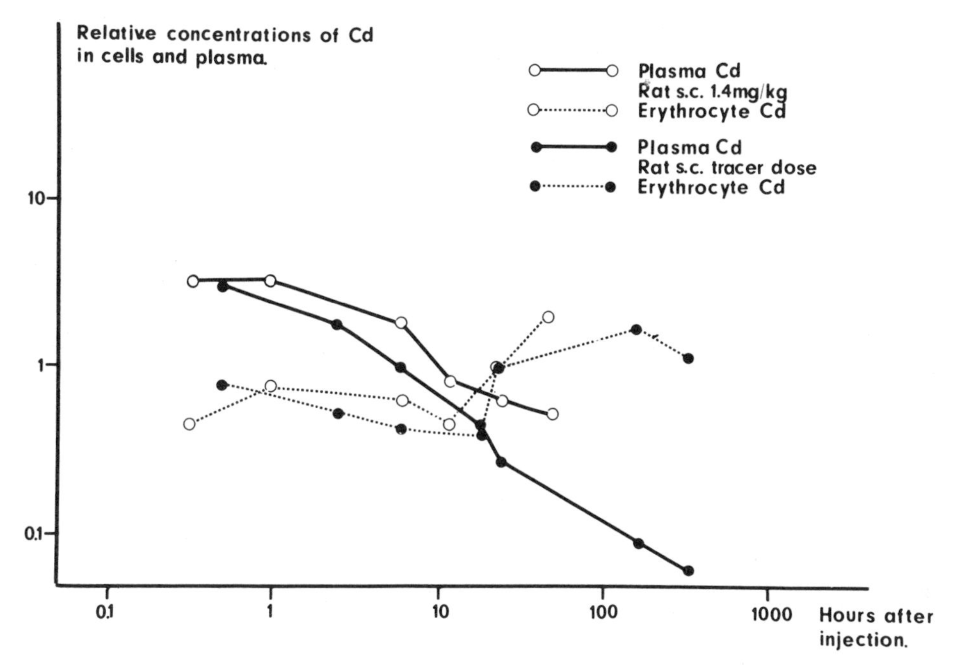

FIGURE 4:5. Concentrations of cadmium in cells and plasma related to time after subcutaneous injections in rats. In each report the concentration in the cells after 24 hr has been set at 1.0 and the concentrations in cells and plasma are expressed in relation to the concentration in cells at 24 hr. (From data by Eybl, Sýkora, and Mertl, 1966a, and Lucis, Lynk, and Lucis, 1969.)

hemoglobin and an insignificant amount in the fraction corresponding to metallothionein. After 96 hr a redistribution had taken place, so that about one third of the total cadmium was found in fractions corresponding to metallothionein. The main part was still in high molecular weight fractions, and no cadmium was observed in the hemoglobin peak. In plasma, at 20 min and 4 hr after injection, the main part of the cadmium was in fractions corresponding to albumin or larger proteins.

4.2.1.1.2 First Weeks After Injection

Whereas there are large differences in distribution during the first hours after injection, depending on how fast cadmium enters the blood, distribution and clearance mechanisms seem to be more similar when some time has elapsed since injection. Blood levels decrease during the first 12 hr. Then a change occurs: the blood levels increase again. In rats and mice, this increase is due to an increase of the cadmium content of the erythrocytes, as shown in Figures 4:5 and 4:6.

Niemeier, 1967, followed rats after an intravenous injection of ^{115}Cd sulfate and found that after 16 days the cadmium level in blood was more than twice the level at day 1. In the goat (Miller, Blackmon, and Martin, 1968) and the hamster (Ferm, Hanlon, and Urban, 1969), a rise in the cadmium level between day 1 and day 7 and day 1 and day 2, respectively, after intravenous injections, has been noted.

These results indicate that after the first rapid clearance of cadmium from blood, a second phase will occur in which cadmium will be released into the blood, probably from the liver, where the main part of the injected cadmium will be found (see Section 4.2.2).

4.2.1.2 Fate of Cadmium in Blood After Repeated Injections

Friberg, 1952, found, in rabbits given subcutaneous injections of ^{115}Cd sulfate (0.65 mg Cd/kg 6 days a week for 4 and 10 weeks), that the cadmium levels in blood, taken about 1 week after cessation of exposure, were on the average 450 and 1,000 ng/g blood, respectively. He could not demonstrate the presence of cadmium in the plasma and concluded that the cadmium was in the blood cells. Truhaut and Boudene, 1954, gave two rabbits subcutaneous injections of cadmium sulfate, 13 injections of 2.1 mg/kg, and 19

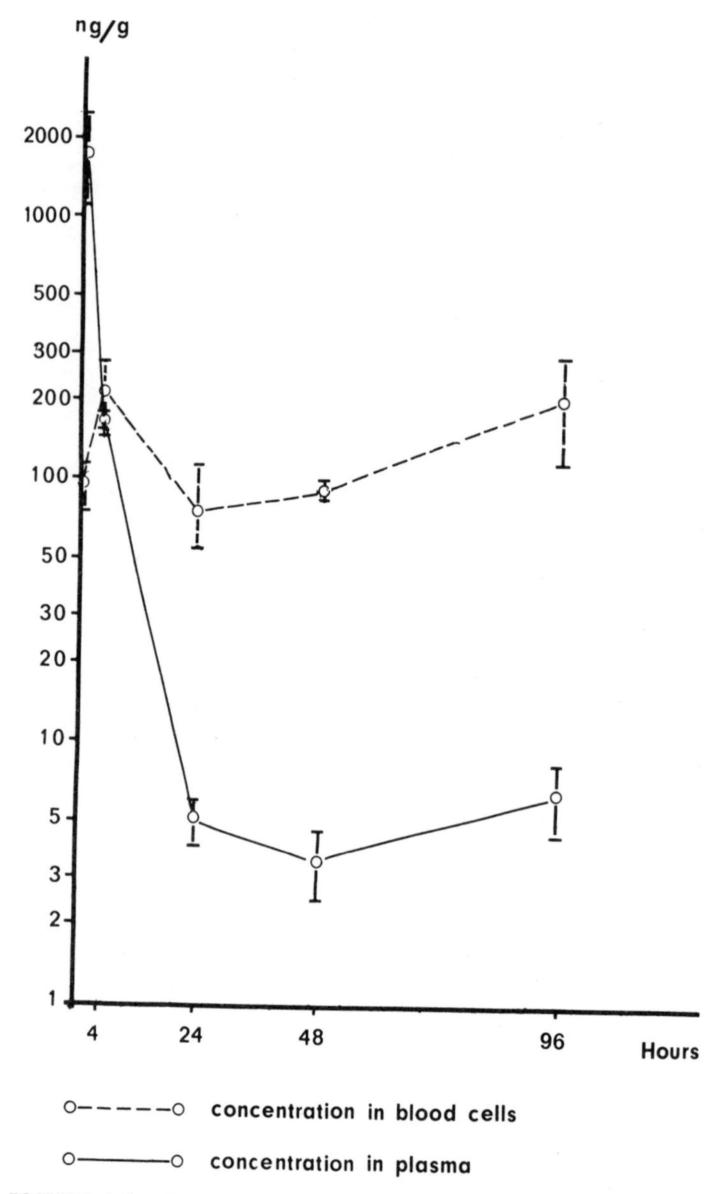

FIGURE 4:6. Concentrations of cadmium in plasma and blood cells, respectively, in mice given a single subcutaneous injection of $^{109}CdCl_2$ (1 mg Cd/kg) and killed various times after injection. Vertical bars indicate ranges of two or three mice and circles are mean values. (From Nordberg, G. F., *Environ. Physiol. Biochem.*, 2, 7, 1972. With permission.)

injections of 1.8 mg/kg. They found that the concentration in the erythrocytes was 18 and 10 times greater, respectively, than in the plasma. Friberg, 1955, made new experiments with similar exposure conditions as in 1952. In one experiment rabbits were given ^{115}Cd during 70 days. In another, they were first given the radioactive isotope for 50 days and then ordinary cadmium for the remaining 20 days. In both experiments cadmium levels rose steadily until the 50th day, at which time they were around 1,750 ng/g. In rabbits given further injections of ^{115}Cd, blood levels of the radioactive isotope remained at the same level after 10 weeks. In rabbits given non-radioactive cadmium, a decrease in ^{115}Cd levels to about 1,250 ng/g in 20 days was found. The results show that at a certain cadmium level in the blood a plateau was reached. A continuous exchange between cadmium accumulated in the blood and newly administered cadmium took place.

In mice repeatedly exposed to cadmium, Nordberg, 1972a, studied the blood concentrations of cadmium and their relationship to whole body and urinary cadmium concentrations. The results from these studies are summarized in Table 4:2, in which it is seen that blood values increase during periods of exposure and decrease during periods without exposure. The table also shows that mean urinary values for the group tend to follow whole body values better than they follow blood values.

TABLE 4:2

Blood Concentrations of Cadmium in Mice Given Repeated s.c. Injections of $^{109}CdCl_2$* and Their Relation to Body Burden and Urinary Excretion of Cadmium

		Whole body	Blood		Urine
Week no.	n	μg	Mean ng/g	S.D.	ng/24 hr
3	4	11.0	23.8	8.6	0.80
6	4	11.9	14.9	1.4	0.80
21	4	61.1	48.4	19.5	4.65
24	8	65.2	25.9	6.4	6.34

*During weeks 1 to 3 and 7 to 21 the mice were exposed by daily s.c. injections of 0.025 mg Cd/kg body weight for 5 days per week.

From Nordberg, G. F., *Environ. Physiol. Biochem.*, 2, 7, 1972. With permission.

Carlson and Friberg, 1957, gave 1 mg ^{115}Cd/kg to two rabbits daily for 1 week. The animals were killed 2 days after cessation of exposure. They also gave 0.65 mg ^{115}Cd/kg to two other rabbits daily for 3 weeks. The animals were killed 3 weeks after cessation of exposure. Of the cadmium in whole blood, 1 to 6% was found in the plasma. When red cells from these rabbits were hemolyzed and centrifuged, all cadmium was recovered in the supernatant. Of the cadmium in the hemolysate, 40 and 37%, respectively, was found to be dialyzable through polyvinyl tubings. Separation by zone electrophoresis in starch of hemolysates from two other rabbits exposed 2 to 4 weeks showed that most of the cadmium in the hemolysate was in the same fractions as hemoglobin, and the authors concluded that part of the cadmium was bound to hemoglobin. Nordberg, Piscator, and Nordberg, 1971, exposed mice for 6 months, 5 days a week, to 0.25 or 0.5 mg ^{109}Cd/kg by subcutaneous injections. They found that part of the erythrocyte cadmium was in a fraction corresponding to the molecular weight of metallothionein. It is possible that this part behaved as dialyzable cadmium in the studies by Carlson and Friberg, 1957.

4.2.2 Tissue Distribution and Retention

4.2.2.1 Single Exposure

Oral route – The amount absorbed after a single dose is small, as has been discussed in Section 4.1.2. The distribution in organs has been studied by several authors. Decker, Byerrum, and Hoppert, 1957, gave rats ^{115}Cd-nitrate (6.6 mg Cd/kg) and determined cadmium distribution from 8 to 360 hr after the dose. After 8 hr the largest total amount of cadmium was in the liver while the highest concentration was in the kidneys. After maximum concentrations in both organs were reached at 72 hr after exposure, a slow decrease seemed to occur in both organs. Miller, Blackmon, and Martin, 1969, gave a single oral dose of ^{115}Cd-chloride to goats (0.04 mg Cd/goat) and found that after 14 days the concentration in the kidneys was nearly twice as high as that in the liver. Miller et al., 1968, gave about 0.06 mg of ^{109}Cd to groups of goats, one group of which had been pretreated 1 week with a diet containing 100 μg/g of nonradioactive cadmium. Determination of ^{109}Cd content in organs was made 14 days after exposure. They did not find any differences between the groups.

Suzuki, Taguchi, and Yokohashi, 1969, put mice on diets with high calcium-low phosphorus and low calcium-high phosphorus content, respectively, for 10 days and then gave them a single dose of ^{109}Cd-chloride (about 0.08 mg Cd/mouse). Liver and kidney levels were measured after 24 and 72 hr, respectively. No difference was found between the groups, each of which was comprised of 3 to 5 mice for each survival time. However, within the groups differences in levels at the two times of measurement could be seen. After 72 hr both kidney and liver levels were only half of the 24 hr values, indicating rapid clearance.

Injection route – The distribution of cadmium will vary considerably depending on the dose and the time. The experiments mentioned in Section 4.2.1.1, in which the fate of cadmium in blood after a single dose was followed, showed in addition that most of the cadmium initially goes to the liver and relatively small amounts to the kidneys. With increasing time after exposure, the kidney levels will increase and will become higher than liver levels.

Gunn and Gould, 1957, gave a single intracardiac injection of ^{115}Cd-nitrate to rats (dose not stated) and followed the animals for up to 8 months. From their data it is seen that during the first month there was a decrease in liver levels of ^{115}Cd and an increase in kidney cortex levels, so that after a month of exposure, the concentration in the cortex was the same as that in the liver (Figure 4:7). During the following months liver levels decreased very slowly while levels in cortex increased so that after 5 months the ratio between cortex and liver levels was 2.4, and after 8 months 4 to 5. The levels in kidney medulla also increased with time and after 8 months they exceeded liver levels. In addition, the authors observed that age could influence uptake in the kidney in rats from 1 to 7 weeks old, so that the younger the rat, the less the uptake. To the contrary, liver levels did not vary with age. This was explained as due to the smaller number of nephrons in the younger rats. Decker, Byerrum, and Hoppert, 1957, injected ^{115}Cd-nitrate into rats by the intravenous route (approximately 0.65 mg Cd/kg). They found that from 4 hr to 5 weeks after injection the liver contained between 62 to 70% of the injected dose. The kidney contained between 1.6 and 2.4% of the dose from 4 hr to 2 weeks. Then, at 5 weeks, 5.1% of the total dose was found in the kidney.

Burch and Walsh, 1959, found that in dogs

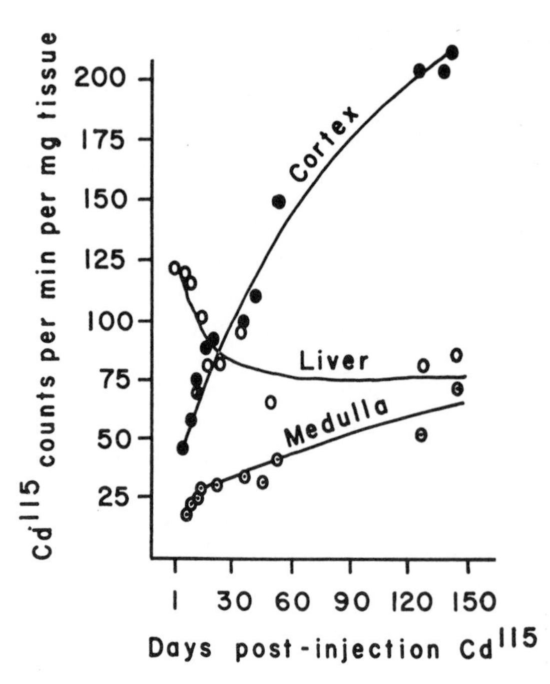

FIGURE 4:7. Accumulation of cadmium in kidney and liver of rats after intracardiac injection. (From Gunn and Gould, 1957.)

given an intravenous injection of ^{115}Cd-nitrate (0.2 to 0.4 mg Cd/kg), the concentration in the liver during the first 24 hr was considerably greater than that in the kidneys. After 20 to 30 days, the concentration in the kidneys was 50 to 100% of the liver levels. Lucis, Lynk, and Lucis, 1969, followed cadmium concentrations for 14 days after giving a tracer dose of ^{109}Cd-chloride to rats by subcutaneous injection. They found that already 2 hr after injection kidney levels were slightly higher than liver levels, and that 10 hr after injection the concentration in the kidneys was about 50% higher than that in the liver. At 2 weeks both the kidney and liver concentrations had increased compared with the concentrations at 30 hr. Miller, Blackmon, and Martin, 1968, investigated the distribution in goats 14 days after a single intravenous administration, using ^{109}Cd (0.04 mg/goat). Concentrations in liver were three times higher than those in kidney, whereas, after an oral dose of the same size, the concentrations in kidney were nearly twice as high as those in liver. Lucis and Lucis, 1969, studied four strains of mice and found considerable differences in organ concentrations 24 hr after subcutaneous injection. Liver levels varied between 37 and 52% of injected dose per gram.

Shaikh and Lucis, 1972a, gave mice and rats single subcutaneous injections of 1.1 and 2.2 μg ^{109}Cd as chloride, respectively. There was a continuous accumulation of cadmium in mouse kidneys during the observation period of 25 days, whereas in the rat kidney accumulation ceased after 10 hr, after which renal levels of cadmium decreased. The results in rats are in sharp contrast to the results obtained by Gunn and Gould, 1957, who found an increase in renal cadmium levels continuing for months after a single intracardiac injection in rats. The drop in renal levels after 10 hr is also the opposite from the earlier results obtained in rats by Lucis, Lynk, and Lucis, 1969.

Suzuki et al., 1971, gave pregnant as well as nonpregnant mice single subcutaneous injections of 2.5 μCi ^{109}Cd as chloride (specific activity not stated) per kilogram body weight. The pregnant rats tended to have higher amounts of cadmium in kidneys and liver, whereas concentration in digestive tract and bones tended to be less than in controls. The authors concluded that pregnancy may alter the metabolism of cadmium.

The liver and the kidney seem to be the two organs of greatest interest with regard to cadmium storage. Cadmium, however, will be found in most compartments of the body. The pancreas and the spleen will store relatively large amounts. Nordberg and Nishiyama, 1972, found that in mice cadmium levels in the pancreas increased with time and exceeded liver levels 110 days after injection (4.5 μg Cd/kg, intravenous).

Figure 4:8 shows the distribution of ^{109}Cd 112 days after a single injection. Berlin and Ullberg, 1963, did not find cadmium in osseous tissue after a single intravenous injection of ^{109}Cd-chloride, but they did observe that cadmium was accumulated and retained in bone marrow and periostium. They detected traces of cadmium in the brain. Their finding of an accumulation in the testis and the hypophysis is also of interest.

4.2.2.2 Repeated Exposure

Oral route – Long-term experiments with exposure via the oral route have been made by Decker et al., 1958, who gave cadmium in drinking water to rats for 6 and 12 months. The concentrations of cadmium in the water were 0, 0.1, 0.5, 2.5, 5, and 10 μg/g. They found that the concentrations in kidneys and liver were roughly proportional to the intake of cadmium. The retention in the kidneys during the period between 6 and 12 months was at least as high as during the first 6 months. It was also found that the larger the dose, the larger the ratio became between concentrations in liver and kidneys, as shown in Table 4:3. Data from a similar experiment on dogs by Anwar et al., 1961, are included in the table.

Piscator, 1971, determined the cadmium concentration (atomic absorption) in liver and kidney cortex of slaughtered horses mainly between 10 to 20 years of age and without known disease. In Figure 4:9, the concentration ratio between liver and renal cortex is given. There is no change in ratio with increasing renal concentration. This seems contradictory to the data displayed in Table 4:3, but might be explained by the longer expo-

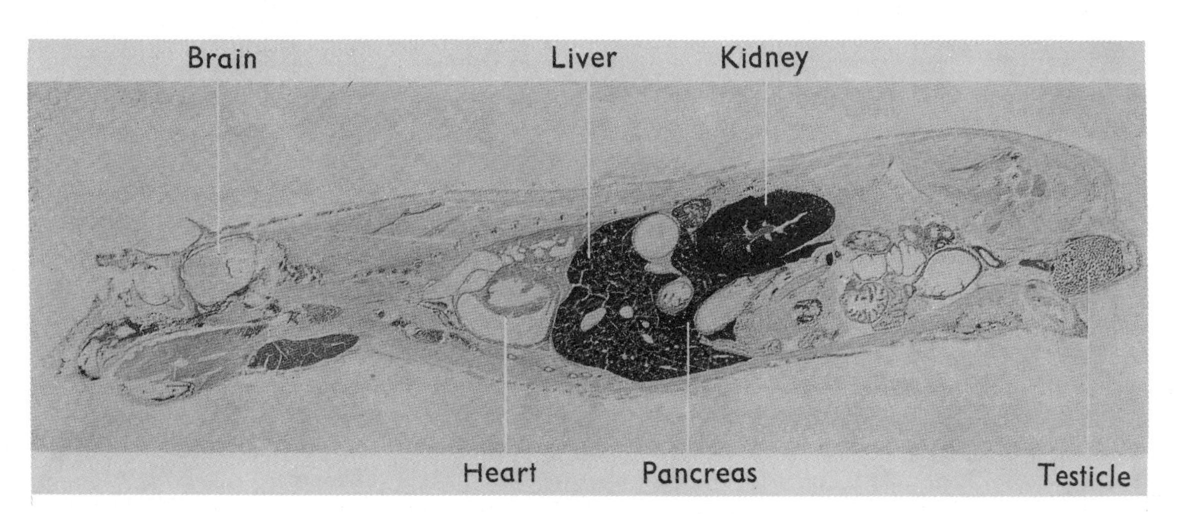

FIGURE 4:8 Autoradiographic distribution of ^{109}Cd in a mouse 112 days after a single intravenous injection. (From Nordberg and Nishiyama, 1972.)

sure period and a lower exposure intensity for the horses studied by Piscator.

The distribution among various tissues after long-term oral exposure has been studied by Stowe, Wilson, and Goyer, 1972. Rabbits were given 160 μg/g as the chloride in drinking water for 6 months. The mean exposure was 15.5 mg Cd/kg body weight per day. Mean cadmium concentrations in liver, kidney, pancreas, and spleen were 188, 170, 29, and 10 μg/g wet weight, respectively. Histopathological examination revealed slight renal tubular damage. At this high exposure level, the liver/kidney ratio was higher. This is in accord with observations from long-term injection experiments to be reviewed next.

TABLE 4:3

Ratios between Cadmium Concentrations in Liver and Kidneys in Rats and Dogs in Relation to Exposure Level

	Ratio liver Cd/renal Cd		
	Rats		Dogs
μg Cd/g H_2O	6 months	12 months	4 years
0.1	–	0.14	–
0.5	0.20	0.19	0.17
2.5	0.38	0.24	0.19
5.0	0.43	0.31	0.24
10.0	0.67	0.47	0.30

Data from Decker et al., 1958, and Anwar et al., 1961.

Injection route – Friberg, 1952, gave subcutaneous injections of ^{115}Cd-sulfate (0.65 mg Cd/kg) to rabbits 6 days a week for 4 and 10 weeks, respectively. Four rabbits exposed for 4 weeks had mean levels of 1,160, 600, 75, and 45 μg/g Cd (dry weight) in liver, kidney, pancreas, and spleen, respectively. After 10 weeks the corresponding figures were 1,480, 1,000, 193, and 180 μg/g. During the period of 4 to 10 weeks of exposure the accumulation rate of liver and kidney decreased, that of spleen increased, and that of the pancreas showed no change. Since cadmium will be excreted during this period (see Section 4.2.4), these results are partly explained. Friberg, 1955, performed a new experiment with similar exposure for 10 weeks, but he gave one group nonradioactive cadmium for the last 20 days. After 10 weeks liver levels of ^{115}Cd were considerably higher than kidney levels. In animals to which nonradioactive cadmium had been given for the last 20 days of the exposure time, the level of radioactive cadmium in the kidneys was only about half of the level in animals given the isotope during the entire period. This indicates that during those 20 days the kidneys had excreted part of the stored cadmium.

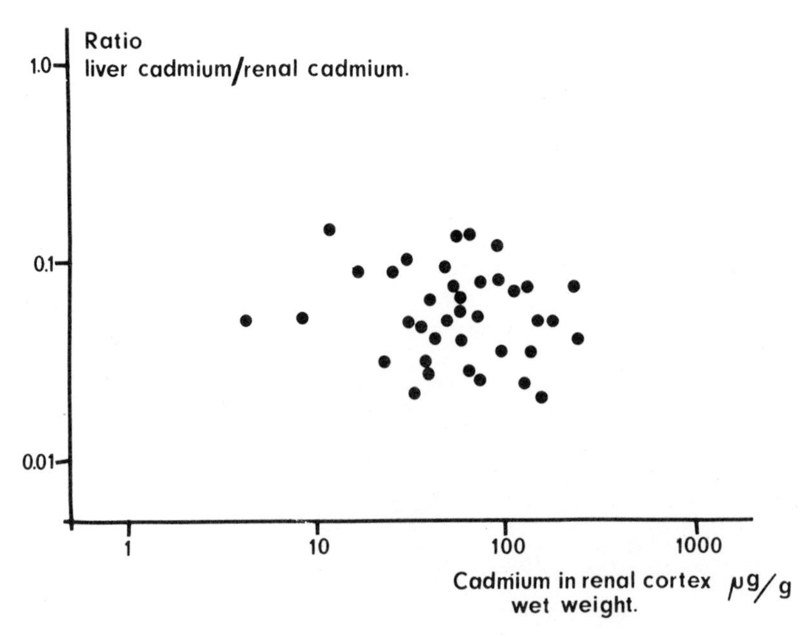

FIGURE 4:9. Ratio between liver and renal cortex concentrations of cadmium in relation to cadmium concentrations in renal cortex in horse. (From Piscator, 1971.)

Axelsson and Piscator, 1966a, gave rabbits cadmium chloride (0.25 mg Cd/kg) by subcutaneous injections 5 days a week for 11 to 29 weeks. There was an increase in liver and renal cortex levels up to 17 weeks. The levels at that time were about 450 and 400 μg/g wet weight, respectively. Further exposure did not increase the concentration in the liver. In the renal cortex there was a decrease so that the concentration was about 275 μg/g wet weight after 23 and 29 weeks. As excretion of cadmium was low until after 17 weeks but then increased considerably (see Section 4.2.4.1), these findings agree with Friberg's results showing first an accumulation stage, followed by an excretion stage. Piscator and Axelsson, 1970, found in a follow-up of a group exposed for 24 weeks under similar conditions (0.25 mg Cd/kg, subcutaneous injection, 5 days a week) and killed 7 months after the last injection that kidney levels had only decreased a little compared with the above mentioned concentrations at 23 or 29 weeks of exposure. Liver levels were about half of the levels at the time exposure ceased.

Bonnell, Ross, and King, 1960, gave intraperitoneal injections of cadmium nitrate to rats (0.75 mg Cd/kg 3 days/week for 5 to 6 months). Exposure was then discontinued for 2 months, after which injections were resumed for some animals (0.25 mg/kg). Other animals received no further exposure. The total time for the experiment was 12 months. A linear increase of cadmium in liver was seen during the first 3 months when a plateau was reached at about 350 μg/g wet weight (see Figure 4:10). It is conceivable that at this time excretion of cadmium began. Urinary cadmium, however, was not determined. The resumed exposure did not increase the levels; a gradual decrease was actually seen. In the kidneys linear increase was also seen during the first 3 months when a plateau was reached at 200 to 300 μg/g wet weight (Figure 4:11). At the end of the experiment the levels were lower than at 3 months. No difference in cadmium levels between animals who received further injections during the last 5 months and those animals who did not could be detected.

The metabolic changes after repeated exposure are further elucidated by Nordberg (unpublished data). The experimental conditions have been described in detail elsewhere (Nordberg and Piscator, 1972). In Table 4:4 are shown organ concentrations of cadmium in groups of mice

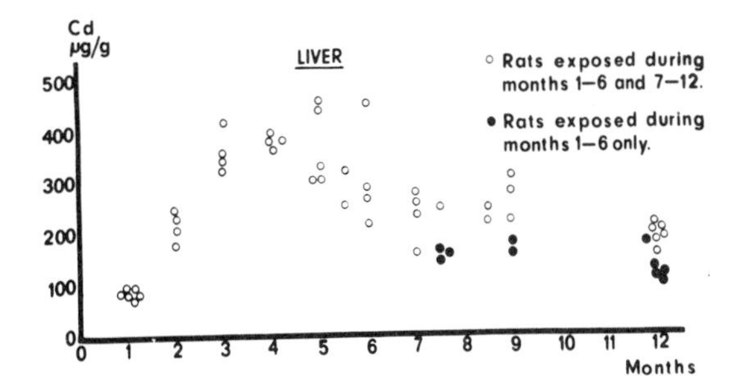

FIGURE 4:10. Cadmium concentrations in liver of rats exposed to cadmium chloride. Intraperitoneal injections of 0.75 mg Cd/kg three times a week, months 1 to 5 or 6; 0.25 mg Cd/kg three times a week, months 7 to 12. (Modified from a figure by Bonnell, Ross, and King, 1960.)

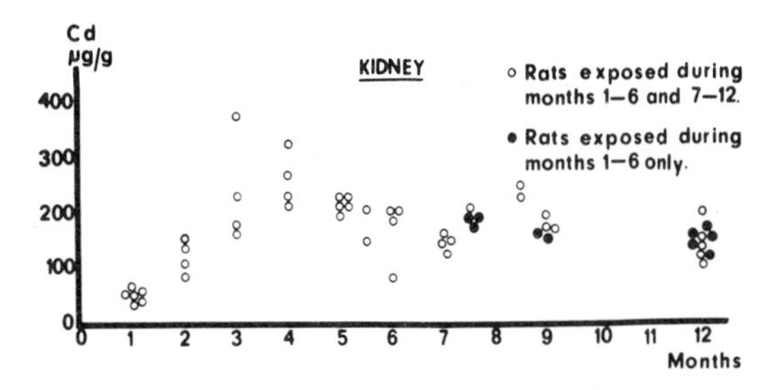

FIGURE 4:11. Cadmium concentrations in kidney of rats exposed to cadmium chloride. Intraperitoneal injections: 0.75 mg Cd/kg, three times a week, months 1 to 5 or 6; 0.25 mg Cd/kg three times a week, months 7 to 12. (Modified from a figure by Bonnell, Ross, and King, 1960.)

exposed to 0.25 mg and 0.5 mg Cd/kg body weight subcutaneously for 6 months. In the group with highest exposure, signs of renal tubular dysfunction, as indicated by tubular proteinuria, and a sharp increase in the excretion of cadmium had become evident 3 weeks before the end of the experiment. It will be seen that in the spleen the cadmium concentration is twice as high in the high exposure group as in the group with lower exposure. This was also the case with other organs such as intestines, testis, eye, etc. Pancreas concentrations are about 50% higher in the high exposure group. In contrast, liver and renal levels of cadmium are about the same in both groups.

The results by Nordberg shown in Table 4:4 also confirm the previously mentioned results by Friberg, 1952, and Bonnell, Ross, and King, 1960, who found that during continuous exposure an accumulation of cadmium in pancreas continued after accumulation in liver and kidneys had

TABLE 4:4

Cadmium Concentrations (μg/g Wet Weight) in Liver, Kidney, Pancreas, and Spleen in Mice Exposed to ^{109}Cd for 6 Months

		Liver		Kidney		Pancreas		Spleen	
Dose	n	Mean	Range	Mean	Range	Mean	Range	Mean	Range
s.c. 0.25 mg Cd/kg body weight, 5 days a week	4	281	271–288	162	141–172	71	67–74	10.1	9.5–10.6
s.c. 0.5 mg Cd/kg body weight, 5 days a week	4	258	190–288	172	158–194	114	111–121	19.5	18.4–21.4

From Nordberg, unpublished data.

stopped. These differences in accumulation of cadmium in liver, kidneys, and pancreas will be of importance when discussing the findings in human beings excessively exposed to cadmium.

With regard to distribution in other organs, it has already been mentioned that only traces of cadmium were found in brain of adult mice after a single injection (Berlin and Ullberg, 1963, Nordberg and Nishiyama, 1972). Some data by Lucis, Lucis, and Shaikh, 1972, may indicate that cadmium will more easily penetrate the blood-brain barrier in rats in the fetal stage. As it was not possible to quantitate the uptake in the brain, more data are needed. In the same work newborns exposed to cadmium via milk from the mothers showed a rapid build-up of cadmium concentrations in the intestines, but only a very slight increase in the liver. It is not clear whether the high concentration in the intestines was due to a local uptake in the intestinal walls or to a slow passage of the intestinal contents. Lucis, Lucis, and Shaikh, 1972, also found that cadmium accumulated in the mammary glands of pregnant rats but only small amounts were excreted via milk.

4.2.3 Distribution within Organs

4.2.3.1 Single Exposure

Friberg and Odeblad, 1957, gave a single subcutaneous injection of ^{115}Cd-sulfate (1.3 mg Cd/kg) to rats and killed the animals after 5.5 and 24 hr, respectively. The cadmium distribution in kidney, liver, and pancreas was studied by an autoradiographical method after fixation in ethanol. The renal cortex accumulated more cadmium than the medulla. In the pancreas, distribution was equal between exocrine and endocrine parts. Gunn and Gould, 1957, gave a single intracardiac injection of ^{115}Cd to rats and found a selective accumulation of cadmium in the renal cortex by means of radioactivity measurements of small tissue parts. They followed the animals for 150 days. At that time the cortex contained four times more cadmium than the medulla.

Berlin and Ullberg, 1963, gave a single intravenous tracer dose of ^{109}Cd-chloride to mice and studied the distribution by whole body autoradiography. The animals were killed at times after injection varying from 5 min to 16 days and were thereafter frozen to allow for whole body sectioning. In the renal cortex the authors found scattered areas with high activity. In the liver the distribution was uniform during the first 24 hr, but after 8 days a higher concentration was found in the periphery of the liver lobules. The whole body autoradiograms prepared by Berlin and Ullberg also revealed a nonuniform distribution in several other organs, which later was shown to be constant over a considerable time period (Nordberg and Nishiyama, 1972); see Figure 4:8. In the testes especially the interstitial tissue contained cadmium.

A special study on the distribution in the kidneys was made by Berlin, Hammarström, and Maunsbach, 1964. They used freeze-drying and dry mounting of the specimens in order to prevent as much as possible the dislocation of the cadmium isotope in the tissues. The radiation properties of ^{109}Cd which they used also allowed a better autoradiographical resolution than that in earlier studies by Friberg and Odeblad, 1957. Mice were killed 24 hours after a single intravenous tracer dose of ^{109}Cd-chloride. In the renal cortex the largest accumulation was found in the outer cortex and corresponded to proximal tubules. Especially the first segment of the proximal tubules showed high activity. The glomeruli did not retain cadmium to the same degree.

The subcellular distribution in the liver has been studied by Kapoor, Agarwala, and Kar, 1961. Rats were given a single subcutaneous injection of cadmium chloride (10 mg/kg), and the livers were fractionated by ultracentrifugation from 6 to 168 hours thereafter. The total concentration of cadmium was the same at the different times. However, there was a change in distribution with time, so that at 6 and 12 hours after injection about 30% was in the supernatant, whereas at 24, 48, and 168 hours, about 60% was in that fraction. This corresponded to a decrease in the nuclear, mitochondrial, and microsomal fractions.

Twenty-four hours after a single subcutaneous injection of ^{109}Cd-chloride to mice and to rats, cadmium was distributed so that 80% was in the liver supernatant (Shaikh and Lucis, 1969, 1970, and 1972b). About 9% was found in mitochondria and microsomes (Shaikh and Lucis, 1972b). Only very small amounts were found in the nuclei. Webb, 1972a, made similar studies on rat kidney cortex at different times after subcutaneous injections of cadmium chloride, 2.4 mg/kg body weight. With time, the cadmium content of the soluble fraction increased, whereas cadmium content of lysosomes and mitochrondria decreased. After 15 days cadmium was practically absent from these organelles. There was some increase with time in the microsomes, paralleling the increase in the supernatant. Nordberg, Piscator and Lind, 1971, found that a change in distribution took place with time after a single injection of cadmium chloride (3 mg Cd/kg) in mice. During the first 24 hr almost more than 80% of the cadmium in the supernatant was in the high molecular weight proteins, as shown by gel filtration, whereas later a change occurred so that more than 50% of the cadmium was in a low molecular weight fraction, probably metallothionein. Shaikh and Lucis, 1970, had found a similar low molecular weight fraction in their experiment.

Lucis, Lucis, and Shaikh, 1972, separated mammary gland soluble proteins from cadmium-exposed pregnant rats, and found that cadmium was bound to high molecular weight protein and was not found in low molecular weight protein fractions. Placenta, however, contained cadmium both in high molecular weight proteins and in a protein fraction with the same molecular weight as metallothionein.

Nordberg, 1971a, studied the distribution of cadmium among cellular organelles, free cytoplasma, and different cytoplasmic proteins in testicles of mice. A homogenate of testicles was ultracentrifuged at 100,000 $\times$ g and the supernatant was separated by G-75 Sephadex gel chromatography. A change with time after single subcutaneous injection of ^{109}Cd-tagged cadmium chloride (1 mg Cd/kg body weight) was observed. Four hours after injection 50% of testicular cadmium was in the supernatant and only 26% of the cadmium in the supernatant was in a small protein, corresponding to the size of metallothionein. The corresponding figures, 4 days after injection, were 74% in the supernatant and 55% in the small protein.

4.2.3.2 Repeated Exposure

Friberg, 1952, gave rabbits subcutaneous injections of ^{115}Cd sulfate 6 days a week for 10 weeks (0.65 mg Cd/kg) and found by autoradiography that in the kidneys the cortex contained five times more cadmium than the medulla. In the liver more activity was in the periphery of the lobules than in the inner halves. In the pancreas the glandular part accumulated cadmium, while no cadmium was found in the connective tissue. Friberg did not note any difference between the exocrine and the endocrine parts. Friberg's use of fixation in ethanol has been subject to some discussion by Berlin, Hammarström, and Maunsbach, 1964, who stated that the technique did not guarantee against the dissolution of cadmium from the specimens. However, the direct measurements on pieces of tissue reported by Axelsson and Piscator, 1966a (see below), show that it is improbable that such dissolution would have been of importance for the distribution between cortex and medulla in Friberg's studies.

Long-term experiments on the rabbit by Axelsson and Piscator, 1966a, revealed that during exposure to subcutaneous injection of 0.25 mg cadmium/kg 5 days a week for 11 to 29 weeks, the ratio between cortex and medulla was 5.1 (2.5 to 9). In rabbits exposed in a similar way for 24 weeks and then followed for another 30 weeks, the ratio was 3.6 (2.4 to 5.5) (Piscator and Axelsson, 1970).

By using a staining method for metals, Axelsson, Dahlgren, and Piscator, 1968, found the main metal accumulation in renal proximal tubules in the above mentioned rabbits. Considerably smaller amounts of deposited metals were found in the glomeruli, collecting tubules, and medulla.

Nordberg (unpublished data) studied the relation between cadmium in kidney cortex and whole kidney in mice given ^{109}Cd subcutaneously 5 days a week for 3 to 6 months. In six mice given 0.5 mg Cd/day the ratio cortex/whole kidney varied between 1.1 and 1.3. In five mice given 0.25 mg Cd/day the ratio varied between 1.0 to 1.48. A mean ratio in eight mice exposed for 6 months to 0.025 mg Cd/day was 1.43 (range: 1.1 to 1.7). The ratios found in mice agree fairly well with ratios found in human beings.

Nordberg, 1972a, demonstrated by autoradiography of mouse testicles that cadmium was located in the interstitial capillaries and their immediately adjacent structures. This distribution pattern was found regardless of whether single or repeated long-term exposure had taken place.

4.2.4 Excretion

4.2.4.1 Urinary Excretion

4.2.4.1.1 Single Exposure

After a single oral dose of ^{109}Cd to goats, Miller, Blackmon, and Martin, 1968, found that on days 1, 2, and 3 to 7, after exposure, 0.5, 1, and 0.1%, respectively, of the absorbed amount was excreted via the urine. The original values have been recalculated to fit an assumed absorption of 2%. A dose of the same amount of cadmium as the oral dose in the previously mentioned experiment was administered by intravenous injection. Excretion values were 0.025, 0.004, and 0.002%. The authors pointed out that these values could be too high as fecal contamination could increase the cadmium levels. A similar recalculation based on data from Moore, Stara, and Crocker, 1973, gives the result that in rats 1.25 and 0.65% of the absorbed dose was excreted on days 1 and 6, respectively. In rats, Lucis, Lynk, and Lucis, 1969, found that 1% of a single subcutaneous injection of ^{109}Cd solution had been excreted with the urine after 1 week. In rats given a single subcutaneous injection of cadmium, Shaikh and Lucis, 1972a, found a daily urinary excretion during 4 days of between only 0.003 to 0.007% of the dose. Burch and Walsh, 1959, calculated the half-time after a single intravenous injection of ^{115}Cd as nitrate (5 dogs, followed for 20 to 30 days). It can be calculated from their data that it would take 3 to 7 years to eliminate 50% of the cadmium via the urine alone. The excretion of cadmium compounds in the rat has been studied by Furst, Cadden, and Firpo, 1972. Cadmium in urine was analyzed by atomic absorption spectrophotometry without any special pretreatment for eliminating interferences. The authors calculated that more than 50% of the dose was excreted in a few days. This must be an error, probably due to influence from sodium chloride in the urine, as has been discussed in Chapter 2. No conclusions can be drawn from the studies with regard to excretion.

Murata, 1971, gave oral doses of 10 to 20 μCi ^{109}Cd as chloride (carrier free) to mice. During the first day about 25% and in 3 weeks about 50% of the total dose were excreted with the urine. The total amount of cadmium in feces during 3 weeks was only about 25% of the dose. The data do not fit at all with earlier experience, since they indicate an extremely high absorption.

4.2.4.1.2 Repeated Exposure

Friberg, 1952, gave subcutaneous injections of cadmium sulfate containing ^{115}Cd in daily doses of 0.65 mg Cd/kg to rabbits. When the excretion of the isotope was followed for 10 weeks, as shown in Figure 4:12, it was found that for a period varying between 6 and 7 weeks, the daily cadmium excretion was very small, less than 1% of the daily injected dose. During the last weeks of the experiment there was a sharp rise in cadmium excretion, up to about 100 times the amount excreted during the first weeks. This rise in cadmium excretion coincided with the appearance of proteinuria in the rabbits.

Further investigations were made by Friberg, 1955, under similar experimental conditions as above. One group of rabbits was given first radioactive cadmium for 50 days and nonradioactive cadmium for the remaining 20 to 50 days. Another group was given radioactive cadmium for 70 days, whereafter cadmium excretion was followed for another 30 days. During the administration of nonradioactive cadmium, the excretion of radioactive cadmium increased markedly. In the group that received no more cadmium whatsoever after the end of the exposure to radioactive cadmium, the concentration of radioactive cadmium in urine decreased rapidly. It was thus

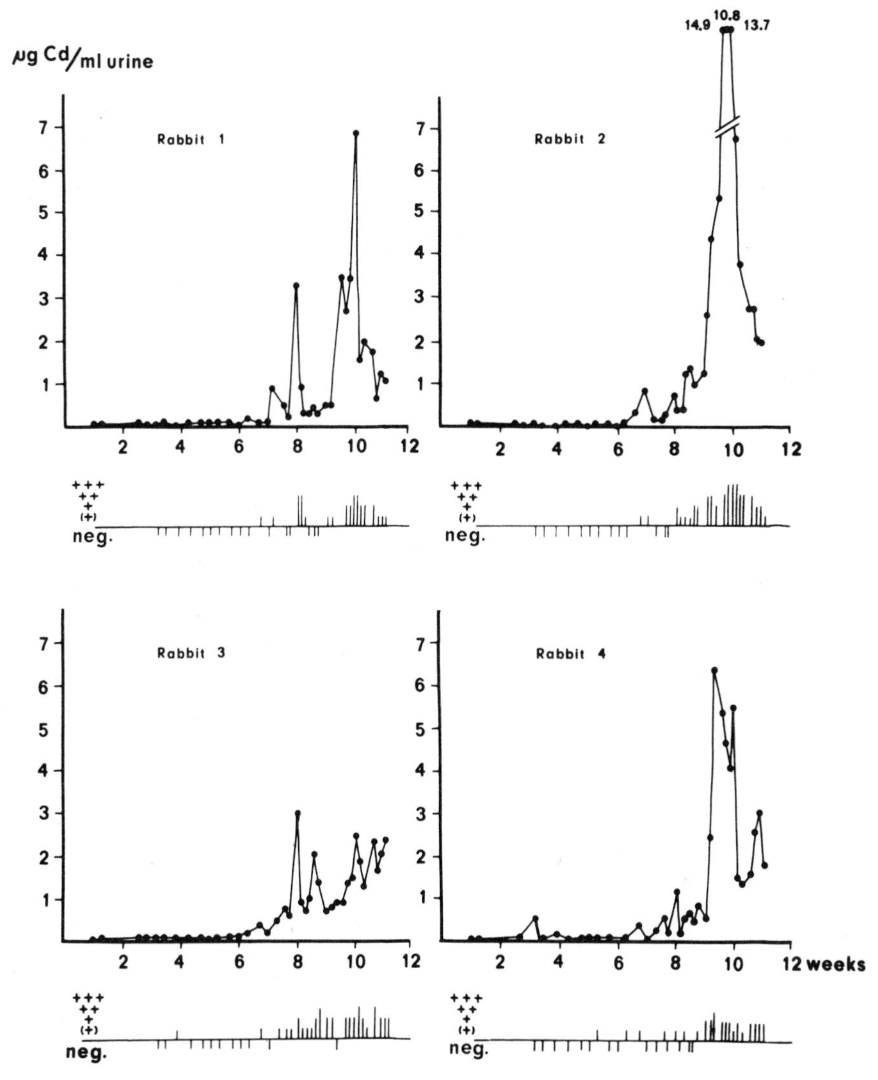

FIGURE 4:12. Cadmium concentrations and occurrence of protein in urine of cadmium-exposed rabbits. (+) = proteinuria demonstrated only with trichloroacetic acid; + – +++ = proteinuria demonstrated also with nitric acid. (Modified from a figure by Friberg, 1952.)

clearly demonstrated that at the time kidney damage existed, the urinary cadmium excreted during exposure came partly from accumulated deposits.

Axelsson and Piscator, 1966a, gave rabbits 0.25 mg Cd/kg as cadmium chloride for 5 days a week for 6 months by subcutaneous injection. During the first 4 months there was an insignificant excretion of cadmium, as determined by polarographic analysis, but in the last 2 months there was such a sudden rise in cadmium excretion that in some animals the daily excretion exceeded the daily dose. The increase in urinary excretion of cadmium paralleled an increase in urinary protein. Six months after a six-month exposure, two rabbits in a group of six still excreted large amounts of cadmium. These two rabbits also had hemolytic anemia, indicating that release of cadmium from red cells might have been partly responsible for the high excretion values (Piscator and Axelsson, 1970).

Nordberg and Piscator, 1972, and Nordberg, 1972a, found that during repeated subcutaneous exposures to 0.025 to 0.5 mg cadmium/kg body weight for up to 6 months the urinary excretion in mice was very low. At the time that pathological

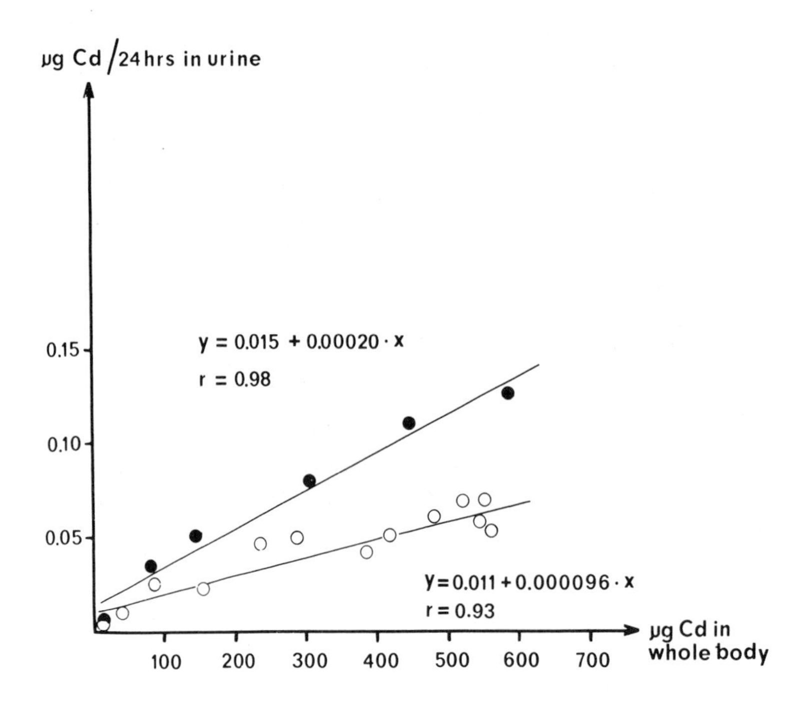

● Mean 24 h excretion in mice given 0.5 mg Cd/kg daily 5 days per week

○ Mean 24 h excretion in mice given 0.25 mg Cd/kg daily 5 days per week

FIGURE 4:13. Relation between urinary excretion and whole body content of cadmium in mice given daily injections of ^{109}Cd (0.25 or 0.5 mg Cd/kg) and followed by whole body and excretion measurements. (From Nordberg, G. F., *Environ. Physiol. Biochem.*, 2, 7, 1972. With permission.)

urinary proteins were observed at electrophoresis (after 21 weeks) in the group given the highest exposure, there was a sharp increase in urinary cadmium excretion. It was shown by Nordberg and Piscator, 1972, by means of gel filtration, that a large part of the cadmium in urine during this phase of prominent excretion was bound to proteins with a molecular weight corresponding to that of metallothionein.

Nordberg, 1972a, 1972b, in the studies referred to above, also showed that the urinary excretion on a group basis was significantly related to total body burden (Figure 4:13). At the highest dose level 0.02% was excreted daily, whereas at the lower dose levels about 0.01% was excreted. However, as seen in Figure 4:14, there is a considerable individual scatter.

In summary, there is evidence that during long-term exposure the daily excretion of cadmium with the urine is very low. Studies in mice have shown that excretion prior to tubular dysfunction constitutes an average of 0.01 to 0.02% (depending on exposure levels) of the total body burden. A wide individual scatter has been shown to exist. Once tubular dysfunction has occurred, the urinary excretion of cadmium increases markedly. In experiments on rabbits, this increase was 50- to 100-fold.

4.2.4.2 Excretion Via the Alimentary Tract

4.2.4.2.1 Single Exposure

Animal experiments have shown that injected cadmium will be partially excreted with the feces. Decker, Byerrum, and Hoppert, 1957, gave the rat a single intravenous dose of ^{115}Cd (0.63 mg/kg). After the first 24 hr, 7.3% was found in the feces.

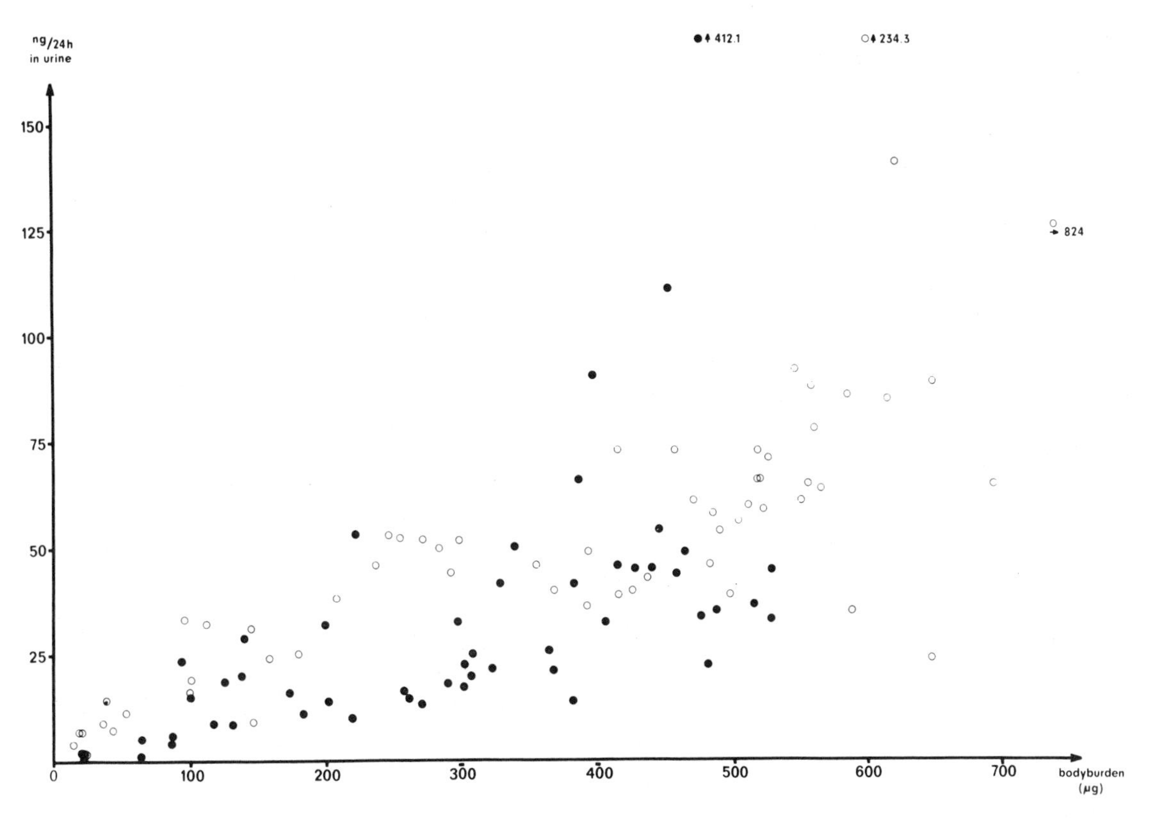

FIGURE 4:14. Individual values on urinary excretion of cadmium related to body burden. All mice given subcutaneous injections of 0.25 mg Cd/kg body weight per day, 5 days per week. • = one group; o = another group, treated in the same way. (From Nordberg, G. F., Urinary, Blood, and Fecal Cadmium Concentrations as Indices of Exposure and Accumulation, XVII Int. Cong. Occup. Health, proceedings to be published 1974.)

After 72 hr 18.5% of the given dose had been excreted via feces. Shaikh and Lucis, 1972a, found that about 5% of a single subcutaneous dose of 2.2 μg ^{109}Cd as chloride was excreted via feces of rats during the first 4 days. Miller, Blackmon, and Martin, 1968, gave a single intravenous dose of ^{109}Cd as chloride to goats and found that 5.6% was excreted via feces within 5 days.

Burch and Walsh, 1956, injected a single dose of ^{115}Cd (0.32 to 0.40 mg/kg) intravenously into dogs. They determined the amount of the isotope in feces from five dogs for 20 to 30 days. They calculated that 419 to 659 days were required for elimination of half of the injected cadmium by fecal excretion alone.

Berlin and Ullberg, 1963, using whole body autoradiography after a tracer dose of ^{109}Cd intravenously injected into mice, found that cadmium rapidly accumulated in the mucous membrane of the intestinal tract and that after 24 hr the cadmium in the mucosa seemed to decrease. Cadmium accumulated mainly in the secretory part of the gastric mucosa and in the colonic mucosa. The investigators could also find radioactivity in the contents of the stomach and colon 20 min after injection. When Lucis, Lynk, and Lucis, 1969, gave a tracer dose of ^{109}Cd by a single subcutaneous injection into rats, they found that cadmium accumulated rapidly in the wall of the stomach and detected only traces in the contents of the stomach. In the intestines the wall of the small intestine had the largest concentration. Radioactivity in the intestinal contents increased during the first 24 hr.

Johnson and Miller, 1970, compared the results of single subcutaneous and intraperitoneal injections of ^{109}Cd (0.02 mg/kg) as chloride given to rats, and with both techniques found cadmium in the duodenum after 2.5 min.

Biliary excretion of cadmium may be of importance (Caujolle, Oustrin, and Silve-Mamy, 1971). Rats were given intraperitoneal injections of ^{109}Cd as sulfate. The dose was 7 μCi, specific activity not stated. A relatively constant excretion of cadmium via bile was found up to 6 hr after injection. When the bile duct of one rat was connected to the small intestine of another rat, cadmium excreted via bile was reabsorbed to a certain extent, indicating an enterohepatic circulation. However, the investigators did not present their data in a way that makes a quantitative evaluation possible. A similar study covering a longer time interval has been performed by Nordberg and Robért (unpublished data). They injected $^{109}CdCl_2$ in a dose corresponding to 0.5 mg Cd/kg body weight s.c. in four rats and recovered 0.005% of it per hour in bile during 3 consecutive days. During the first few hours of the experiment some rats excreted up to 0.15% per hour of the dose in the bile. In all instances, however, the total daily biliary excretion constituted only a small fraction (on the average, 5%) of the fecal excretion measured in the same animals. It can be concluded that routes other than via the bile are mainly responsible for gastrointestinal excretion of cadmium, at least under the described experimental conditions in rats.

4.2.4.2.2 Repeated Exposure

Ceresa, 1945, determined the daily excretion of cadmium in five rabbits. The animals received about 5.5 mg Cd/kg daily as the sulfate by the subcutaneous route, which resulted in the deaths of the animals after 7 to 9 days. During the period of exposure, the mean fecal excretion rate was 1.8% of the injected amount and slightly higher than the urinary excretion.

Long-term experiments have been performed by Axelsson and Piscator, 1966a, on rabbits given 0.25 mg Cd/kg as chloride 5 days a week for 29 weeks by subcutaneous injections. Fecal excretion was slightly higher than urinary excretion after 11 weeks but corresponded to only about 1.6% of the daily dose. After 17 weeks the daily fecal excretion corresponded to 2.8% of the daily dose. After 23 and 29 weeks, when the urinary excretion of cadmium was greater, 11.2 and 6.6%, respectively, were found in the feces.

Nordberg, 1972b, found a certain correlation between total body burden and fecal excretion of cadmium. In Figure 4:15 (where individual data

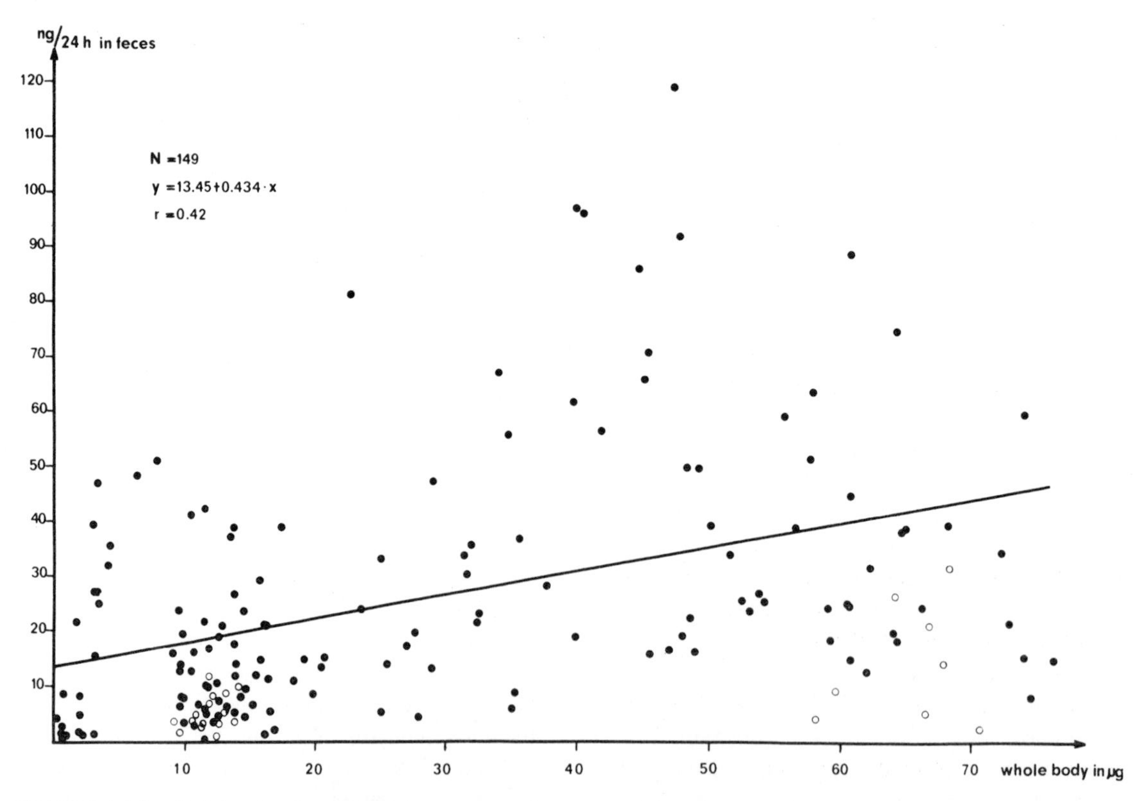

FIGURE 4:15. Individual values on fecal excretion of cadmium related to body burden in mice given subcutaneous injections of 0.025 mg Cd/kg body weight per day, 5 days per week. • = during or immediately after exposure; o = 3 weeks after exposure. The line refers to filled circles. (From Nordberg, G. F., Urinary, Blood, and Fecal Cadmium Concentrations as Indices of Exposure and Accumulation, XVII Int. Cong. Occup. Health, proceedings to be published 1974.)

are given), the results are seen from a group of mice exposed to 0.025 mg Cd/kg body weight 5 days per week for about 5 months. It is seen from the figure that a considerable individual scatter exists. If a linear regression analysis is carried out, a significant correlation is found ($r = 0.42$, $n = 149$). The regression coefficient is 0.39 and differs significantly ($p < 0.001$) from zero. This indicates that some part of the excretion is dependent on body burden of cadmium. On the other hand, in another experiment in which the daily dose was ten times higher (Figure 4:16), it was shown that the fecal excretion at a certain body burden also was about ten times higher. This means that at the dose levels studied the main part of the excretion must be related to the daily dose. Upon two occasions during the experiment with the lower doses, exposure was withheld for 3 weeks. Observations made after the exposure-free weeks are shown in Figure 4:15 (unfilled circles). These values consistently fall below the regression line, further supporting a relation between fecal excretion and daily exposure. The fact that the average of these fecal excretion values was higher at the end of the experiment (14.2 μg/24 hr S.D. = 11.0) than in the beginning (5.1 μg/24 hr S.D. = 3.1) also is consistent with part of the fecal excretion's having a relation to body burden.

Thus, there is evidence that injected cadmium is excreted via some parts of the alimentary tract. On the other hand, the daily excretion is low and mainly exposure related. Less than 5% of the daily absorbed dose is excreted this way, as judged from long-term experiments on mice and rabbits. That both the gastric mucosa and intestinal mucosa excrete cadmium is clear, but, as the parotic gland was shown by Berlin and Ullberg, 1963, to accumulate relatively large amounts and the pancreas has been shown in several of the reports to store large amounts of cadmium, the possibility of excretion from these two organs should be remem-

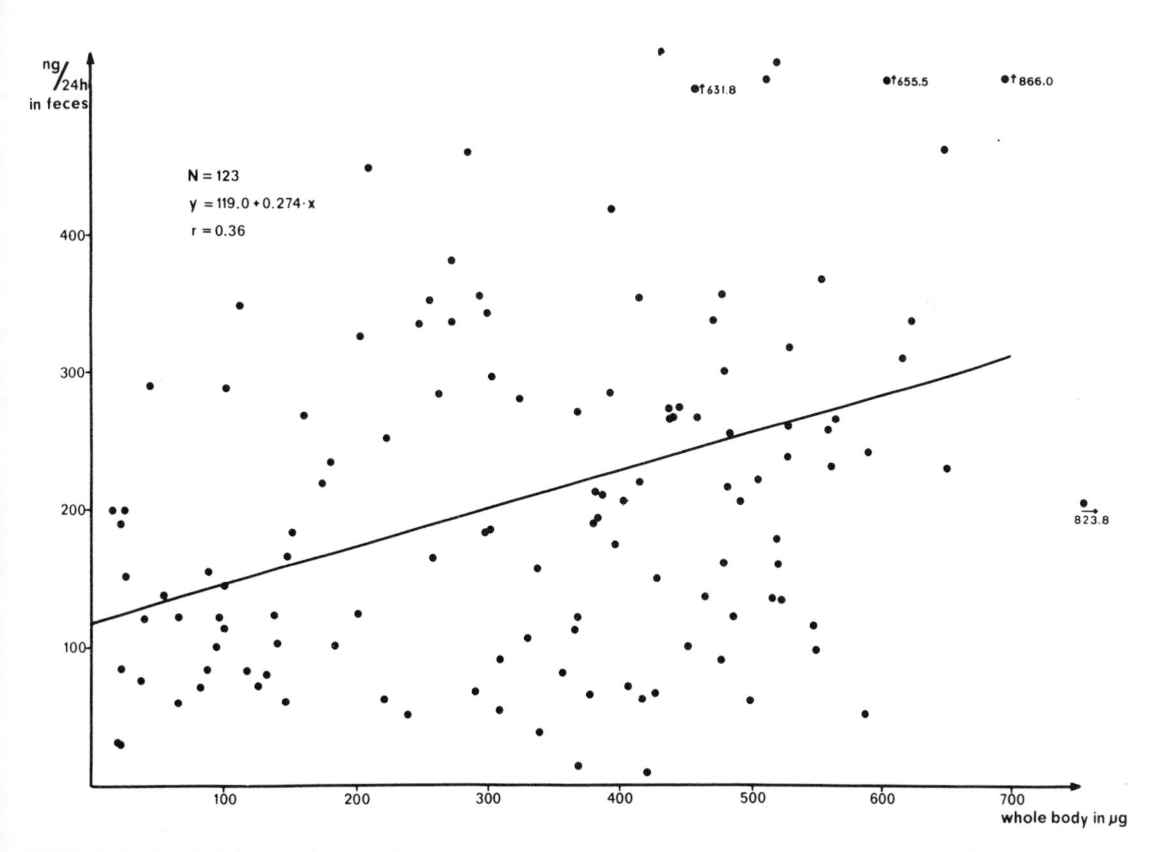

FIGURE 4:16. Individual values on fecal excretion of cadmium related to body burden in mice given subcutaneous injections of 0.25 mg Cd/kg body weight per day, 5 days per week. (From Nordberg, G. F., Urinary, Blood, and Fecal Cadmium Concentrations as Indices of Exposure and Accumulation, XVII Int. Cong. Occup. Health, proceedings to be published 1974.)

bered. The role of the bile also has to be taken into account, as discussed above.

4.2.4.3 Other Excretion Routes

Truhaut and Boudene, 1954, found high concentrations of cadmium in hair from rats and rabbits injected with cadmium and suggested that hair analysis could be of value for determining cadmium levels in exposed workers. Berlin and Ullberg, 1963, found that cadmium accumulated in hair of mice given a single dose of ^{109}Cd. Miller, Blackmon, and Martin, 1968, found that the concentration in goat hair was about 20% of the concentration in the liver 15 days after a single oral dose of ^{109}Cd as chloride. After an intravenous dose the corresponding figure was about 1%. When this experiment was repeated by Miller et al., 1969, the concentrations in hair after oral exposure were only 3% of the concentrations in the liver. Nordberg and Nishiyama, 1972, found that in mice given a single intravenous injection of ^{109}Cd-chloride (4.5 μg Cd/kg), the decrease in cadmium levels in hair paralleled the decrease in whole body retention. Less than 0.5% of the total excretion between day 41 and 105 was excreted via the hair.

There are other excretion routes that may be worth comment. From the data of Lucis, Lucis, and Shaikh, 1972, it can be calculated that excretion was less than 0.05% per gram milk per day after a single subcutaneous postpartum injection to female rats. Tanaka et al., 1972, gave pregnant mice 50 μCi of ^{115m}Cd (15 μg Cd) in a single intravenous injection 24 to 36 hr before delivery. In the newborns only 0.09% of the dose was found, i.e., the amount excreted via the placenta. The cadmium concentration in milk was not measured. Sucklings born to female mice not given radioactive cadmium were suckled by the female mice that had been given radioactive cadmium before delivery. Whole body measurements of the sucklings showed that after 14 days 0.3% of the dose given the sucklers was found in the sucklings. As some cadmium must have been passed out via the feces of the sucklings, the total amount excreted via the sucklers' milk must have been more than 0.3%.

Cadmium may be excreted through saliva. Using multielement emission spectrometry, Dreizen et al., 1970, studied the concentration of cadmium in marmoset saliva after stimulation with pilocarpine and mecholyl. The average concentrations were 0.044 and 0.078 μg/g, respectively, with a maximum of 0.344 μg/g.

4.2.5 Biological Half-time

One of the more important questions with regard to cadmium metabolism is the biological half-time. Animal experiments have shown highly varying values for biological half-time of cadmium, depending on dose, administration, single or repeated exposure, length of observation period, etc. A general discussion of questions on evaluating the biological half-time can be found in a paper by Nordberg, 1972a, and in a report by the Task Group on Metal Accumulation, 1973.

After a single oral exposure to ^{109}Cd (intubation, compound not stated), Richmond, Findlay, and London, 1966, estimated the biological half-time in female mice to be about 200 days. After a rapid clearance during the first 5 days, representing the passage of nonabsorbed cadmium through the intestines, between 0.5 and 3% of the dose was retained. The retained amount then decreased slowly during an observation period of nearly 1 year.

A similar experiment was undertaken on squirrel monkeys (*Saimiri sciureus*) by Nordberg, Friberg, and Piscator, 1971. Four weeks after oral exposure (for details see Section 4.1.2.1.1) the retention in two animals given high enough amounts of cadmium for long-term studies of biological half-time was for both animals 2.7% of the dose given. The corresponding figures were 2.5 and 2.4%, respectively, 4½ months after dosing. These figures indicate that the half-time in the squirrel monkey after the first 4 weeks is long, more than 2 years. Whether there exist components with still longer half-times than that has not been possible to demonstrate as yet. They may well exist in view of the continuous redistribution that takes place within the body for several months after exposure (see, for example, Figure 4:7).

Cotzias, Borg, and Selleck, 1961, gave mice ^{109}Cd as chloride (carrier free, 17 μCi) by a single intraperitoneal injection and determined whole body retention during 20 days. After 5 days about 95% of the injected dose remained and after 20 days, about 90%. The slow decrease during the latter part of the observation period indicates a half-time of at least 100 days.

After single injections via the intraperitoneal or intravenous route, biological half-time was studied in female mice by Richmond, Findlay, and London, 1966. They gave ^{109}Cd (compound and dose not stated) and made repeated determinations of whole body retention for periods of more than 400 days. After an initial rapid clearance during the first day, they estimated half-times from about 40 days and increasing to 200 to 300 days during the latter part of the observation period. Other investigators (Eybl, Sýkora, and Mertl, 1970, Tomita, 1971, and Nordberg and Nishiyama, 1972) have studied whole body retention during 1 to 4 months after single intravenous or subcutaneous injections of cadmium. From their data it can be calculated that if the first rapid clearance is excluded, biological half-time in mice will vary between 25 to 100 days, generally confirming the findings of Cotzias, Borg, and Selleck, 1961, and Richmond, Findlay, and London, 1966, for that period of time.

In rats, Durbin, Scott, and Hamilton, 1957, estimated the biological half-time to be about 200 days after an intramuscular injection of ^{109}Cd (2 μCi per rat, compound not stated). They did not determine whole body retention but determined cadmium in organs after 1, 8, and 64 days and estimated also the excretion via feces and urine for 64 days. If an estimate is made using their data after 8 days compared with 64 days, thus eliminating the initial excretion phase, a biological half-time of about 300 days will be obtained.

Burch and Walsh, 1959, estimated that in dogs it would take 260 to 500 days to eliminate 50% of an intravenously injected dose of ^{115}Cd as nitrate. Also experiments on goats by Miller, Blackmon, and Martin, 1968, indicate a long biological half-time of cadmium.

In four male CBA mice, the elimination rate of a single s.c. dose of $^{109}CdCl_2$ (0.25 mg Cd/kg) has been measured by whole body measurements (Nordberg, 1972a) when repeated doses of non-radioactive Cd (0.25 mg Cd/kg 5 days/week) were given during the 25 weeks following the s.c. administration of radioactive cadmium. The decrease in the whole body value during that time (corrected for radioactive decay) was only about 10% (Figure 4:17B), which corresponds to a biological half-time of a couple of years. The method used for whole body measurements gave rise to some variation in the values so that it is not possible (based on whole body measurements alone) to judge whether there was an early excretion phase corresponding to a few percent of the dose during the first week. The total 24-hr combined urinary and fecal excretion was measured from time to time in these mice. A high diuresis was obtained by adding glucose to the drinking water. The values (expressed as excretion per week per mouse) are seen in Figure 4:17D. During the first week about 4% of the dose was excreted but during the subsequent weeks, only about 0.5% per week was excreted.

The reason that the biological half-time is so long compared with other studies on mice is not clear. However, Nordberg set up considerably different experimental conditions, basing the evaluation of the biological half-time upon the single injection of radioactive cadmium at the beginning of the experiment followed by continuous exposure to fairly high amounts of nonradioactive cadmium. It might be that the last mentioned cadmium exposure prolonged the half-time of the first dose of radioactive cadmium. The seeming lack of any exchange whatsoever between radioactive and nonradioactive cadmium is in sharp contrast to what has been shown by Friberg, 1955 (Sections 4.2.2.2 and 4.2.4.1.2). The explanation is probably that in the earlier studies on rabbits by Friberg, the exchange was observed when toxic effects were already manifest. In fact, Nordberg did find a tendency toward an increased excretion of radioactive cadmium in the very end of the experiment when it is not possible to rule out the occurrence of some toxic manifestations.

Other groups of mice were exposed continuously to radioactive cadmium ^{109}Cd, 5 days per week for 25 weeks, and the whole body retention and the excretion in urine and feces were determined at weekly intervals. One group of four mice received 0.25-mg/kg doses and another group of four mice 0.5-mg/kg doses. The results of the whole body measurements are seen in Figure 4:17A. In the group receiving 0.25 mg/kg a linear increase in the whole body retention throughout the experiment can be seen. In the group receiving 0.5 mg/kg, there is a deviation from the linearity at 15 to 19 weeks and onward. During this latter part of the study, these animals showed a changed pattern of urinary proteins and a sharply increased excretion of cadmium, most marked in urine. These features were not evident in the first group, which received 0.25 mg Cd/kg daily. The excretion values for the two groups can be seen in Figure 4:17C.

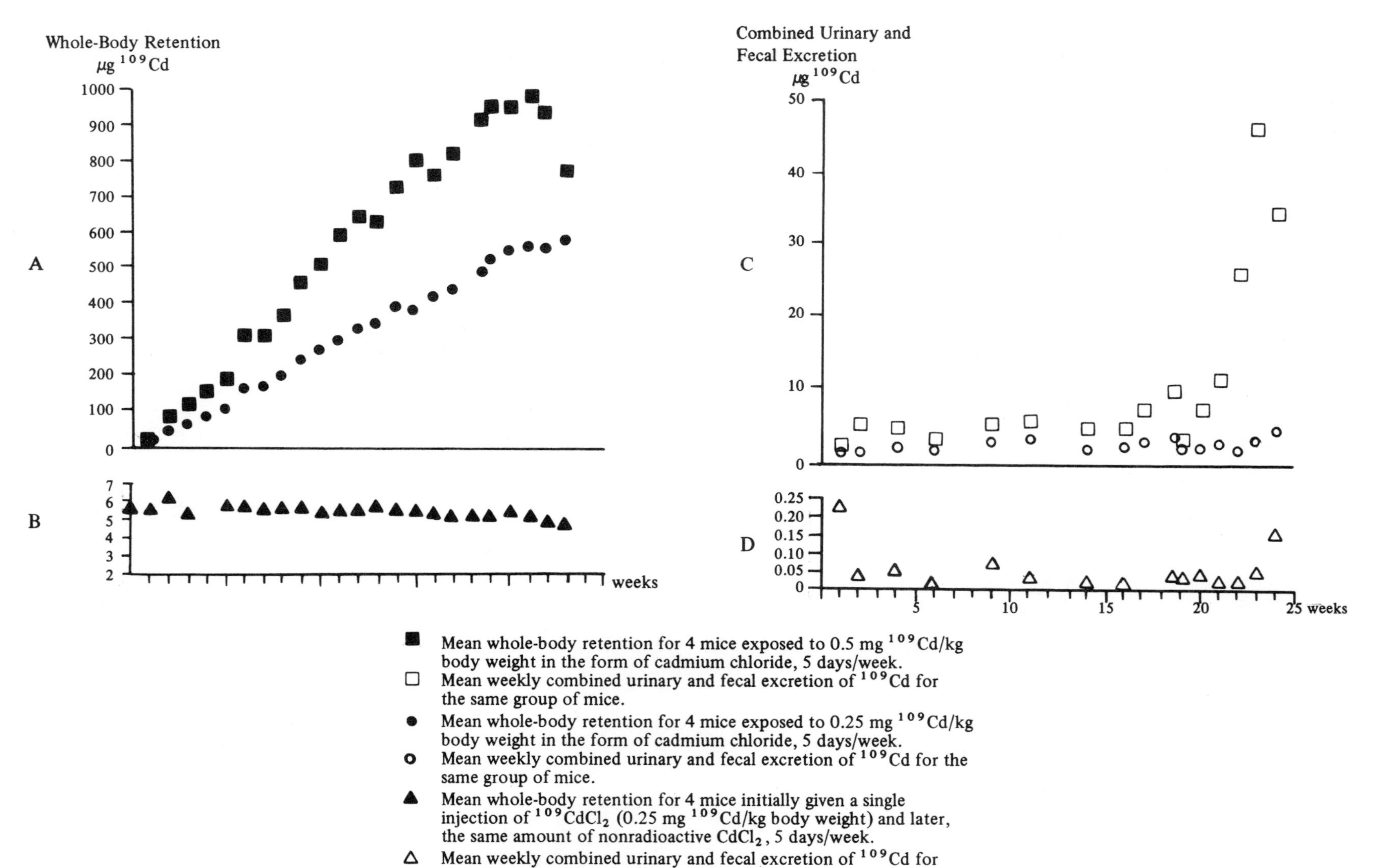
Whole-Body Retention
μg 109Cd
1000
900
800
700
600
500
400
300
200
100
0
A
7
6
5
4
3
2
B
weeks
Combined Urinary and
Fecal Excretion
μg 109Cd
50
40
30
20
10
0
C
0.25
0.20
0.15
0.10
0.05
0
D
5
10
15
20
25 weeks
Mean whole-body retention for 4 mice exposed to 0.5 mg 109Cd/kg body weight in the form of cadmium chloride, 5 days/week.
Mean weekly combined urinary and fecal excretion of 109Cd for the same group of mice.
Mean whole-body retention for 4 mice exposed to 0.25 mg 109Cd/kg body weight in the form of cadmium chloride, 5 days/week.
Mean weekly combined urinary and fecal excretion of 109Cd for the same group of mice.
Mean whole-body retention for 4 mice initially given a single injection of 109CdCl2 (0.25 mg 109Cd/kg body weight) and later, the same amount of nonradioactive CdCl2, 5 days/week.
Mean weekly combined urinary and fecal excretion of 109Cd for the same group of mice.

FIGURE 4:17 Whole-body retention (A and B) and combined urinary and fecal excretion (C and D) with time in animals exposed subcutaneously to $^{109}CdCl_2$. (From

There was no evident increase in total cadmium excretion prior to the appearance of some form of functional changes. Thus, total cadmium excretion is not directly proportional to the total body burden. In explanation, the total excretion is dominated by fecal excretion which is mainly dependent on daily dose (Section 4.2.4.2). This is also in accord with the above mentioned studies involving a single dose of radioactive cadmium (Figure 4:17D) of which a few percent were excreted during the first week and thereafter much smaller amounts. As mentioned above, the slow excretion phase corresponded to a biological half-time of a couple years. With such a long biological half-time, it is reasonable to expect an approximation of the accumulation curve to a straight line during the life span of mice (2 years).

Webb, 1972a, followed rats for up to 82 weeks after a single subcutaneous injection of about 2.5 mg Cd as chloride per kilogram body weight. Up to 21 weeks two animals were killed each week, and then two animals at 35 and 82 weeks, respectively. Maximum levels in liver were reached at about 10 weeks after exposure and slightly later in kidneys. Values at 35 and 82 weeks were only slightly lower than at 21 weeks. The interpretation of the results is difficult, as the number of animals at each time is only two. If the data are taken at face value, they indicate a long biological half-time in rat kidney. In the liver there was a slightly faster clearance.

Webb also studied cadmium concentrations in testis and epididymis. In the former organ there was an increase in cadmium levels up to at least 18 weeks. This may partly be explained by the necrosis and atrophy of the organ, which are induced by the injected dose (2.5 mg Cd/kg). In epididymis maximum levels were obtained after 5 weeks, the concentration had been halved after about 12 weeks, and at 82 weeks no cadmium could be demonstrated there. Also in this organ some pathological alterations could be expected from the dose injected.

The results by Webb, 1972a, given above for testis and epididymis differ from those reported by Gunn, Gould, and Anderson, 1968b, who also used rats but gave a noninjurious dose of ^{109}Cd. They found no time-related changes in concentration of cadmium in testis and epididymis from 1 hr up to 6 weeks after injection. The strikingly slow accumulation in all organs studied by Webb is difficult to account for; one reason may be a retention of the high dose at the injection site.

The above mentioned values for biological half-times of cadmium will only be valid before renal damage has occurred. If there is renal tubular dysfunction, the excretion of cadmium will increase, and this will change the biological half-time considerably.

4.2.6 Cadmium Metabolism in Relation to Zinc Metabolism

It was mentioned in Chapter 3 that cadmium will appear together with zinc in nature, and it could be expected that uptake of zinc could give rise to an uptake of cadmium. Zinc is an essential metal for animals and human beings (see, for example, Underwood, 1962, Bowen, 1966, and Schroeder et al., 1967), and zinc deficiency will give rise to disease with symptoms from skin, gonads, and hematopoietic system. As it was known that zinc deficiency in animals would give rise to effects on the gonads, Parizek, 1957, studied the influence of a simultaneous administration of zinc on the toxicity of cadmium. He found that in rats a large dose of zinc salt could prevent the action of cadmium on the testes (see also Section 6.6.1.5). Similar findings have since been reported by Kar, Das, and Mukerji, 1960, Gunn, Gould, and Anderson, 1961, 1963a and 1963b, Kar and Kamboj, 1965, Mason and Young, 1967, and Gunn, Gould, and Anderson, 1968a and 1968b. Not only the acute toxic effects of cadmium on the testes but also some long-term effects of cadmium may be prevented by simultaneous administration of zinc. Vigliani, 1969, gave rabbits injections of cadmium chloride for several months and found that the degree of proteinuria was less in rabbits treated with zinc. Powell et al., 1964, found that oral exposure to cadmium produced parakeratosis in calves similar to that caused by zinc deficiency. Petering, Johnson, and Stemmer, 1969, reported that typical symptoms of zinc deficiency appeared in rats given a diet low in zinc and in rats given a sufficient amount of zinc and an equimolar amount of cadmium. Hill et al., 1963, Bunn and Matrone, 1966, and Banis et al., 1969, found that some of the effects of cadmium, especially anemia and weight loss, will be corrected by the administration of zinc in rats and mice.

Sporn et al., 1969, gave groups of rats 10 μg/g cadmium, 80 μg/g zinc, and 10 μg/g cadmium + 80 μg/g zinc in food for 60 days. They determined the activities of nine different enzymes in the liver. They found, for example, that cadmium caused decreased GOT (glutamic-oxaloacetictransaminase) activity and that zinc given together with cadmium did not prevent this decrease. Cadmium caused a significant decrease in the oxidative phosphorylation in the liver mitochondria and zinc prevented this action of cadmium. Exposure to cadmium will tend to increase zinc levels in organs (Gunn, Gould, and Anderson, 1962, Bunn and Matrone, 1966, and Banis et al., 1969). This is probably due to the greater burden placed upon zinc to counteract the action of cadmium. Part of this zinc will probably be found in metallothionein together with cadmium, as this protein will normally contain equimolar amounts of cadmium and zinc.

Thus, there is abundant evidence that cadmium and zinc oppose each other in animals. As many enzymes are zinc dependent, it is conceivable that part of the toxic action of cadmium will be caused by an exchange with zinc in some enzymes.

4.2.7 Influence of Other Compounds on the Metabolism of Cadmium

Kar, Das, and Mukerji, 1960, and Mason and Young, 1967, found that selenium could protect the testes against damage by large doses of cadmium (see Section 6.6.1.5). Gunn, Gould, and Anderson, 1968a and 1968b, showed that though the uptake of cadmium in the testes increased after the administration of selenium dioxide, the testes were not damaged. Gunn, Gould, and Anderson, 1968a, gave mice ^{109}Cd as chloride (1.4 mg/kg), subcutaneous injection. One group was also given a subcutaneous injection of selenium dioxide. One hour after injection, the latter group had a blood level of cadmium 11 times higher than that of the group given cadmium alone. Organ distribution was also influenced. The group given selenium dioxide showed lower concentrations in liver and pancreas. There was no difference in kidney levels. Parizek et al., 1969, found a similar increase in blood levels of cadmium in rats given 2.2 mg Cd/kg and sodium selenite. Selenium given in the diet had a similar effect. Selenite was more effective than selenate. By separating plasma proteins by gel filtration, they found a large amount of cadmium in macromolecular fraction in the selenite-treated animals.

Injection of cysteine has also been shown to change the transport and distribution of cadmium. Gunn, Gould, and Anderson, 1968a, found that cysteine prevented testicular damage by cadmium chloride in mice while kidney damage was made more prominent. One hour after a single injection of ^{109}Cd given simultaneously with cysteine, kidney levels of Cd were 30 times higher than in a group given Cd alone. At first the blood level of cadmium was also higher, but 7 days after injection the cysteine-treated animals had lower values than the group given Cd alone.

Various chelating agents can influence the distribution and toxicity of cadmium in the body. A discussion of the possible relationship between distribution changes and observed alterations of toxicity will be found in Section 6.1.1.3.

As an introduction to the following rundown of some chelating agents, it should be mentioned that BAL (2,3-dimercaptopropanol) can increase the cadmium uptake in the kidney (Tepperman, 1947, Niemeier, 1967), whereas HEDTA (hydroxyethylenediaminetriacetic acid) and DTPA (diethylenetriaminepentaacetic acid) decrease the uptake of cadmium in the kidneys (Eybl, Sýkora, and Mertl, 1966a). Certain chelating agents, e.g., NTA (nitrilotriacetic acid) and EDTA (ethylenediaminetetraacetic acid) are of special interest since they are used as components of widely used detergents.

Friberg, 1956, in studies of the action of EDTA on the excretion and distribution of cadmium, found that when EDTA was mixed with cadmium and injected into rabbits, cadmium excretion in the urine increased markedly, about 1,000-fold. Liver, kidney, spleen, and pancreas values in these rabbits were considerably lower than corresponding values in controls injected with the same amount of cadmium. In rabbits given an injection of cadmium, after which a 4-hr waiting period took place before the administration of EDTA, the urinary cadmium excretion was increased by a factor of about 100 in relation to controls. Cadmium concentrations in the liver were lower than in the controls whereas kidney, spleen, and pancreas values were higher than in the controls. In animals treated for a long time with EDTA and cadmium, there was also a marked increase in urinary cadmium, but on the other hand, organ values did not differ from those of controls.

Scharpf, Ramos, and Hill, 1972, gave female rats single oral doses of $CdCl_2$ and NTA. There is reason to believe that serious methodological errors were introduced because cadmium in urine was determined with atomic absorption spectrophotometry without previous extraction or other treatment. Because of salt effects (Chapter 2) the urinary values will be too high under such circumstances. One group of animals was given 64 mg of cadmium chloride per kilogram body weight, which corresponds to about 10 mg of cadmium to a 250-g rat. The excretion of cadmium in urine after 96 hr was 1.97% of the dose in rats given cadmium and NTA and 0.76% of the dose in rats given only cadmium. It has repeatedly been demonstrated that in normal rats absorption of cadmium after oral exposure is only about 2%, indicating that in this case not more than about 200 μg cadmium per rat was absorbed. An excretion of 0.76% of the dose would mean about 76 μg, which is more than one third of the absorbed amount, an extremely unlikely figure.

In the same report it was found that in liver and kidneys, where the results of cadmium analysis probably are more reliable than in urine, cadmium concentrations were considerably lower in animals given cadmium chloride and NTA than in those given cadmium chloride alone. This has also been shown by Scharpf et al., 1972, in another experiment on rats in which cadmium chloride was given without NTA to one group and with NTA, 200 mg/kg body weight, to another group. Liver levels of cadmium were more than three times higher in the group given cadmium chloride alone. Whether this is due to a lower absorption or to a higher excretion when cadmium is given together with NTA is not known.

Nolen et al., 1972, and Nolen, Bohne, and Buehler, 1972, studied the effects of NTA on cadmium metabolism in rats. Nolen et al. gave groups of pregnant rats cadmium as the chloride in drinking water so that the exposure was 0, 0.01, 1, and 4 mg/kg body weight per day. Sodium-NTA was given in water in doses corresponding to 0, 0.1, and 20 mg/kg body weight per day. Exposure time was between days 6 to 14 of the gestation period. On day 21 the animals were killed. At all cadmium exposure levels there were no differences in cadmium content of liver and kidneys between animals without or with NTA. In the study by Nolen, Bohne, and Buehler, pregnant rats were studied under similar conditions as above. Cadmium was given in oral doses of 4 mg/kg/day. Na_3NTA was given in a dose of 20 mg/kg/day and also in that dose together with $FeCl_3$ (7 mg/kg/day). One group was given sodium citrate (20 mg/kg/day). This time, rats given NTA or NTA + Fe had significantly higher cadmium levels in liver than animals given only cadmium. Citrate also caused higher liver levels. In kidneys both NTA and citrate caused higher cadmium levels.

Nordberg, 1971b, gave 0.1% NTA in the drinking water for 3 to 4 weeks to one group of four mice and ordinary tap water to another group of four mice. Both groups were given 7 μg Cd/kg in the form of $^{109}CdCl_2$ via stomach tube on the third day of the experiment. The body burden of cadmium as well as the excretion of cadmium in urine and feces was followed by measurements of radioactivity. The NTA group had retained 9.7% of the administered dose and the other group 18% 3 days after the cadmium exposure. The corresponding figures were 1.9 and 1.8%, respectively, 7 days after dosing, and 1.6 and 1.6%, respectively, 14 to 18 days after dosing. The mice were killed 24 to 29 days after dosing. The mean liver concentration of cadmium in the NTA-exposed group was 0.53 ng/g (range: 0.27 to 1.1) and in the nonexposed group 0.76 ng/g (range: 0.24 to 1.5). Kidney concentrations were 1.4 ng/g (range: 0.92 to 2.7) in the NTA group and 1.6 ng/g (range: 0.76 to 2.34) in the control group. In short, no large difference in organ retention was found.

Nordberg, 1972c, also studied liver and kidney retention of radioactive cadmium in mice that had been given a single subcutaneous injection of $^{109}CdCl_2$ (0.1 mg Cd/kg body weight). They then drank water containing 50 μg/g nonradioactive cadmium for several months. Half of the 38 mice in the study were also exposed to 500 μg/g NTA in the drinking water. At the end of the observation period (4 months) 14 mice in the cadmium-NTA group were alive vs. 11 in the control group. The mean concentration of radioactive cadmium in liver and kidney of the surviving mice was as follows: group Cd + NTA: liver 1.18 (S.D. 0.23) μg/g (wet weight), kidney 0.98 (S.D. 0.23) μg/g; group Cd: liver 1.33 (S.D. 0.27) μg/g, kidney 1.21 (S.D. 0.44) μg/g. If a statistical analysis is carried out (Student's t test), there appear to be no statistically significant differences between the NTA-treated group compared to the group exposed to only cadmium ($p < 0.1$). In spite of this observation, it may be noted that the mean values in both kidney and liver in this study and in the short-term study described in the preceding paragraph were lower in the NTA-treated group compared with the group given only cadmium.

In summary, the experiments with NTA have shown that there is usually a tendency towards lower concentrations of cadmium in liver and kidney of animals given a combination of Cd + NTA compared to rats and mice given only Cd.

4.2.8 Discussion of Mechanisms for Transport, Distribution, and Excretion

In this section an attempt will be made to explain some of the mechanisms for transport, distribution, and excretion of cadmium against the background of the data presented in the previous sections.

It is not known in what form cadmium is transported from the lungs, the intestines, or injection sites to various organs. The studies by Perry et al., 1970, and Perry and Erlanger, 1971, showed that, in the first hours after intravenous and intraperitoneal injections, respectively, cadmium in blood was mainly in the plasma and partly dialyzable. This may mean that the initial uptake in the kidneys after injection is the result of glomerular filtration and reabsorption in the tubules of low molecular weight compounds. As will be discussed later, a similar mechanism probably holds true later when cadmium in plasma is bound to metallothionein.

There will be a rapid decrease in plasma levels of cadmium, whereas blood cell levels will decrease slowly. About 24 hr after injection cadmium in blood will be mainly in the cells. At that time there is a change in metabolism and cadmium concentrations begin to increase in the red cells. In the red cells and in the liver, cadmium appears in a low molecular weight protein fraction, probably metallothionein, the cadmium and zinc binding protein described by Kägi and Vallee in 1960 and 1961. The many unique properties of this protein are given in Appendix 4:1. Metallothionein must play an important role for cadmium metabolism, both in normal and in chronically exposed animals, but as it probably does not exist in amounts large enough to handle injected doses in acute exposure, it will be involved only to a minor degree in the transport of cadmium during the first hours after injection.

After oral exposure only a relatively small percentage of the given dose is absorbed. The capability of the intestinal walls to produce metallothionein is not clear, but Starcher, 1969, has found a low molecular weight protein (mol. wt. around 10,000) in the duodenal mucosa of the chicken. This protein had the ability to bind copper and was thought to be important for the transport of this metal. Cadmium and zinc could displace the copper. A similar protein was found in the duodenal mucosa of the rat (Evans and Cornatzer, 1970) and in the duodenal mucosa of the bovine (Evans, Majors, and Cornatzer, 1970). This protein might be identical with metallothionein. However, it is not clear whether it is synthesized in the duodenal mucosa or if it originates in the liver. Piscator, 1964, and Nordberg et al., 1972, found large amounts of the protein in liver from exposed rabbits, but other organs might also contribute, as Lucis, Shaikh, and Embil, 1970, found that human fibroblasts could produce a similar protein.

That pretreatment with small doses of cadmium will prevent some of the effects of a large dose has been shown in several experiments (Terhaar et al., 1965, Ito and Sawauchi, 1966, the National Institute of Industrial Health, Tokyo, 1969, Nordberg, Piscator, and Lind, 1971, and Nordberg, 1971a, 1972a. This finding is consistent with an induction of cadmium-binding protein by the small dose which can then bind a large cadmium dose. This theory has received some verification from the experiments by Nordberg, 1971a (Section 6.6.1.5).

Metallothionein is probably one of the proteins responsible for cadmium transport in the blood, but other proteins must also be involved. Carlson and Friberg, 1957, found that hemoglobin probably plays a role in cadmium metabolism. This was discussed by Piscator, 1963, and Axelsson and Piscator, 1966b, in connection with the finding that rabbits exposed to cadmium for longer periods of time developed hemolytic anemia, and that the release of cadmium-containing hemoglobin from the cells could be important for metabolism of cadmium.

As Carlson and Friberg, 1957, found that about one third of the cadmium in hemolysate was dialyzable, cadmium could also be in a lower molecular weight form. Data from Nordberg, Piscator, and Nordberg, 1971, indicate that a part of the erythrocyte cadmium in exposed mice corresponded to metallothionein. As metallothionein has a low molecular weight (about 7,000), it is possible that it enters the erythrocytes via the erythrocyte membrane. However, it could also be synthesized in the cells. In the plasma a small amount of cadmium was found both in high and low molecular weight fractions, the latter corresponding to metallothionein.

It is thus possible for cadmium to appear in at least two forms, bound to metallothionein or to hemoglobin. Then it can also be expected that albumin and zinc-dependent enzymes may bind it, probably depending upon the degree of exposure. By using a chelating agent (CaDTPA), Eybl, Sýkora, and Mertl, 1966a, found that the bond of Cd to hemoglobin was weaker than that of Cd to albumin.

The selective accumulation in the kidneys has been thought to be due to reabsorption of metallothionein in the renal tubules (Piscator, 1964 and 1966c). The possibility that cadmium bound to hemoglobin released through hemolysis can reach the kidneys has been pointed out by Piscator and Axelsson, 1966b. Recent data (Nordberg, Piscator, and Nordberg, 1971) suggest that the release of methallothionein from red cells is of greater importance. As metallothionein is of low molecular weight, the small amount of free metallothionein in plasma can be expected to be cleared completely. As proteins in normal kidneys are almost completely reabsorbed, only trace amounts of the protein can be expected in the urine. When the kidney has become saturated with cadmium, the reabsorption decreases and tubular proteinuria appears (Section 6.1.2.2). If metallothionein is still filtered, less will be reabsorbed and it could be expected to appear in urine. Nordberg and Piscator, 1972, have recently shown that this might happen. They found a low molecular weight protein fraction containing cadmium in urine from mice given injections of cadmium chloride for 5 months.

4.3 TRANSPORT, DISTRIBUTION, AND EXCRETION OF CADMIUM IN "NORMAL" AND EXPOSED HUMAN BEINGS

In Section 4.2 it has been shown that cadmium will be retained to a high degree in animals, especially in the kidneys and liver. Accumulation of cadmium in man is unavoidable, as even in nonpolluted areas, cadmium will be found in food, air, and water often appearing together with zinc in the environment. For practical reasons, "normal" concentrations of cadmium are here defined as those found in human beings without *known* exposure to cadmium in industry or to excessive amounts in food, water, and ambient air.

4.3.1 Transport and Distribution in Blood

Little is known about transport of cadmium in blood in human beings. Data from Szadkowski, 1972, indicate that more cadmium is in the plasma than in the cells in persons without known exposure to cadmium. He found a ratio between plasma and cell levels of 1.9 with a mean level in whole blood of 3.5 ng/g (n = 18). Cadmium was determined by atomic absorption spectrophotometry after extraction. Vens and Lauwerys, 1972, found a mean concentration of 9.5 ng/g in whole blood (S.D. 10) and 21.6 ng/g (S.D. 26.9) in red cells taken from 24 normal subjects. These values are higher than usually reported, and it is not quite clear from the paper as to whether actual determinations of cadmium in red cells were made or hematocrit values used instead. The analytical method was atomic absorption spectrophotometry (for earlier pertinent remarks, see Section 2.4).

As has been discussed in Chapter 2, it is difficult to determine accurately low concentrations of cadmium in biological material. Results from several investigations during the last 10 years, in which methods such as neutron activation, atomic absorption spectrophotometry after extraction of metal, or spectrographic analysis with modern equipment have been used, have indicated that the average normal level, both in whole blood (Table 4:5) and in serum (Table 4:6), is well below 10 ng/g. Recently, a mean "normal" level of 0.6 ng/g was reported (Ediger and Coleman, 1973). But the true value may be still lower. In these studies there are still relatively large discrepancies among values reported by different authors. To what extent this reflects differences in analytical methods or in actual cadmium levels is not known. To date, no accurate studies on analytical methods for determination of cadmium in blood (see Chapter 2) have been carried out.

Apart from those data listed in Tables 4:5 and 4:6, there are some reports in which considerably higher values have been stated. Butt et al., 1964, reported an average level of cadmium in whole blood of 2,090 ng/g and in serum, 410 ng/g. Mertz

TABLE 4:5

Cadmium Concentration (ng/g) in Whole Blood from "Normals" and "Exposed Workers"

Country	n	Mean	Median	Range	Method	References
"Normals"						
Sweden	6	5		2–8	Neutron activation	Brune et al., 1961
Sweden	150	<2	<2		Spectrographic	Piscator, 1971
W. Germany	18	3.5			Atomic absorption after extraction	Szadkowski, 1972
Belgium	24	9.5 (S.E. = 2.1)			Atomic absorption after extraction	Vens and Lauwerys, 1972
U.S.A.	153	8.5	7.0	3.4–5.4	Spectrographic after extraction	Imbus et al., 1963
U.S.A.	243	<5	<5	<5–142	Atomic absorption after extraction	Kubota, Lazar, and Losee, 1968
U.S.A.	50		0.6	<0.2–2	Atomic absorption (Delves Cup technique)	Ediger and Coleman, 1973
U.K. (children)	89	4.9		0–19	Atomic absorption after extraction	Delves, Bicknell, and Clayton, 1972
U.K. (hospitalized children)	204	5.7		0–79	Atomic absorption after extraction	Delves, Bicknell, and Clayton, 1972
"Exposed workers"						
Sweden (presently exposed)	22	27		4–58	Spectrographic	Piscator, 1971
Sweden (presently exposed)	26		20	<4–50	Spectrographic	Piscator, 1972
Sweden (not exposed for 10–20 years)	35	8.6		3–22	Spectrographic	Piscator, 1971
W. Germany (presently exposed)	22	41			Atomic absorption after extraction	Lehnert (pers. commun.)

TABLE 4:6

Cadmium Concentration (ng/g) in Serum from "Normals" and "Exposed Workers"

Country	n	Mean	Median	Range	Method	References
"Normals"						
W. Germany	15	3.3		1.2–9.0	Atomic absorption after extraction	Lehnert, Schaller and Haas, 1968
U.S.A.	27	12			Atomic absorption	Morgan, 1970
W. Germany	18	2.3			Atomic absorption after extraction	Lehnert (pers. commun.)
"Exposed workers"						
W. Germany (presently exposed)	17	1.9		0–9.0	Atomic absorption after extraction	Lehnert et al., 1969
W. Germany (presently exposed)	22	13.7			Atomic absorption after extraction	Lehnert (pers. commun.)

et al., 1968, reported average normal levels of 250 ng/g in serum.

In a paper by Mertz, Koschnik, and Wilk, 1972, the mean cadmium concentration reported in serum was 230 ng/g, with a maximum value of 800 ng/g. They used spectrographic methods, and it seems obvious that analytical errors were involved (see also Section 4.3.3.1 where this method gave values for normal excretion of cadmium in urine of almost 1 mg/day). It can be mentioned that the values are in the same range as the one reported by Friberg in 1952 and 1955 in the blood from rabbits given large doses of cadmium (^{115}Cd) by injections. The rabbits had symptoms of intoxication (Section 6.1.2.2). They are also considerably higher than the values given in Table 4:5 for workers exposed in industry.

In exposed workers, Lehnert (personal communication) found more cadmium in cells than in plasma. At a mean blood level of 41 ng/g, the ratio of plasma to cells was 0.5 (n = 22), which is in accordance with the findings in animal experiments: cadmium will be bound mainly to the blood cells during exposure.

In Tables 4:5 and 4:6 some data on cadmium concentrations in blood or serum of exposed workers are shown, but as exposure conditions are not fully known, they can only be used for demonstrating that cadmium concentrations in exposed workers are considerably higher than in "normals."

There will be fluctuations in blood levels of cadmium. Rogenfelt (personal communication) has determined cadmium concentrations three to six times a year for 3 years in blood of 20 workers exposed to cadmium oxide fumes in a Swedish factory producing a copper cadmium alloy. Production has varied to a large extent. When levels of cadmium in air have been measured, they have varied between 10 to 360 $\mu g/m^3$. In Figure 4:18 it is seen that during exposure cadmium concentrations in blood from one worker varied between 53 to 91 ng/g (spectrographic method, "normal" concentration < 2 ng/g). After the worker was removed from exposure, the cadmium levels slowly decreased. The concentration in blood was still high, about 30 ng/g, 32 months after exposure had ceased. In some other workers similar fluctuations were seen during exposure. In others a more steady level was seen. It was not possible to relate cadmium concentrations to actual exposure or to duration of work.

Tsuchiya, 1969, determined cadmium in blood with an interval of 1 year in four workers exposed to cadmium oxide fumes. In two workers no change was seen, in one there was a decrease, and in another one there was an increase, so that the mean level for the group was about the same (190 ng/g) at both determinations. Piscator, 1972, studied a group of 26 workers in an alkaline battery factory and related cadmium concentrations in blood (spectrographic method, "normal" concentration < 2 ng/g) to exposure time but found no relationship (Figure 4:19). Short exposures could be combined with high concentrations in blood and long-term exposure with low concentrations of cadmium in blood. In another study (Piscator, unpublished data) it was possible to follow the decline in blood levels of cadmium in a group of about 20 workers who for 1 to 2 years

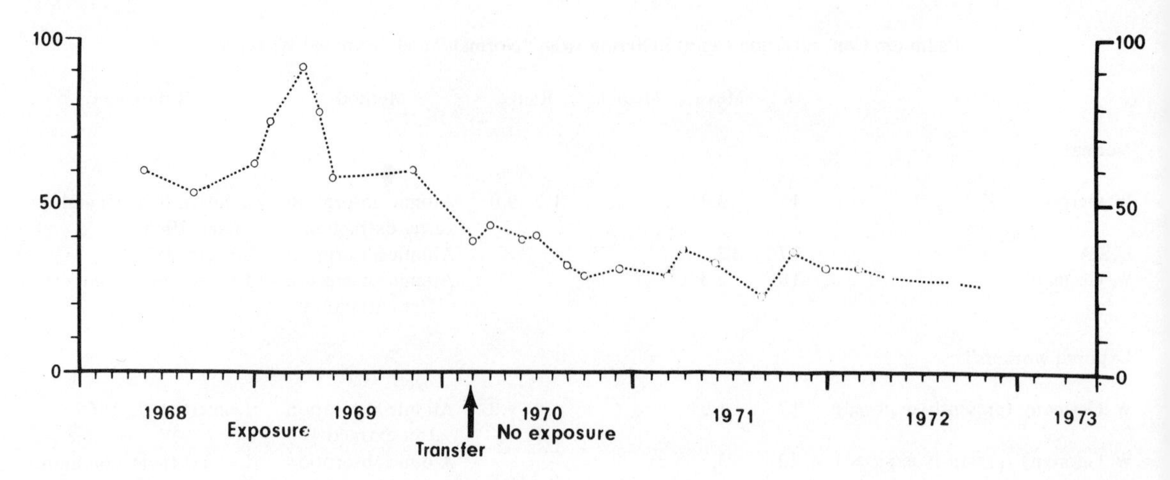

FIGURE 4:18. Cadmium concentrations in blood from a worker during and after exposure to cadmium oxide fumes (spectrographic method). (From Rogenfelt, personal communication.)

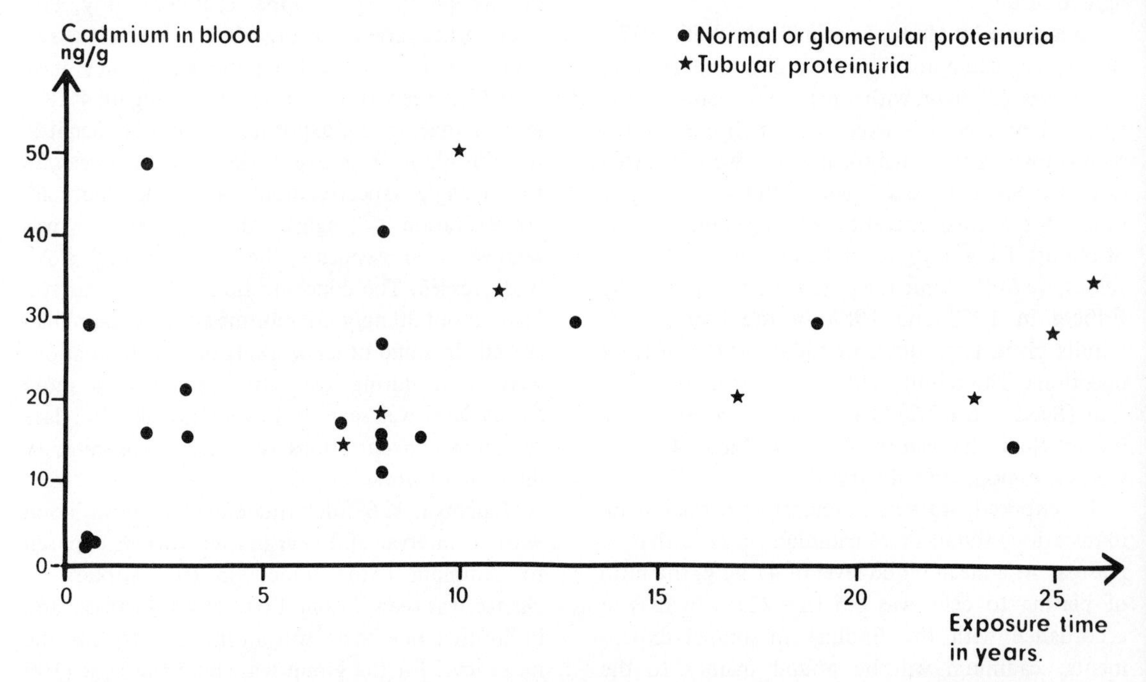

FIGURE 4:19. Relationship between exposure time and cadmium concentrations in blood among workers in an alkaline battery factory. (From Piscator, M., Cadmium Toxicity – Industrial and Environmental Experience, XVII Int. Cong. Occup. Health, proceedings to be published 1974.)

must have been very heavily exposed to a cadmium aerosol during polishing of cadmiated metal. Extremely high levels of cadmium were found in blood, after which the operations were closed down and the workers moved to another workshop. Cadmium in blood was determined 3 and 9 months after cessation of exposure (Figure 4:20), and during that period the half-time of cadmium in blood was about half a year.

After exposure has ceased, blood concentrations of cadmium will usually decrease, as was illustrated in Figures 4:18 and 4:20. The individual variations that exist are also evidenced by these two figures. In addition, it may be mentioned that Tsuchiya, 1969, reported a decrease from about 280 to 140 ng/g in 1 year in two workers removed from exposure to cadmium oxide fumes. In a third worker removed from the exposure there was no marked decrease. On both occasions his blood concentrations were about 140 ng/g. Piscator, 1971, found that in a group of workers who had not been exposed for 20 years to cadmium oxide dust (this group included workers examined by Friberg, 1950) the mean cadmium concentration in blood was 9 ng/g (n = 18; spectrographic method).

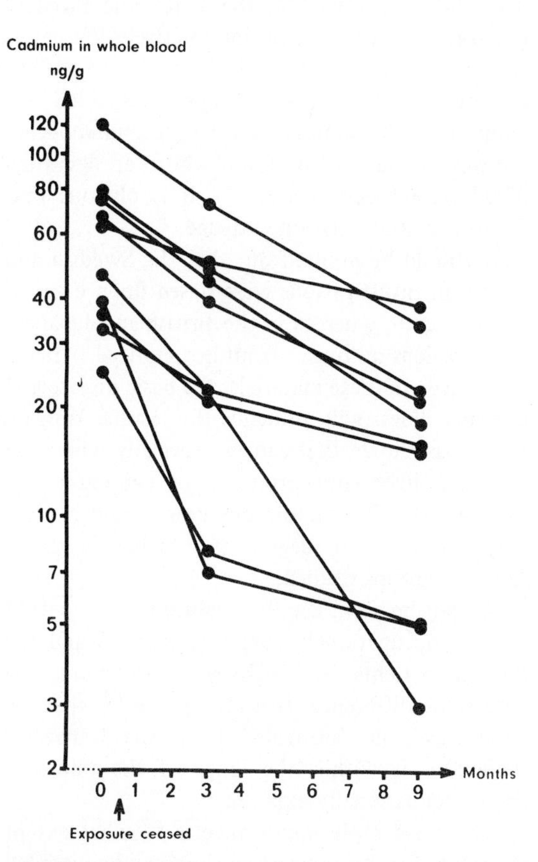

FIGURE 4:20. Cadmium concentrations in blood in workers during and after exposure. (From Piscator, unpublished data.)

4.3.2 In Organs

4.3.2.1 In Liver and Kidney

In contrast to blood and urine, analysis of cadmium in liver and kidney from adult humans will give accurate results with most methods, as the concentrations usually are relatively high; thus, interference from other compounds will be negligible. The basic work in this field has been done by Tipton, Schroeder, Perry, and co-workers. In a series of papers they presented normal values for cadmium and many other metals in organs of subjects from the United States and from other parts of the world (Tipton and Cook, 1963; Tipton, 1960; Perry et al., 1961; Schroeder and Balassa, 1961; Tipton et al., 1965; and Schroeder et al., 1967). During recent years, other investigators have also studied this problem (in Sweden: Piscator, 1964, and Piscator and Lind, 1971; in Germany: Geldmacher-v. Mallinckrodt and Opitz, 1968, Henke, Sachs, and Bohn, 1970, and Anke and Schneider, 1971; in Japan: Inhizawa et al., 1967, Kitamura, Sumino, and Kamatani, 1970, and Ishizaki, Fukushima, and Sakamoto, 1970b, and Tsuchiya, Seki, and Sugita, 1972a; in the United States: Morgan, 1969, Hammer and Finklea, 1972, and Gross, Cholak, and Pfitzer, unpublished data; and in the United Kingdom: Curry and Knott, 1970). Data on exposed workers have been reported by Friberg, 1950, 1957; Bonnell, 1955; Smith, Smith, and McCall, 1960; Kazantzis et al., 1963; and Piscator, 1971.

As the tissue concentrations have been expressed in different ways, i.e., μg/g wet weight, dry weight, or ash, and as with regard to renal concentrations some authors have determined cadmium in whole kidney and some in renal cortex, conversion factors have been used. For liver, the ash values have been multiplied by 0.013 (Tipton and Cook, 1963), and the dry weight values have been multiplied by 0.29 to obtain wet weight values (Piscator, 1971). For kidney the ash values have been multiplied by 0.011 (Tipton and Cook, 1963), and the dry weight values by 0.21 (Piscator, 1971) to obtain wet weight values. Whole kidney values have been multiplied by 1.5

(Geldmacher-v. Mallinckrodt and Opitz, 1968) to obtain cortex concentrations.

Data from organs stored for long periods in, for example, formalin, cannot be used if corrections for the cadmium leached out into the formalin are not made. For this reason, the data by Inhizawa et al., 1967, are not cited.

From data in the above mentioned papers the ratios between cadmium concentrations in kidneys and liver can be calculated. In the U.S. the mean ratios were between 11 and 15 for age ranges 20 to 29, 30 to 39, 40 to 49, and 50 to 59 (Schroeder and Balassa, 1961) and between 9 and 14 in seven groups with mean ages around 60 years (Morgan, 1969, 1970, unpublished data). In Sweden the mean ratio in subjects in the age range 40 to 59 was 15.5 (range 5 to 43) (Piscator, 1971), in the U.K. 5.8 (1.3 to 10) and 8.1 (4.3 to 13) in the age ranges 20 to 29 and 40 to 59 (Curry and Knott, 1970), and in West Germany 6.2 (2.6 to 12) in age range 20 to 59 (Geldmacher-v. Mallinckrodt and Opitz, 1968). With regard to Japanese ratios the original data used in the papers by Ishizaki, Fukushima, and Sakamoto, 1970b, and Kitamura, Sumino, and Kamatani, 1970, have been obtained by the courtesy of the authors, and it was found that in the Kanazawa area the ratios were 6.9 (3.3 to 11.6) and 12.6 (2 to 61) in age ranges 25 to 39 and 40 to 59, respectively, whereas in the Kobe area the ratio was 8.7 (2 to 18) in age range 40 to 59. In exposed workers, on the other hand, lower ratios have been seen, and sometimes liver levels have even exceeded kidney levels (Smith, Smith, and McCall, 1960).

With regard to distribution within organs, Geldmacher-v. Mallinckrodt and Opitz, 1968, found at autopsy in nine normal human beings a mean ratio of 2.2 (1.9 to 3.0) for cadmium in renal cortex to cadmium in medulla. Smith, Smith, and McCall, 1960, found ratios of 1.1 to 1.4 in three controls. From the data of Ishizaki, Fukushima, and Sakamoto, 1970b, and Kitamura, Sumino, and Kamatani, 1970, it has been calculated that the ratios between cadmium concentrations in cortex and medulla were 2.3 (0.6 to 9.6) and 2.9 (1.3 to 6.3) in adults. In four exposed workers, who died 2 to 18 years after the last exposure, the ratios were 1.2 to 1.6 (Smith, Smith, and McCall, 1960). Further information on the distribution of cadmium in kidneys has been obtained by analysis of serial sections (Livingston, 1972). From the outer layer of the cortex and inward, eight horizontal layer sections, four cortical and four medullary, were cut and analyzed for cadmium, zinc, and mercury by neutron activation analysis. An example is shown in Figure 4:21. The concentration of cadmium was found to decrease from the outer layer inward. In the illustrated case, the content in the outer cortex was about twice as high as in the inner cortex nearest the medulla. These data clearly show the necessity of standardizing sampling methods.

In Figure 4:22 the data from investigations of cadmium in liver (μg/g wet weight) in relation to age have been assembled for "normals" in different countries, occupationally exposed workers, and Itai-itai patients. As can be seen, the values for "normals" in the U.S., the U.K., and Sweden are comparatively low, and the average values do not exceed 2 μg/g wet weight. The normal values from three Japanese studies are considerably higher, particularly the values reported by Ishizaki, Fukushima, and Sakamoto, 1970b, from the Kanazawa Prefecture, which are five to ten times the values for the U.S., the U.K., and Sweden. Kanazawa, regarded as having relatively low levels of cadmium in rice, is not included in the present investigation of areas in Japan suspected for pollution with cadmium. As the cadmium concentrations in the liver of the newborn are less than 0.002 μg/g (Section 4.1.3.2), it is obvious that there is an accumulation with age.

It should be pointed out that the Swedish and American investigations were based upon cases of sudden death, whereas in the British and Japanese investigations autopsies from hospitalized patients were used. In these materials, the basic disease and terminal illness will influence the results. Morgan, 1969, has shown that cancer especially will cause increased liver concentrations of cadmium (see Chapter 7). This factor, however, could explain only to a minor degree the higher levels in Japanese groups studied.

As can be seen, the liver values in Figure 4:22 for the occupationally exposed workers and the Itai-itai patients are high with no substantial observable difference. It is also noticeable that the upper level in "normals" from the Kanazawa Prefecture is within the range of the values in those occupationally exposed.

Of considerable importance is to what extent the values for the exposed workers can be regarded as representative for the actual values *during* exposure, as all of the workers included had been

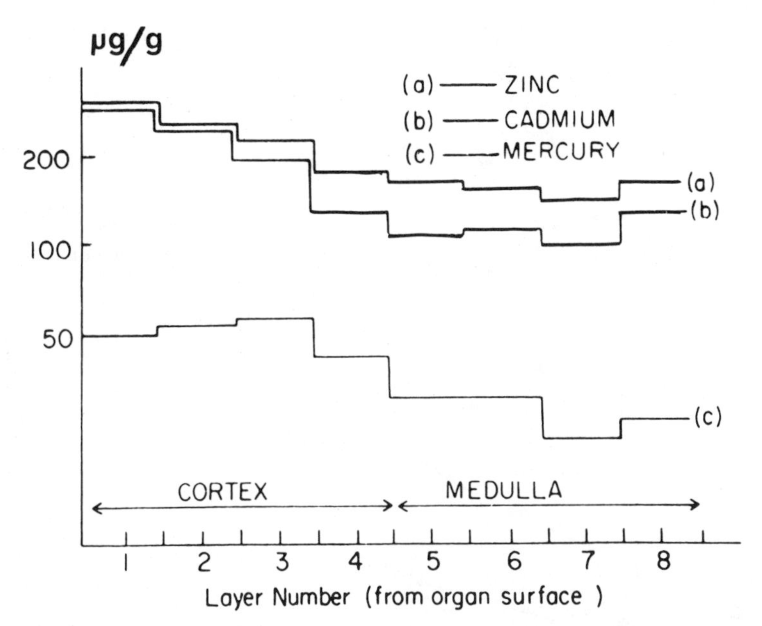

FIGURE 4:21. Cadmiun, zinc, and mercury concentrations (µg/g dry weight) in different layers of the kidney cortex and medulla from a 42-year-old woman. (Cause of death: gunshot wound to head.) (From Livingston, H. D., *Clin. Chem.*, 18, 67, 1972. With permission.)

without exposure for a considerable time before death. In Figure 4:23 the liver values have been plotted against the passage of time after the end of the exposure. In this figure the actual exposure times for the workers are given. Workers exposed to cadmium dust are separated from workers exposed to cadmium oxide fumes. As can be seen, there is no tendency for the liver values to decrease substantially with the time from the end of exposure. This would mean that the biological half-time in the liver of human beings subjected to this type of exposure is very long, since there is no evidence that those without exposure for only a short period had had a different exposure than those without exposure for a longer period. The liver values in Figure 4:23 should be fairly representative for the values during exposure, provided that the clearance during the first year did not differ considerably from the later clearance.

In Figures 4:24 and 4:25, values for renal cortex (µg/g wet weight) in relation to age are given. It can be seen that the European values are the lowest, but even these show a considerable accumulation with time. The values from Tokyo and the Kanazawa Prefecture are the highest among the "normals." In one U.S. study, the values are comparable with the European ones, while in another study the values are considerably higher than the European data but still low compared with the said Japanese values.

It is evident from Figure 4:25, where the data are plotted on a linear scale, that the increase in renal cadmium concentrations is almost linear in the age interval 5 to 50 years. As judged from general biological experience, a certain amount of curvilinearity would be expected. Possible reasons for the linear increase instead will be discussed in Section 4:5.

Those occupationally exposed are scattered over a wide range, with about two thirds of the values over 100 µg/g and the rest between 20 and 100 µg/g. The reason that the concentrations in some occupationally exposed workers and in the Itai-itai patients are even somewhat lower than in the "normals" is discussed in detail in Sections 6.1.2.1.4 and 8.2.4.2. If renal damage is present, cadmium excretion will increase and renal levels of cadmium will decrease.

Figures 4:24 and 4:25 show that cadmium has been accumulated with age in all groups studied, thus supporting the findings of Schroeder and Balassa in 1961.

It will also be seen from Figures 4:24 and 4:25 that after 50 to 60 years of age, a decrease in cadmium levels has taken place. At present it is very difficult to explain these changes. They might

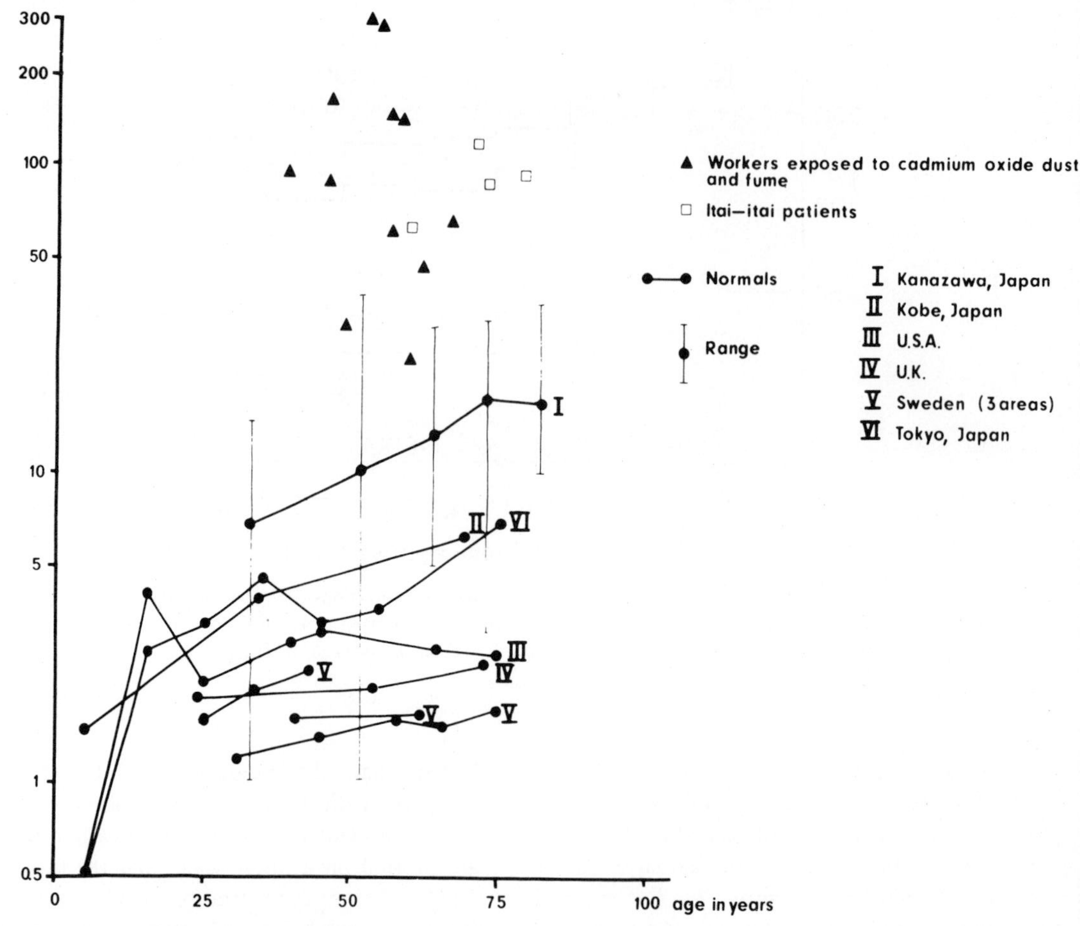

FIGURE 4:22. Cadmium concentrations in liver from normal human beings in different age groups (mean values), exposed workers (single values), and Itai-itai patients (single values). (Data on normal human beings from Kanazawa, Japan – Ishizaki, Fukushima, and Sakamoto, 1970b; Kobe, Japan – Kitamura, Sumino, and Kamatani, 1970; U.S.A. – Schroeder and Balassa, 1961, Schroeder et al., 1967; U.K. – Curry and Knott, 1970; Sweden – Piscator, 1971; Tokyo, Japan – Tsuchiya, Seki, and Sugita, 1972a; data on exposed workers, see Figure 4:23; data on Itai-itai patients – Ishizaki, Fukushima, and Sakamoto, 1970b.) Logarithmic scale.

be connected with factors related to a low cadmium intake a long time ago and with different food habits in older age groups. To what extent metabolic changes in kidneys or other organs in old age are of importance is not known (Section 4.5.3).

In the studies by Anke and Schneider, 1971, the cortex levels of men in the age group of 50 years were about 25 to 30 µg/g while corresponding values for women were about 15 µg/g.

The transport to, and metabolism of, cadmium in renal cortex are closely related to zinc metabolism, as pointed out by Schroeder, 1967. Analysis of human renal cortex from 36 normal persons, aged 6 to 50 years, showed an equimolar increase in zinc content with increasing cadmium levels (regression equation: $Zn = 1.13 \cdot Cd + 0.5$, µmol/g wet weight) (Piscator and Lind, 1971). The basic level of zinc in the kidneys was not affected. The molar relationship between cadmium and zinc under normal circumstances is 1:1 in metallothionein, which is found in human kidneys (Pulido, Kägi, and Vallee, 1966). It is possible that the equimolar relationship between increases in zinc and cadmium in normal renal cortex reflects the metallothionein content.

4.3.2.2 Other Organs

In the pancreas the levels of cadmium in "normals" are usually below 2 µg/g wet weight.

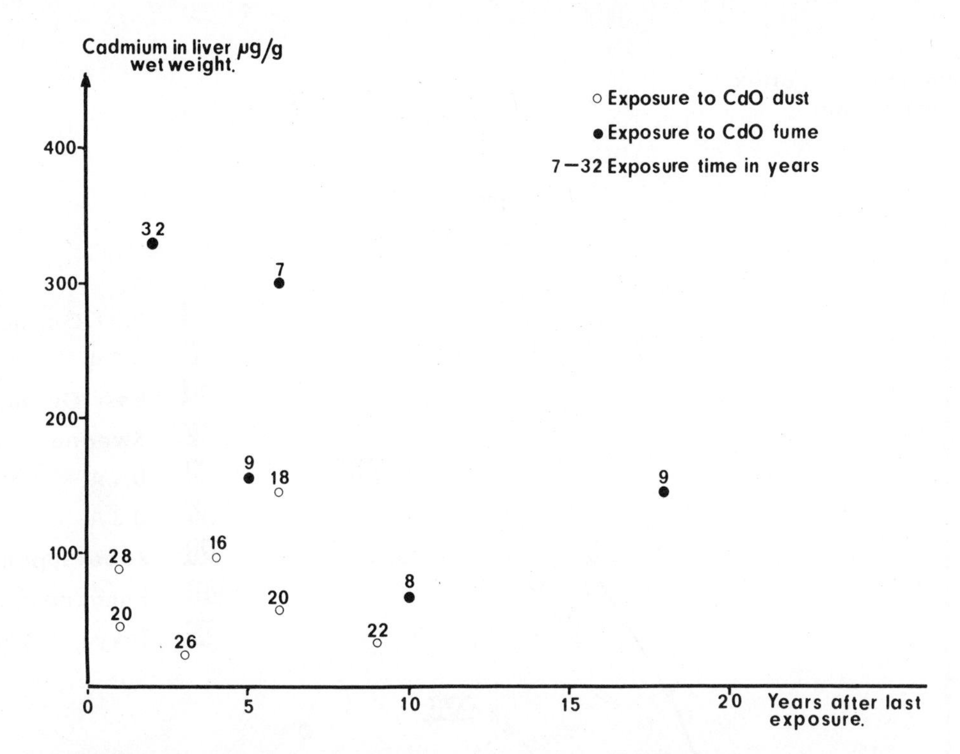

FIGURE 4:23. Cadmium concentrations in livers of exposed workers in relation to time since exposure ceased. (From data by Bonnell, 1955, Friberg, 1957, Smith, Smith, and McCall, 1960, Kazantzis et al., 1963, and Piscator, 1971.)

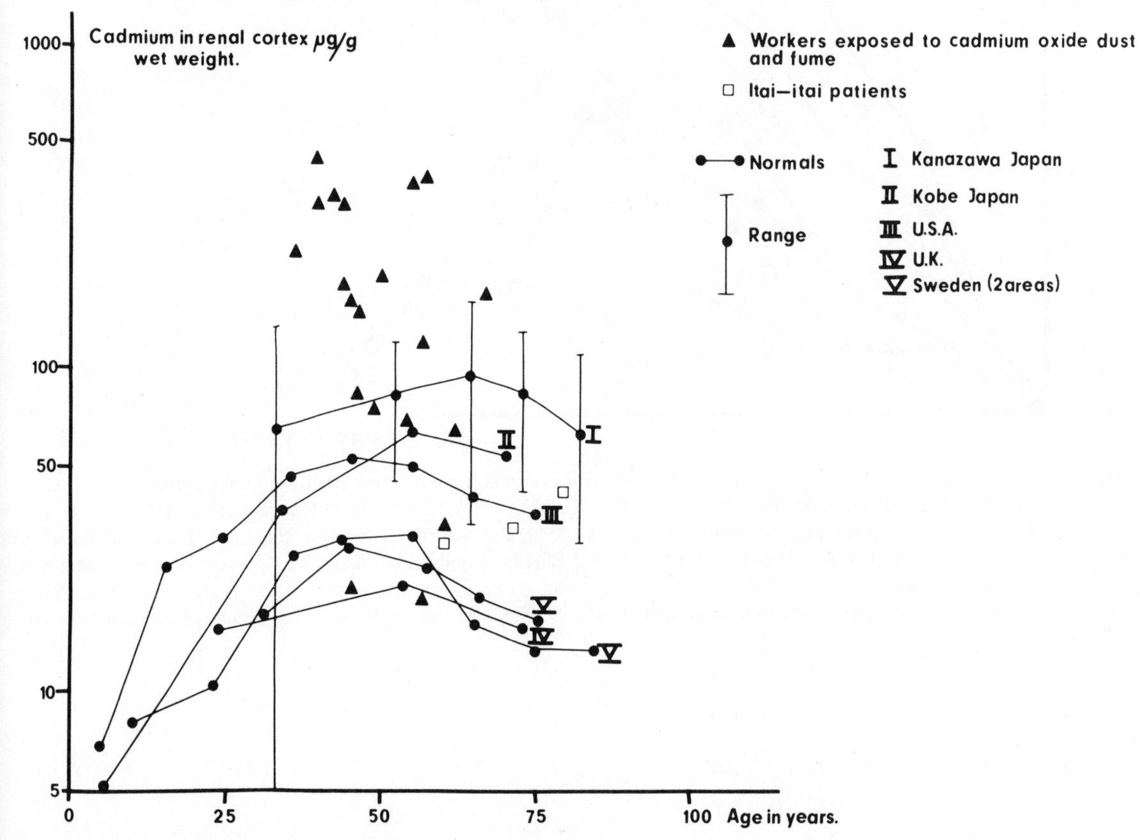

FIGURE 4:24. Cadmium concentrations in renal cortex from normal human beings in different age groups (mean values), exposed workers (single values), and Itai-itai patients (single values). Logarithmic scale. (Data from reports cited in Figures 4:22 and 4:23.)

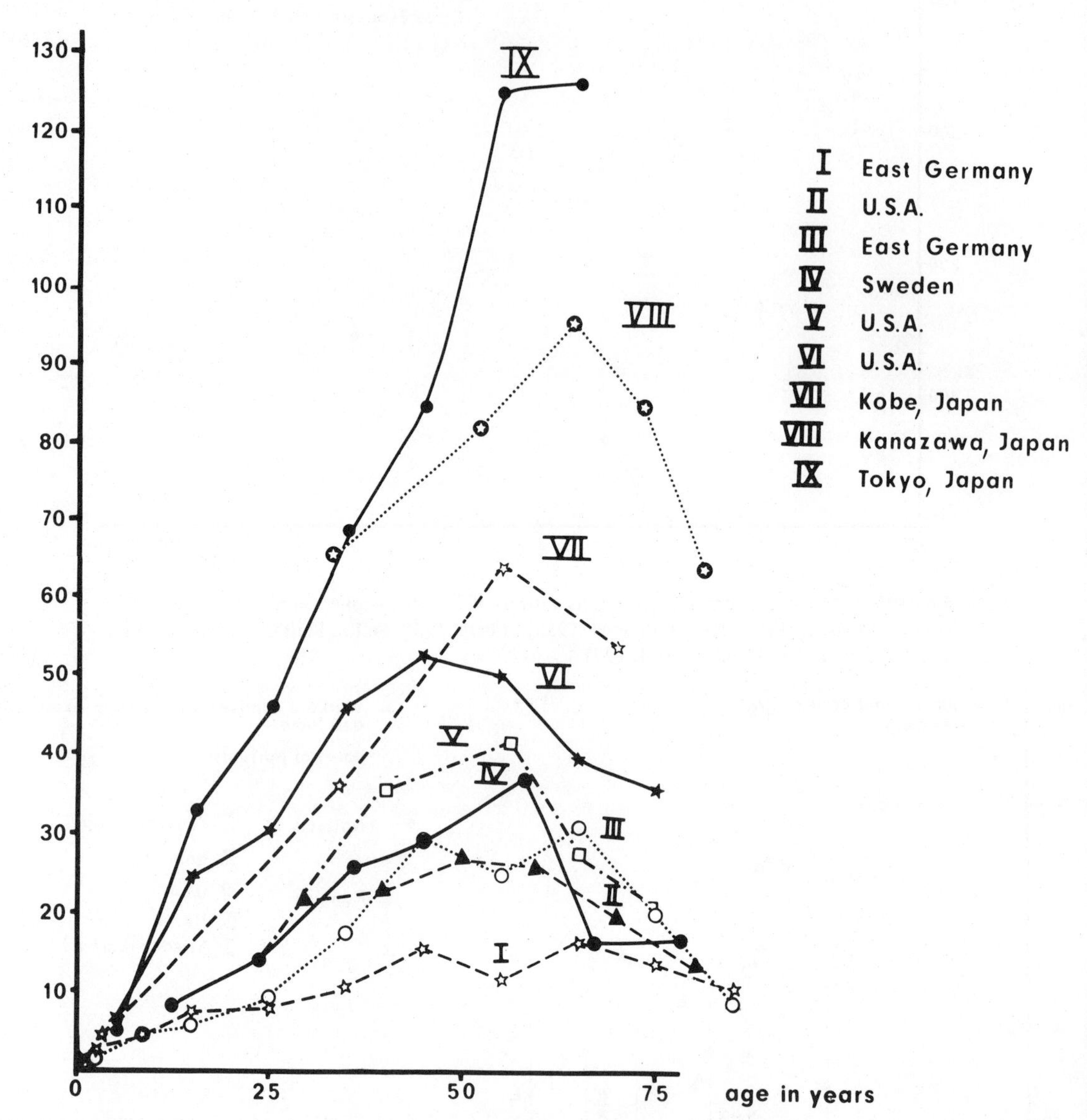

FIGURE 4:25. Cadmium concentrations in renal cortex from normal human beings in different age groups. Linear scale. I: women, East Germany (from Anke and Schneider, 1971). II: both sexes, U.S.A. (from Hammer and Finklea, 1972). III: men, East Germany (from Anke and Schneider, 1971). IV: men, Sweden (from Piscator and Lind, 1971, and Piscator, unpublished data). V: men, U.S.A. (from Gross, Cholak, and Pfitzer, unpublished data). VI: U.S.A. (from Schroeder and Balassa, 1961, and Schroeder et al., 1967). VII: Kobe, Japan (from Kitamura, Sumino, and Kamatani, 1970). VIII: Kanazawa, Japan (from Ishizaki, Fukushima, and Sakamoto, 1970b). IX: Tokyo, Japan (from Tsuchiya, Seki, and Sugita, 1972a).

Tipton and Cook in 1963 found a median value of about 1 μg/g in U.S. citizens, whereas Tipton et al. in 1965 found a median value of 2.8 μg/g wet weight in subjects from the Far East. None of the presently available data has related cadmium in pancreas to age. In exposed workers, 39 to 80 μg/g wet weight have been recorded by Friberg in 1957 and Smith, Smith, and McCall in 1960, and in Itai-itai patients, 45 to 65 μg/g, by Ishizaki, Fukushima, and Sakamoto in 1970b.

"Normal" concentrations of cadmium in the lungs have been reported by Tipton and Shafer in 1964. In subjects from San Francisco, they found a mean level of 0.34 μg/g wet weight. Molokhia and Smith in 1967 found in 21 Scottish subjects (40 to 70 years old) a mean level of 0.26 μg/g wet weight and Smith, Smith, and McCall in 1960 found in three British subjects (mean age: 54) a mean level of 0.71 μg/g wet weight. In Germany Geldmacher-v. Mallinckrodt and Opitz in 1968 found a mean level of 0.11 μg/g wet weight in subjects who were 3 to 62 years of age. Cadmium levels in the lungs seem to increase with age. In most other organs cadmium will be found, as Smith, Smith, and McCall, 1960, and Tipton and Cook, 1963, have attested, but there are not enough data on accumulation with regard to age to draw any conclusions. Concerning cadmium concentrations in other organs, it can be mentioned that Friberg in 1957 and Smith, Smith, and McCall in 1960 found relatively large concentrations of cadmium (25 to 83 μg/g wet weight) in the thyroid in cases of chronic cadmium poisoning. In bone, cadmium levels normally are low, below 1 μg/g with accumulation with age, as found by Kitamura, Sumino, and Kamatani, 1970.

4.3.3 Excretion

4.3.3.1 Urinary Excretion

Just as in the case of blood values, big differences in reported normal urinary concentrations of cadmium have sometimes become manifest. Even so, most reports during the last 10 years have stated an average normal excretion of 1 to 2 μg/day (Table 4:7 and Figures 4:26 and 4:27). In some reports, such as those by Schroeder et al., 1967, Tipton and Stewart, 1970, McKenzie, 1972a, 1972b, and Mertz, Koschnik, and Wilk, 1972, considerably higher figures, i.e., 30 to 900 μg, have been stated. There are good grounds for believing that serious methodological errors were involved (Chapter 2).

The urinary excretion of cadmium in relation to age has been charted in some instances. Szadkowski, Schaller, and Lehnert, 1969, in an extensive study, found practically no increase with age in the excretion of cadmium calculated as micrograms per gram creatinine (Figure 4:26). Suzuki and Taguchi, 1970, studied 38 men (18 to 44 years old) and 49 women (16 to 49 years old) and did not find any relationship between age and excretion of cadmium. The mean excretion was 2.4 μg/l in the men and 2.0 μg/l in the women. The age range was rather narrow, though.

Katagiri et al., 1971, determined cadmium in urine in 303 persons (males and females, ages 4 to 59) from Gifu in Japan (Table 4:7). At ages 4 to 6 the average urinary concentration was 0.47 μg/l, in age group 30 to 35 1.1 μg/l, and in age groups 40 to 49 as well as 50 to 59 1.8 μg/l. Similar results were obtained by Tsuchiya and Sugita, 1971, and Tsuchiya, Seki, and Sugita, 1972a, who studied 609 inhabitants of Tokyo (Figure 4:27). From age 5 to 35 years there is an increase in cadmium excretion from about 0.5 to 2 μg/l. Then the excretion is about the same to age 55, after which it falls off slowly.

These last mentioned data fit with recent experience from studies on mice (Nordberg, 1972a) showing that the urinary excretion of cadmium on a group basis was correlated with the total body burden. The animal data likewise brought out a large individual scatter.

Piscator, 1972, found in ten men aged 34 to 63 years and living in a polluted area of Sweden a mean concentration of 2.1 μg/l. In urines from ten persons in Stockholm, Sweden, 20 to 40 years of age, the average cadmium excretion was 0.39 μg/day (Table 4:7). This corresponded to a concentration of 0.34 μg/l (0.07 to 0.63).

Determinations reported by Szadkowski, Schaller, and Lehnert, 1969, Suzuki and Taguchi, 1970, Katagiri et al., 1971, Tsuchiya and Sugita, 1971, Tsuchiya, Seki, and Sugita, 1972a, and Piscator, 1972, have been made by atomic absorption spectrophotometry after extracting with a chelating agent into an organic solution.

Cadmium excretion in exposed workers has been studied extensively and documented by Truhaut and Boudene, 1954, Bonnell, 1955, Smith, Kench, and Lane, 1955, Smith and Kench, 1957, Kazantzis et al., 1963, Suzuki, Suzuki, and

TABLE 4:7

Urinary Excretion of Cadmium in "Normal" Subjects*

Country	n	Age group	Mean	S.D.	Range	Unit	Method	References
U.S.A.	154		1.59		<0.5–10.8	μg/l	Spectrographic after dithizone extraction	Imbus et al., 1963
Japan	30		3.1		0–15.9	μg/l	Dithizone	Suzuki, Suzuki, and Ashizawa, 1965
Japan (Gifu)	46	4–6	0.47	0.25		μg/l	Atomic absorption after extraction	Katagiri et al., 1971
	40	9–10	0.65	0.45				
	41	14–15	0.72	0.50				
	56	20–29	0.99	0.63				
	37	30–39	1.13	1.06				
	40	40–49	1.76	1.33				
	43	50–59	1.75	1.38				
W. Germany	14		1.0			μg/l	Dithizone	Mappes, 1969
W. Germany	15		0.98		0.34–1.57	μg/24 hr	Atomic absorption after MIBK-APDC extraction	Lehnert, Schaller, and Haas, 1968
W. Germany†	169		1.25		0–5	μg/g creatinine	Atomic absorption after MIBK-APDC extraction	Szadkowski, Schaller, and Lehnert, 1969
Belgium	44		0.95	0.8		μg/g creatinine	Atomic absorption after extraction	Vens and Lauwerys, 1972
Sweden	10	20–47	0.39		0.05–0.77	μg/24 hr	Atomic absorption after MIBK-APDC extraction	Linnman and Lind, unpublished data

*Adults if not otherwise stated.

†In this case children are also included since their values did not differ on an average from those of the adults.

TABLE 4:7 (Continued)

Urinary Excretion of Cadmium in "Normal" Subjects*

Country	n	Age group	Mean	S.D.	Range	Unit	Method	References
Sweden	88	50–59	0.62		0.1–2.0	μg/g creatinine	Atomic absorption after extraction	Kjellström, 1973
Sweden	10	15–16	0.25		0.2–0.5	μg/l	Atomic absorption after extraction	Nordberg, unpublished data
Sweden	10	15–16	0.21		0.1–0.3	μg/g creatinine	Atomic absorption after extraction	Nordberg, unpublished data
Sweden[‡]	10	34–63	1.7		0.4–3.7	μg/g creatinine	Atomic absorption after extraction	Piscator, 1972

[‡]Workers without known industrial exposure to cadmium but living in a cadmium-contaminated area.

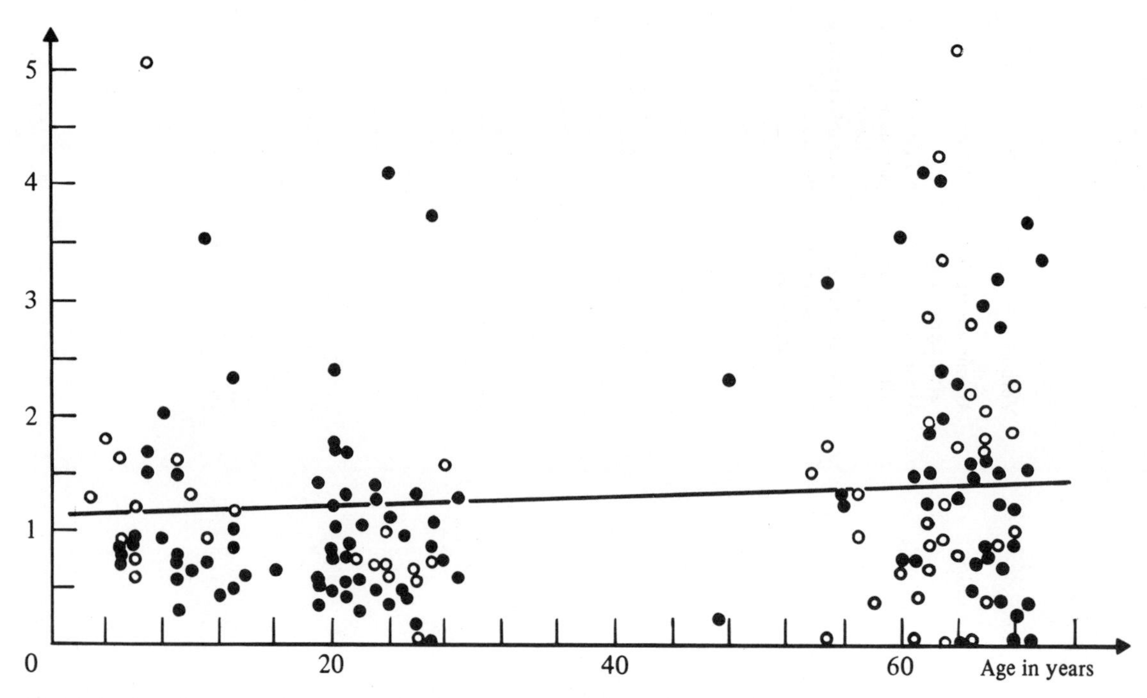

FIGURE 4:26. Urinary cadmium excretion in relation to age. (From Szadkowski, Schaller, and Lehnert, 1969.)

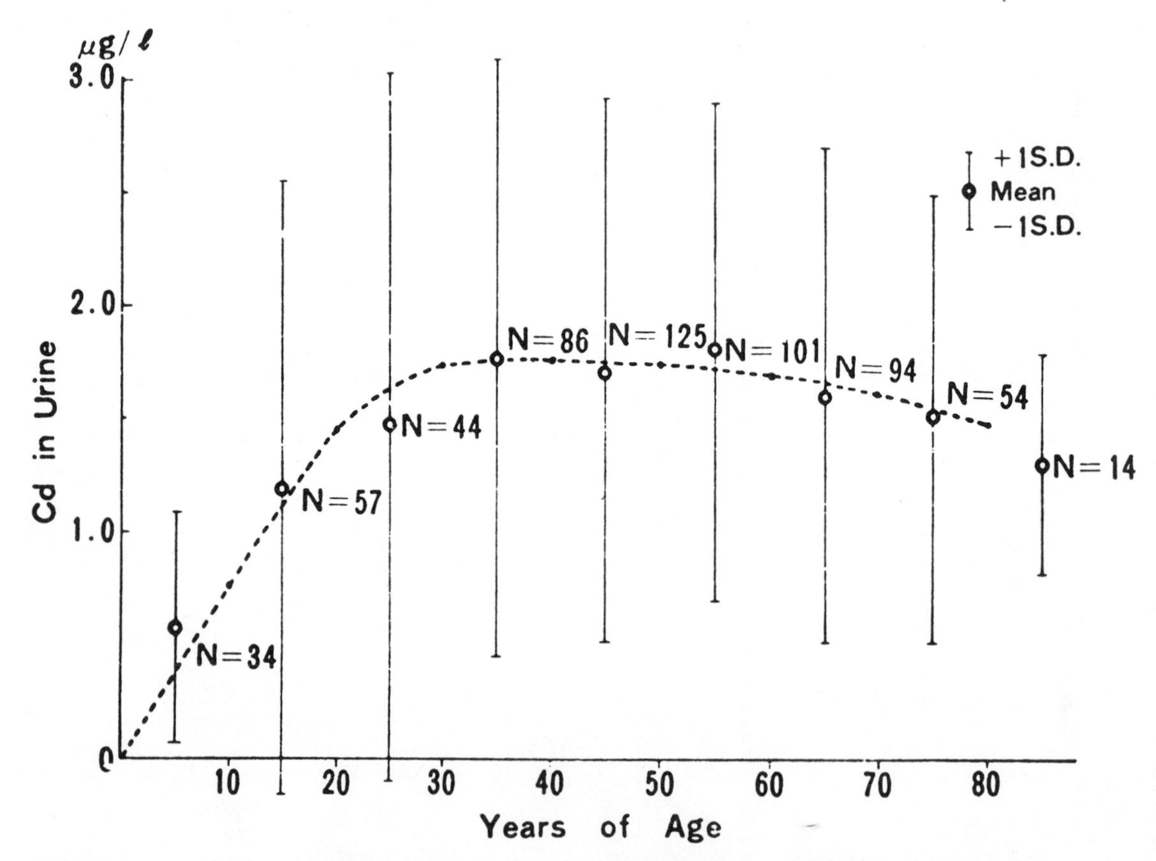

FIGURE 4:27. Average cadmium concentration in urine within different age groups in Tokyo. (From Tsuchiya, K., Seki, Y., and Sugita, M., Organ and Tissue Cadmium Concentrations of Cadavers from Accidental Deaths, XVII Int. Cong. Occup. Health, proceedings to be published 1974.)

Ashizawa, 1965, Tsuchiya, 1967 and 1969, Adams, Harrison, and Scott, 1969, and Lehnert et al., 1969. It will be seen in Table 4:8 that during exposure, cadmium excretion will vary from near zero to around 1,000 μg/day.

In several of the above mentioned reports it can be seen that some workers were exposed to cadmium without this resulting in a marked increase of urinary excretion of cadmium. This is in accord with the now well-documented large individual scatter with regard to urinary cadmium excretion among members of an exposed group. Ten workers studied by Adams, Harrison, and Scott, 1969, had proteinuria and excreted 30 to 170 μg/l. Other workers in the same study who were free of proteinuria excreted less cadmium. Some had been exposed for several years to cadmium concentrations known to cause considerable accumulation of cadmium yet had cadmium concentrations in urine of less than 2 μg/l. It is conceivable that their renal levels of cadmium had not yet reached the point at which renal damage occurs and a pronounced cadmium excretion begins.

An increase in cadmium excretion has been reported in workers without detected proteinuria by Smith and Kench, 1957, and Adams, Harrison, and Scott, 1969, but it should be remembered that for the determination of cadmium, quantitative methods were used, whereas semiquantitative methods were used for the determination of urinary protein.

When, during exposure, proteinuria is present, there is always an increase in the excretion of cadmium, which is also in accordance with the results from animal experiments (Section 4.2.4.1.2). It has not been possible to relate excretion of cadmium to excretion of protein (Smith, Kench, and Lane, 1955, and Smith and Kench, 1957).

After exposure has ceased, cadmium excretion will decrease, whereas proteinuria will persist (see also Section 6.1.2.1.1). Adams, Harrison and Scott, 1969, found that in a group of workers under exposure and with proteinuria, the mean excretion of cadmium was 94 μg/l, whereas in a group of retired workers (not exposed for 0.5 to 5

TABLE 4:8

Urinary Excretion of Cadmium in Exposed Workers

Country	n	Range	Unit	Type of exposure	Method	References
France	2	530–1120	μg/24 hr	CdO dust	Nephelometric	Truhaut and Boudene, 1954
England	12	12–487	μg/l	CdO dust	Dithizone	Smith, Kench, and Lane, 1955
England	10	9–36*	μg/l	CdO dust	Dithizone	Smith and Kench, 1957
England	26	14–428	μg/l	CdO dust	Dithizone	Smith and Kench, 1957
England	27	16–425	μg/l	CdO fume	Dithizone	Smith and Kench, 1957
Japan	34	0–15.9	μg/l	Cd stearate	Dithizone	Suzuki, Suzuki, and Ashizawa, 1965
Japan	13	3–140	μg/l	CdO fume	Dithizone	Tusuchiya, 1967
England	56	0–168	μg/l	CdO dust	Dithizone	Adams, Harrison, and Scott, 1969
England	9	6–74**	μg/l	CdO dust	Dithizone	Adams, Harrison, and Scott, 1969
W. Germany	18	0.8–6.2	μg/g creatinine	CdO fume CdS	Atomic absorption after MIBK-APDC extraction	Lehnert et al., 1969

*Workers in alkaline accumulator factory not directly exposed to cadmium.
**Workers with proteinuria without present exposure.

years) with proteinuria, the mean concentration of cadmium in urine was 28 μg/l.

Tsuchiya, 1969, determined urinary cadmium in three workers at the time they were removed from exposure to cadmium and 1 year later. In one worker there was a decrease from about 250 to 125 μg/l, whereas in the other two workers, cadmium excretion was at both times about 100 μg/l.

Bonnell, Kazantzis, and King, 1959, found that urinary excretion of cadmium could be affected by general health. Acute illnesses such as pneumothorax and pneumonia caused considerable increases in excretion of cadmium. Terminal illnesses will also increase excretion, as pointed out by Smith, Smith, and McCall, 1960.

One main difficulty in evaluating urinary excretion of cadmium in exposed workers as described in much of the data at hand is the fact that it has not always been made clear whether tubular dysfunction, which could have caused an increased excretion, also existed. Piscator, 1972, studied urinary excretion of cadmium in workers exposed to cadmium oxide dust in an alkaline battery factory. In 27 workers and 10 controls urinary cadmium, total urinary protein, and urinary β_2-microglobulin were determined and electrophoretic separations of urinary proteins performed. All analyses were blind, without knowledge of exposure conditions. Even in cases with long exposure time (> 5 years) but without signs of renal dysfunction, urinary cadmium generally was below 10 μg/g creatinine. In workers with signs of renal dysfunction as indicated by changes in electrophoretic patterns and increases in β_2-microglobulin, urinary excretion of cadmium was higher (Figure 4:28). From these data, it seems that cadmium excretion was not related to body burden, whereas renal dysfunction caused an increase in the excretion. There are many difficulties in interpreting these data. For example, the cadmium in air during the last 10 years has usually

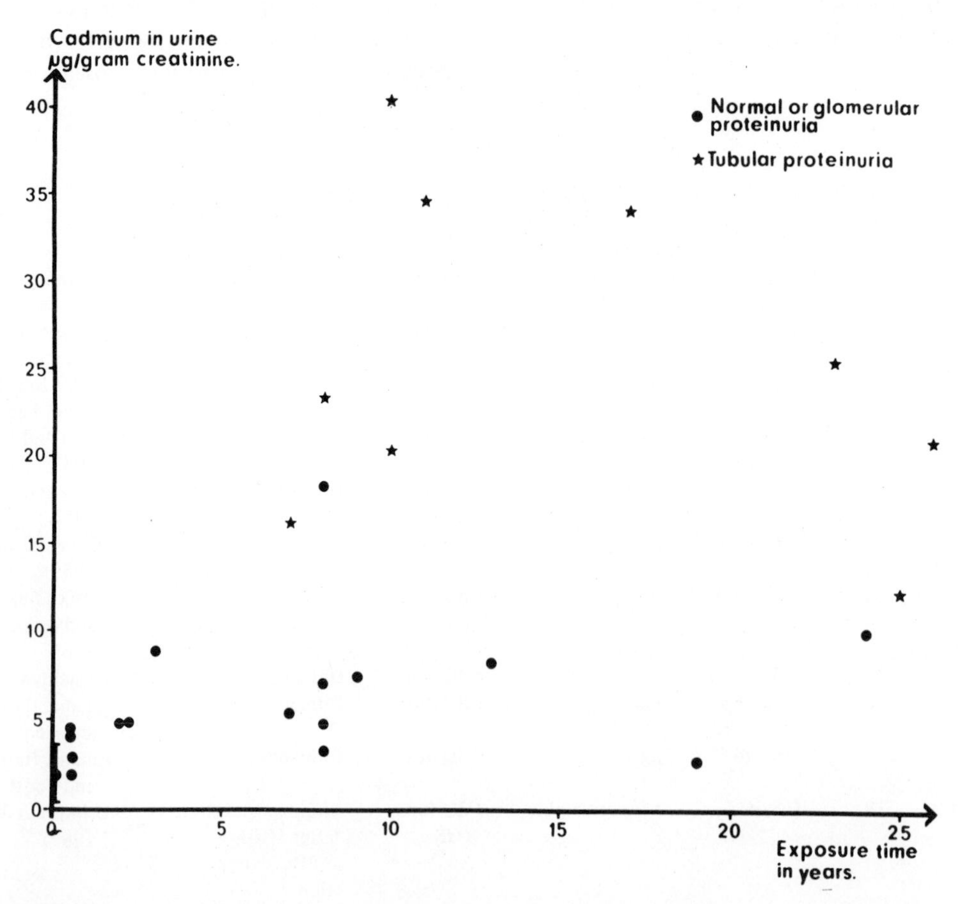

FIGURE 4:28. Relationship between exposure time and cadmium excretion in urine among workers in an alkaline battery factory. (From Piscator, M., Cadmium Toxicity – Industrial and Environmental Experience, XVII Int. Cong. Occup. Health, proceedings to be published 1974.)

been around 50 μg/m^3 but it is difficult to estimate the differences in individual exposure.

In the study by Piscator (unpublished data) referred to in Section 4.3.1, high concentrations of cadmium in blood as well as in urine were observed in workers after 1 to 2 years of exposure to high concentrations of cadmium. After exposure had ceased, large individual variations were found and no regular patterns were seen (Table 4:9). In some cases there were no important changes with time while in other cases considerable decrease or increase in urinary cadmium excretion was observed. In most cases protein patterns were normal. Total protein excretion was in all cases within the normal limits, but as seen in Table 4:9, the relation between different proteins varied. Pathological patterns were sometimes seen on paper electrophoresis.

TABLE 4:9

Change in Urinary Cadmium Excretion with Time after Cessation of Exposure

Worker no.		Urinary cadmium μg/g creatinine at different times after last exposure		
Males	Age	2 weeks	10 weeks	8 months
1	47	32.3 (N)	33.7 (N)	18.1 (N)
2	25	25.4 (G)	39.9 (G-T)	22.3 (G)
3	58	21.9 (N)	18.3 (N)	14.1 (N)
4	63	20.6 (N-T)	34.3 (N)	19.1 (T)
5	53	17.2 (T)	55.0 (T)	22.4 (T)
6	51	1.3 (N)	3.4 (N)	1.9 (N-T)
7	26	10.5 (G)	2.7 (N)	2.9 (N-G)
8	52	4.3 (T)	3.8 (N)	3.7 (N)
9	54	3.7 (N)	3.6 (N)	3.8 (N-T)
10	54	3.6 (T)	11.4 (T)	2.9 (G-T)
11	33	1.2 (N)	1.1 (N)	0.8 (N)
Females				
1	47	72 (N)	55 (N-T)	33 (N)
2	43	17.6 (N)	16.7 (N)	13.0 (N)
3	45	6.3 (N)	4.3 (N)	5.0 (T)
4	43	5.4 (N)	–	6.2 (N)
5	47	2.2 (N)	10.0 (N-T)	3.0 (N)

Letters within parentheses denote patterns on paper electrophoresis:

N	=	normal
G	=	glomerular
T	=	tubular
N-T	=	suspected tubular
N-G	=	suspected glomerular
G-T	=	mixed tubular-glomerular

From Piscator, unpublished data.

According to Tsuchiya, Seki, and Sugita, 1972b, protein excretion will increase first at a cadmium concentration of about 50 μg/l of urine. Such a conclusion does not seem to fit with the data presented above by Piscator, 1972. Tsuchiya and collaborators studied only total protein content. It should be pointed out that protein concentrations in urine may well be within normal limits coincident with slight tubular dysfunction, as shown by increases in β_2-microglobulin and changes in electrophoretic patterns. Piscator studied such early signs of tubular dysfunction. Furthermore, in the Swedish study the workers were exposed to cadmium oxide dust, while in the Japanese study the workers were exposed to cadmium oxide fumes. The Japanese workers had very high blood levels, highest around 400 ng/g. The study by Tsuchiya, Seki, and Sugita, 1972b, would be more comparable with the other study by Piscator (unpublished data) in which the exposure was heavy but over a fairly short period of time. The blood values were high and high amounts of cadmium were excreted in the absence of signs of renal dysfunction. Nomiyama, 1971a, found 5.8 (1.9 to 13.7) μg Cd/l in the urines of nine persons without proteinuria (trichloracetic acid method) compared with 19.8 (17.3 to 21.2) μg Cd/l in three persons with proteinuria.

Sudo and Nomiyama, 1972, studied urinary excretion of cadmium in a 61-year-old male cadmium worker with proteinuria varying between – and ++. Cadmium excretion as measured in three daily urine specimens was studied during 1 month and showed marked variation (10 to 50 μg Cd/24 hr). Spot samples were also studied for 6 months. The authors pointed out the necessity for expressing excretion values on a per-day basis rather than on a per-volume basis.

The relationship between cadmium excretion and zinc metabolism has been illustrated by the findings of Suzuki, Suzuki, and Ashizawa, 1965. They observed a decrease in the excretion of zinc with an increasing excretion of cadmium.

Summing up, there are data that may seem to contradict each other. Japanese studies on people not occupationally exposed to cadmium show an increase in cadmium excretion on a group basis that seems to be related to body burden. A similar association with exposure could not be found in Swedish workers exposed for several years to cadmium oxide dust and without signs of tubular dysfunction. Likewise, in workers having undergone a short but very heavy cadmium exposure and having high blood and urinary values of cadmium, the urinary excretion after cessation of exposure did not follow body burden in a predictable way based on the Japanese data. Exposure situations thus seem to be of paramount importance.

4.3.3.2 Fecal Excretion

Fecal excretion of cadmium has been determined in normal human beings by Essing et al., 1969, who found a mean level of 0.16 μg/g (n = 23). They estimated the daily amount in feces to be 31 μg. Szadkowski, 1972, reached the same result, 31 μg Cd/day, as an average of 16 samples. Tipton and Stewart, 1970, found a mean amount of 42 μg/day in three persons in a long-term balance study. Tsuchiya, 1969, referred to Japanese data showing a mean daily amount of 57 μg in feces from four Japanese men. No known excessive exposure was stated. It is not known how much cadmium will come from cadmium not absorbed from food and how much will have been excreted via the intestines. It is also not known if cadmium excreted into the gastrointestinal tract will be reabsorbed. Tsuchiya, Sugita, and Seki, 1972, found high concentrations of cadmium in bile from autopsy samples and claimed a considerable enterohepatic circulation of cadmium in man. This is in contrast to what has come forth from animal experiments (Section 4.2.4.2.1) on biliary excretion of cadmium. It is desirable that more data on this question be obtained. At present, no definite answer concerning the mechanisms for fecal excretion of absorbed cadmium in human beings can be given. In the studies by Rahola, Aaran, and Miettinen, 1971 (referred to in detail in Section 4.1.2.2) on humans given radioactive cadmium orally, it was found that less than 0.1% of the retained dose of cadmium was excreted in feces. Data from injection experiments on animals suggest that less than 5% of the absorbed cadmium from a daily dose will be excreted daily via feces.

4.3.3.3 Other Excretion Routes

During the last years, the possibility of using hair as an indicator of exposure has been investigated, as seen in Table 4:10.

Analysis of metals in hair is difficult, as external contamination both from dust and from metals in hair lotions, hair sprays, etc., must be taken in account.

In the study by Schroeder and Nason, 1969, the influences of sex, age, and color of hair on the levels of cadmium in hair were investigated. The investigators found that male black hair contained less cadmium than brown, blonde, or red hair. In women, grey hair contained less cadmium than hair with natural color. Their data indicate that cadmium levels in hair are higher in the particular area investigated (Brattleboro, Vermont) than in Stockholm, Sweden.

Ishizaki, Fukushima, and Sakamoto, 1969, determined cadmium in hair from men and women above 55 years of age from the endemic district of the Itai-itai disease. They found mean concentrations of about 0.4 and 0.7 μg/g, respectively. There were, however, no controls for the men. There were only three controls for the women and they had a mean concentration in hair of about 0.4 μg/g.

Hammer et al., 1971, investigated 10-year-old children living in areas with varying degrees of pollution. They found good agreement between exposure levels and hair levels of cadmium.

In exposed workers hair analysis will be of little value, as external contamination will be great. Nishiyama and Nordberg, 1972, showed that cadmium absorbed on the hair due to such contamination was virtually impossible to remove by various washing procedures and could not be distinguished from endogenous cadmium. Analysis of metals in hair is a relatively new undertaking. More information is needed about the chemistry of hair, preanalysis treatment, and significance of levels in hair in relation to organ levels before the value of cadmium analysis of hair can be ascertained.

Cadmium may also be excreted in human saliva. Concentrations of up to 0.1 μg/g have been reported by Dreizen et al., 1970.

4.3.4 Total Body Burden and Renal Burden

High concentrations of cadmium will be found in kidneys and liver and the total amount in these two organs can be calculated quite accurately. In most other organs concentrations are low and accurate determinations have not always been possible. The "Standard American Man" has been reported to contain about 30 mg of cadmium (Report of Committee II on Permissible Dose for

TABLE 4:10

"Normal" Concentrations (μg/g) of Cadmium in Hair

Country	Sex	n	Mean ± S.D.	Median	Range	Method	References
U.S.A.	Male	82	2.77 ± 4.37			Atomic absorption. Hair not treated with detergent solution.	Schroeder and Nason, 1969
	Female	47	1.77 ± 1.64			Atomic absorption. Hair not treated with detergent solution.	Schroeder and Nason, 1969
U.S.A.	Male children*	45	3.5 ± 4.94	2.1		Atomic absorption. Hair pretreated with EDTA.	Hammer et al., 1971
		25	2.0 ± 1.54	1.6			
		37	1.3 ± 0.99	1.0			
		21	1.3 ± 1.30	0.9			
		37	0.9 ± 0.8	0.8			
Sweden	Male	7	0.44 ± 0.14	0.43	0.24–0.60	Atomic absorption. Hair pretreated with detergent.	Nishiyama, 1971
	Female	8	0.87 ± 0.26	0.92	0.41–1.27	Atomic absorption. Hair pretreated with detergent.	Nishiyama, 1971
Yugoslavia rural area	Male	17	0.54 ± 0.27	0.45	0.20–1.48	Atomic absorption. Hair pretreated with detergent.	Nishiyama, 1971

*Ten years of age from five areas with differing degrees of cadmium contamination.

Internal Radiation, 1960, Schroeder and Balassa, 1961, Schroeder et al., 1967), and it has been estimated that about one third of the total cadmium in the body is in the kidneys and about half in liver and kidneys together (Schroeder et al., 1967). However, it is not clear how these figures have been obtained. Recent data from studies in other regions of the United States (Section 4.3.2.1) have shown renal cortex levels of only about two thirds, respectively, one half of those earlier reported. It seems as though regional differences may be of considerable importance.

The "Standard Man" has a weight of 70 kg, of which 30 kg will consist of muscle and 10 kg of fat. The cadmium content of these two components must thus be of great importance for the total body burden. Tipton and Cook, 1963, reported that in most tissues, including muscles, the concentration of cadmium was less than 50 μg/g in ash (less than about 0.5 μg/g wet weight). They did not write about cadmium concentrations in fat but Schroeder et al. later reported that the mean concentration in fat for 28 persons was 39 μg/g in ash.

We have estimated the amount of cadmium in kidneys and liver in relation to total body burden by using data from three studies, in the U.K. (Smith, Smith, and McCall, 1960, dithizone method), in Sweden (Piscator, 1971, atomic absorption after extraction with MIBK-APDC), and in West Germany (Geldmacher-v. Mallinckrodt and Opitz, 1968, spectrographic method). With regard to the results by the last mentioned authors it should be stated that they also determined cadmium in blood and urine. These values were high and have not been included in the normal values for blood (Table 4:5) and urine (Table 4:7) for the same reasons as results by some other authors have been discarded (Sections 4.3.1 and 4.3.3.1). However, there is no reason to believe that their values for cadmium concentrations in muscles and fat are in error.

In Table 4:11 the data on cadmium concentrations in kidney, liver, muscle, and fat are shown. The concentration in the "rest" of the body has been assumed to be an average of the concentration in muscles and fat (from the U.K. study only the concentrations in muscle are included as fat was

TABLE 4:11

Concentrations of Cadmium (μg/g wet weight) in Kidney, Liver, Muscle, and Fat in Unexposed People from the U.K., Sweden, and West Germany

Country	n	Age	Kidney	Liver	Muscle	Fat
U.K.	3	54* (49–56)**	13.5 (10.5–16.5)	2.3 (2.1–2.8)	0.11 (0.10–0.13)	Not determined
Sweden	4	52 (41–62)	22.5 (15–30)	1.4 (1.1–1.6)	0.13 (0.08–0.18)	0.18 (0.13–0.23)
West Germany	11	40 (19–62)	10 (3.5–12)	2.2 (0.4–3.9)	0.06 (0.01–0.10)	0.04 (0.01–0.10)

*Mean
**Range

not determined). This means that the average concentration in the "rest" of the body was estimated at 0.11 (U.K.), 0.15 (Sweden), and 0.05 (West Germany) μg/g wet weight. The following weights have been used in calculating the amount of cadmium in the organs: kidneys, 300 g; liver, 1,700 g; muscles, 30 kg; fat, 10 kg; blood, 6 kg; "rest," 22 kg. In each series the individual values for renal and liver concentrations have been used to estimate the total amounts of cadmium in these two organs, whereas the amounts in muscles and fat have been estimated by using the mean values shown in Table 4:11. For the remaining 22 kg the estimated mean concentrations have been used. Blood was not included as it has been shown (Section 4.3.1) that the concentration of cadmium in blood is too small to be of any importance for these calculations. Mean values have been used for muscles and fat in order to diminish the influence of analytical errors. The estimates, carried out in the manner described, gave the following values for the total body burden: in the U.K. material 15 mg (13.5 to 16.6), in the Swedish material 18 mg (15.4 to 20.2), and in the German material 10 mg (5.1 to 12.9). In the liver and kidneys together there were 54 (50 to 59), 49 (42 to 55), and 65 (34 to 74) % of the total body burden, respectively. The kidneys alone contained 28 (23 to 29), 36 (26 to 45), and 30 (22 to 36) % of the total body burden, respectively. These values thus agree fairly well with the figures given by Schroeder et al., 1967, and in the following we consider that one third of the body burden of cadmium is in the kidneys, while at the same time we are aware that a considerable individual variation might exist. As will be discussed below, in industrial exposure to high concentrations of cadmium, the ratio between liver and kidney concentrations will increase. It seems, however, from data by Piscator on horses (Section 4.2.2.2, Figure 4:9) that no increase in ratio occurs during normal long-term low-level exposure giving rise to kidney cortex concentrations up to about 200 μg/g.

The figure of one third of the total body cadmium in the kidney can be used for estimating total body burden in people in two Japanese areas. Using data on renal concentrations of cadmium in the Kanazawa area (Ishizaki, Fukushima, and Sakamoto, 1970b) the average total body burden will be about 80 mg in a Japanese person of 70 kg and 50 years of age in that area. If data from Kobe, Japan, are used instead (Kitamura, Sumino, and Kamatani, 1970), the average body burden will be lower, about 40 mg. It should be remembered that body weights and thus also organ weights in Japan are lower than in Europe and the United States. As a result, the total body burden, but *not* the body burden expressed as mg Cd/kg body weight, will probably be lower than the values given.

Tsuchiya, Seki, and Sugita, 1972a, calculated total body burden and renal burden in autopsy studies from Tokyo. They incorporated accurate data on organ and body weights. The total body burden was assumed to constitute double the amount found in liver and kidney. At 10 to 30 years of age, total body burden was about 30 mg, at 30 to 50 years about 45 mg, and at 50 to 59 years about 55 mg. The renal burden at age 50 was about 20 mg.

Friberg, Piscator, and Nordberg, 1971, calculated that smoking a pack of cigarettes a day for 35 years might result in 30% higher renal cortex levels of cadmium in smokers than in nonsmokers. Lewis et al., 1972a, determined cadmium in kidney, liver, and lung in autopsy material from Boston, Massachusetts, and Providence, Rhode Island. There were 45 male smokers (mean age at death

60, S.D. 11 years) and 22 male nonsmokers (mean age 60, S.D. 12 years). It was possible to calculate the number of "cigarette pack years," i.e., cigarette packs per day multiplied by years of smoking, for each smoker. In nonsmokers, cadmium in kidney, liver, and lung was on an average 6.6 mg while in smokers the corresponding figure was 15.8 mg. An obvious association was found between number of pack years and cadmium accumulation (Figure 4:29) also when age was controlled for.

In a later paper (Lewis et al., 1972b), a larger number of cases were reported. The same detailed smoking history was not given but the subjects were divided into different smoking categories without any mention of the time of exposure. The mean age at death was about the same (60 to 65 years) in the different groups. The main results are seen in Figure 4:30. The effect of smoking on cadmium accumulation is seen when nonsmokers are compared with light smokers and moderate smokers. It is astonishing that heavy smokers do not have higher values than moderate smokers; this finding is not discussed by the author. Age, cause of death, or occupation could not explain the results.

If organ values are to be used to calculate total body burden in human beings during exposure to large concentrations of cadmium via inhalation or ingestion, the above mentioned proportions can not be used. During or immediately after exposure, the liver contains considerably more cadmium than the kidneys. This is shown by the animal experiments of Decker et al., 1958.

Animal experiments have further shown that during repeated exposure to large doses about 75% of the total administered dose will be found in liver and kidneys before renal damage occurs, as can be estimated from data by Friberg, 1952, and Axelsson and Piscator, 1966a. The main part will be in the liver and only 5 to 10% of the total body burden will be in the kidneys.

There is no reason that human beings should differ much from animals in this respect. For calculating the total body burden in a worker during exposure to high concentrations for fairly

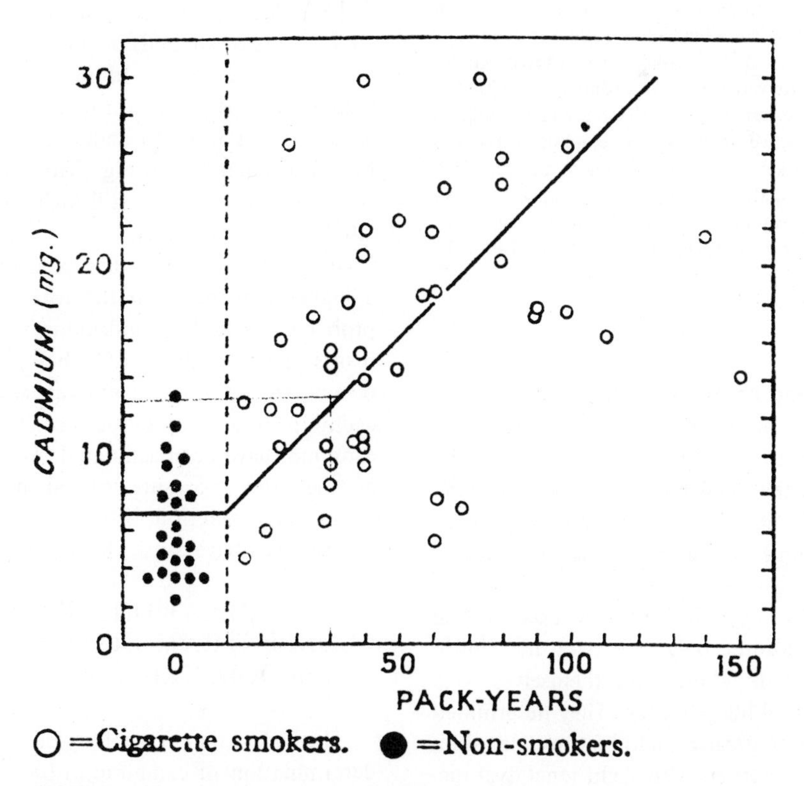

FIGURE 4:29. Calculated sum of renal, liver, and lung burdens of cadmium in relation to smoking. (From Lewis, G.P., Coughlin, L., Jusko, W., and Hartz, S., *Lancet,* i, 291, 1972. With permission.)

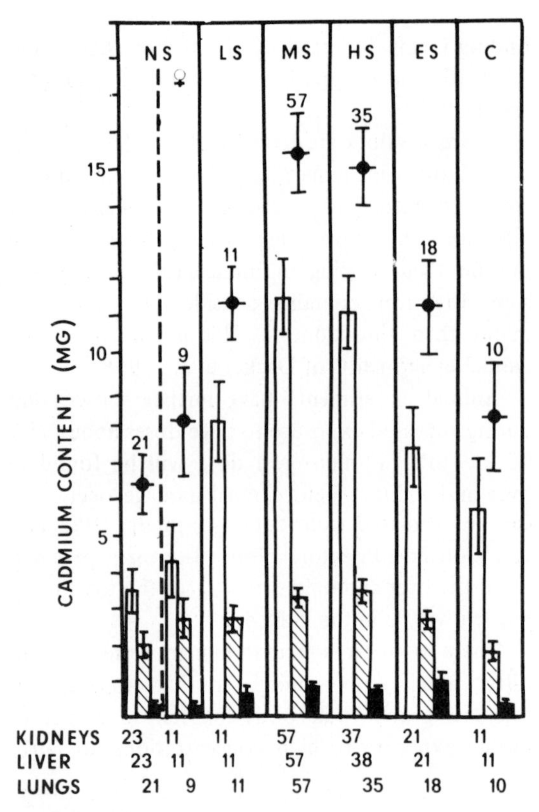

FIGURE 4:30. Cadmium contents of various organs of patients classified by smoking habits. Kidneys: open bars; liver: hatched bars; lungs: solid bars. Closed circles represent composite values (sum of cadmium in kidneys + liver + lungs). The numerals are the number of subjects used in the respective analyses. The smoking categories are NS – nonsmokers (divided into males and females); LS – light smokers; MS – moderate smokers; HS – heavy smokers; ES – exsmokers; C – cigar and pipe smokers. Vertical lines show ± SEM. (From Lewis, G. P., Jusko, W. J., Coughlin, L. L., and Hartz, S., *J. Chronic Dis.*, 25, 717, 1972. With permission.)

short periods and before renal damage has occurred, it can thus be assumed that the amount of cadmium in liver and kidneys together will be about 75% of the total body burden. The same would probably hold true for people exposed to excessive amounts of cadmium via the peroral route.

With regard to body burden in workers several years after exposure, only the data by Smith, Smith, and McCall, 1960, lend themselves to a calculation of total body burden. They determined cadmium in most organs, including muscle, skin, and fat. In three workers with slight renal dysfunction, autopsied 5 to 10 years after the last exposure, it was estimated that between 40 and 70% of the total cadmium accumulation was in liver and kidneys together, with about 10% in the kidneys.

In Figures 4:22 and 4:24 liver and renal concentrations of cadmium at autopsy of exposed workers are shown (see also Table 6:1). With assumed liver and kidney weights of 1,700 and 300 g, respectively, the total amount of cadmium in these workers in liver and kidney would have been between 50 and 600 mg. With 50 to 75% of the cadmium in liver and kidneys, the total body burden at time of death would have been 70 to 1,200 mg. Total body burden at the time exposure ceased must have been higher as excretion of cadmium due to renal disease must have caused losses of cadmium.

In summary it can be stated that in a "standard man" about 50% of the total body burden is in liver and kidney and about 33% in the kidneys alone. In three studies in which smoking habits were not taken into account, the total body burden in a "standard man" in the United States has been calculated to be 30, 20, and 15 mg, respectively. The total body burden in a "standard man," i.e., a 70-kg man, in the United Kingdom and Sweden is probably about 15 to 20 mg. In two control areas studied in Japan the total body burdens for a "standard man" are considerably higher; in one area considered not to have an excessive exposure to cadmium it could be estimated to be about 80 mg. Data recently reported from Tokyo indicate a still higher value if recalculated to "standard man."

For workers during high exposure to cadmium, a higher percentage of the total body burden is probably in the liver and kidneys. Animal experiments indicate about 75% before renal damage occurs. In autopsies of workers with exposure to cadmium total body burdens of 70 to 1,200 mg of cadmium have been estimated. Total body burden at the time exposure ceased must have been higher, as excretion of cadmium due to renal disease must have caused losses of cadmium.

4.4 RELATIONSHIPS AMONG CONCENTRATIONS OF CADMIUM IN BLOOD, URINE, AND ORGANS

The important question of to what extent determination of cadmium in blood or urine may assist in making an estimate of cadmium concentrations in organs and total body burden will be discussed in the following sections. A discussion on the usefulness of blood and urine concentrations of cadmium as indices of exposure and retention will also be found in a paper by the Task Group on Metal Accumulation, 1973.

4.4.1 Relationship between Concentrations of Cadmium in Blood and Organs

4.4.1.1 In Animals

As mentioned in Section 4.2.1.1, the concentration of cadmium in the blood of experimental animals given a single injection of the metal salt displays a two-phase course during the first 2 days after injection. Since this course is not reflected in other organs, it is evident that blood values do not depict the situation in any body organ during this first period. Even if the elimination phase counted from the second maximum seems to be slower, available data indicate that it is considerably faster than the elimination, both from whole body and critical organs. However, more data on these relationships are necessary before any definite conclusions can be drawn.

Data from repeated exposure are scarce. Available evidence has been given in Section 4.2.1.2, and shows that blood concentrations increase continuously during continuous exposure to cadmium – but only to a certain level. Nordberg, 1972a (Section 4.2.1.2), showed that in animals given daily subcutaneous injections of cadmium for 3 weeks, after which they were no longer given cadmium, blood levels decreased about 40% during 3 consecutive weeks. Whole body burden did not decrease noticeably, indicating that blood levels and body burden hold no constant relationship. In general, available animal studies seem to indicate that blood levels partly reflect recent exposure and partly organ concentrations of cadmium. Since none of the investigations mentioned provides estimations of blood/organ concentrations at different exposure levels, definite conclusions are here likewise pending more information.

4.4.1.2 In Human Beings

The data that have been presented do show that blood levels of cadmium in exposed workers may be considerably higher than in people without known exposure to cadmium. There will be fluctuations in blood levels of cadmium, possibly reflecting fluctuations in recent exposure.

Among a group of seven carpenters and repairmen employed at a Swedish factory for more than 20 years but not directly exposed to cadmium, 2 men had renal damage at blood levels of 21 and 31 ng/g blood, whereas two workers in the cadmium department with 3-to-4-years' exposure had cadmium concentrations in blood of 48 and 58 ng/g blood (Piscator, 1971). These cadmium workers did not have any renal damage. A relatively low value for cadmium in blood thus does not exclude a considerable accumulation in the kidneys and a high value does not necessarily mean that renal damage has occurred.

In the investigations by Piscator, 1972, mentioned in Section 4.3.1, no relationship was found between cadmium levels in blood and exposure time, nor was there any relationship between degree of proteinuria and blood levels. On the other hand, Piscator, 1971, found that urinary excretion of protein was related to cadmium concentrations in blood in a group of workers not exposed for 20 years (Figure 4:31). It had earlier been shown (Piscator, 1962a) that in these workers proteinuria was related to exposure time, i.e., the larger the absorbed and accumulated amount, the greater the renal dysfunction. That a correlation could be found between cadmium in blood and proteinuria indicates that the blood levels reflected organ levels in these workers when they were without exposure for several years.

4.4.2 Relationships between Concentrations of Cadmium in Urine or Feces and Organ or Blood Concentrations

4.4.2.1 In Animals

After a single injection of cadmium not more than a few percent of the dose will be excreted in urine during the first weeks. During that period levels of cadmium in blood will both increase and decrease. Liver concentrations will slowly decrease and renal levels will increase. It has not been shown whether urinary excretion of cadmium is related to any of the above mentioned changes in blood or organ concentrations.

After repeated exposure urinary excretion will be related on a group basis to body burden of cadmium, as shown by Nordberg, 1972a and 1972b, in mice. This holds true for the accumulation phase prior to tubular proteinuria. Prolonged exposure in these mice brought a sudden increase in the urinary cadmium excretion simultaneously with the occurrence of tubular proteinuria (Nordberg and Piscator, 1972). The ample evidence on the simultaneity of these two events

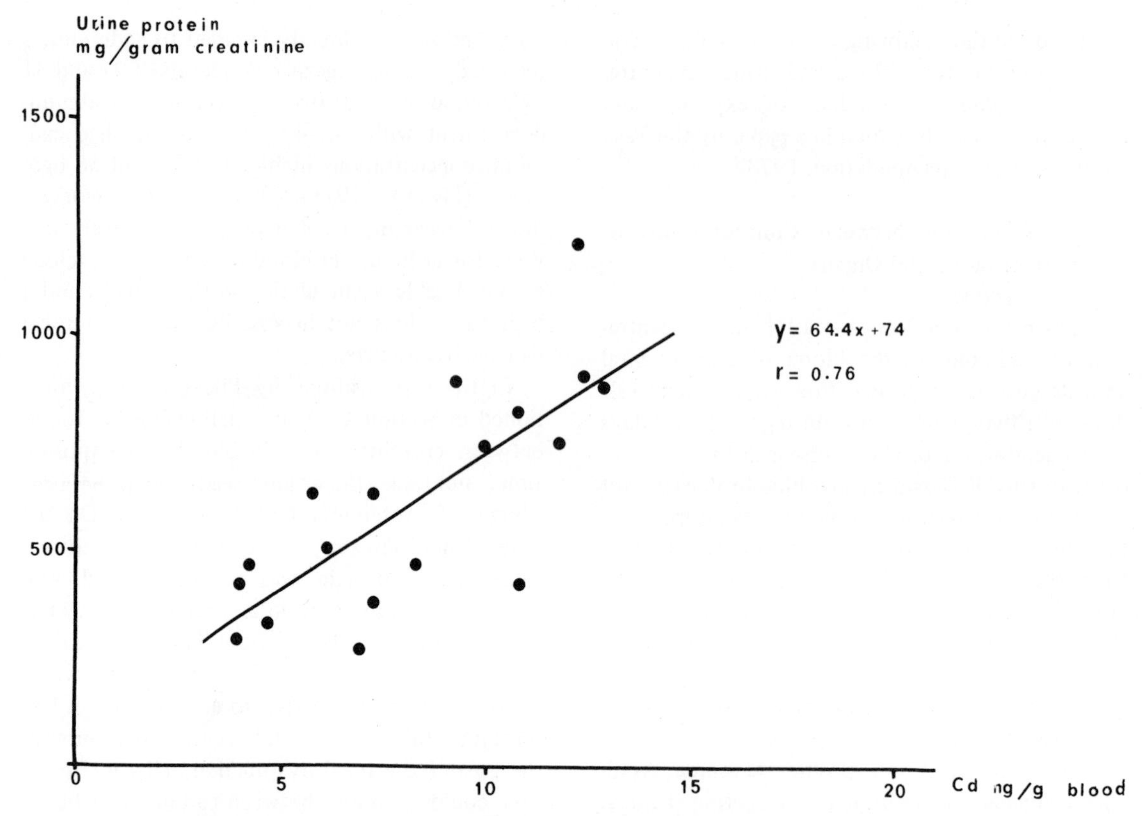

FIGURE 4:31. Urinary protein excretion in relation to blood concentrations of cadmium in workers with earlier exposure to cadmium but not exposed during the last 20 years. (From Piscator, 1971.)

(also, for example, Friberg, 1952, and Axelsson and Piscator, 1966a) indicates a common mechanism. Proteinuria during exposure to cadmium is a sign of renal tubular dysfunction (see Section 6.1.2.3).

Axelsson and Piscator, 1966a, found significant negative correlations between excretion of cadmium and renal tubular capacity to reabsorb glucose and renal activity of alkaline phosphatase in rabbits subjected to a prolonged, proteinuria-inducing cadmium exposure. These results and the relationship between proteinuria and cadmium excretion indicate that once renal injury is present and the amount of excreted cadmium increases, this amount is mainly a reflection of the functional state of the kidney. Hence, at the stage at which tubular proteinuria is present and cadmium exposure continues, the cadmium excretion in itself is a measure of the extent of renal dysfunction rather than a measure of organ concentrations. The mechanisms for this will be further discussed in Section 6.1.2.3.

It has been shown in mice (see Section 4.2.4.2) that the fecal excretion of cadmium is mainly dependent on the daily dose, but a small part of it also may be dependent on body burden.

4.4.2.2 In Human Beings

Concerning fecal excretion there are no data that elucidate its relationship to whole body – or organ – concentrations of cadmium in human beings. The following discussion will be devoted entirely to urinary excretion of cadmium.

The Japanese studies referred to in Section 4.3.3.1 (Katagiri et al., 1971, and Tsuchiya, Seki, and Sugita, 1972a) show a relation between urinary cadmium excretion and total body burden on a group basis during long-term low exposure to cadmium. Due to the wide individual scatter the predictive value of urinary cadmium as an index of total body burden on an individual basis is low.

Data on exposed workers (Section 4.3.3.1) show that in other exposure situations a relation between urinary cadmium excretion and body burden may be absent.

The findings in exposed workers are not easy to interpret, as data on possible renal dysfunction are often not available. Tsuchiya, 1970, determined cadmium in blood and in urine in three groups of workers. One control group (n = 9) in an alloy factory, not directly exposed to cadmium, one group (n = 11) exposed to cadmium oxide fumes in the alloy factory, and one group (n = 20) in an alkaline accumulator factory exposed to cadmium oxide dust were examined. In the control group and in the group exposed to cadmium oxide dust, blood concentrations of cadmium varied between 2 and 40 ng/g and between 30 and 70 ng/g, respectively. Corresponding values for urinary excretion were 2 to 10 and 1 to 40 μg/l. In these two groups there was no correlation between cadmium concentrations in urine and blood. In the group exposed to cadmium oxide fume, blood levels were between 10 to 35 μg/100 ml and urinary cadmium between 10 to 230 μg/l. In that group a correlation ($p < 0.05$) was found between blood levels and urinary levels of cadmium.

There is a great need to study, under different exposure situations, urinary excretion in relation to concentrations in different organs. Under certain conditions and on a group basis it may be that urinary cadmium is a useful index of accumulation of cadmium.

4.4.3 Conclusions

There are no data that show that concentrations of cadmium in blood can be used for estimating organ levels of cadmium in human beings. Available data do not allow any recommendation of a fixed level in blood to be used for control of exposed workers or populations exposed to cadmium in food or ambient air.

The investigations, both on animals and normal human beings, discussed in the foregoing sections indicate that in human beings having a normal renal function and not excessively exposed to cadmium there may exist a relationship between urinary excretion of cadmium and total body burden on a group basis. Due to a wide scatter, the predictive value is low on an individual basis. It is not known to what extent the excretion is related to the concentration in the critical organ. During exposure situations which may occur in factories, relations between body burden and urinary cadmium concentration may be quite different.

4.5 THEORETICAL MODELS OF UPTAKE AND RETENTION OF CADMIUM IN HUMAN BEINGS

It is obvious from preceding sections of this chapter that cadmium accumulates in several "body compartments." Among these compartments a complicated redistribution of cadmium takes place. With our present knowledge about the metabolism of cadmium far from complete, it is not possible to set up an absolutely valid model for cadmium accumulation in various organs or in the body as a whole. The problems involved in accomplishing this task have recently been pointed out by Nordberg, 1972d, and by the Task Group on Metal Accumulation, 1973. Nevertheless, in the following sections an attempt will be made to describe and discuss metabolic models that have been used for cadmium. This is done since these models make possible a discussion of empirical accumulation curves in relation to retention rates and biological half-times. Reservations are necessarily entailed and more or less arbitrary assumptions concerning some of the parameters are unavoidable. Results of calculations of biological half-time are presented in Section 4.6.

4.5.1 Mathematical Model for Accumulation of Cadmium in Kidney Cortex

The kidney, considered the critical organ in chronic cadmium poisoning (Section 6.1), was the first organ for which the modeling technique was employed. Tsuchiya and Sugita, 1971, suggested a model in which the accumulation would follow the differential equation

$$dx = -a \cdot x \cdot dt + b \cdot dt$$

where x = organ (or body) burden of cadmium, a = excretion rate, b = cadmium absorption per time unit, and t = time. The general form of the equation would be

$$x = k_1 \ (1 - e^{-k_2 t})$$

The accumulation curve using this model is a simple logarithmic curve. Tsuchiya and Sugita applied the model to kidney cortex accumulation of cadmium absorbed from the gut. The calorie intake is strongly age related and thus also the cadmium intake via food, which means that b in the equation is a function of time (b = b [t]). The function is such that the equation for organ burden cannot be solved mathematically without approximations.

Calculations based on similar assumptions as those of Tsuchiya and Sugita have been performed by Kjellström et al., 1971, and Kjellström, 1971. They assumed that one third of the body burden was in the kidneys (concentration in cortex 50% higher than the renal mean value). Of the absorbed cadmium, 10% was considered to be rapidly excreted and this fraction was excluded from the calculation because it ultimately could contribute less than 1% to the total body burden of cadmium. Estimations were made using a 5%-absorption of ingested and a 25%-absorption of inhaled cadmium.

For the mathematical description of the model the following symbols were used:

a = 90% of cadmium absorbed daily
e = excretion constant, i.e., fraction of body burden excreted per day via the slow excretion phase
k = fraction of body burden remaining after one day (1 −e)
B = body burden, i.e., total amount of cadmium accumulated in the body (micrograms)
t = time of exposure to cadmium (days)
C = cadmium concentration in kidney cortex (μg/g)
K = weight of kidneys (grams)

Excretion via the slow phase on the first day cadmium is absorbed reduces the total body burden of cadmium by a trivial amount (less than 0.05%). Ignoring this part of the slow phase excretion, body burden can be calculated using an equation for the sum of a geometrical series as follows:

$$B_t = a \cdot \frac{1 - k^t}{1 - k} \qquad t = 0, 1, 2, 3 \ldots \qquad a = \text{constant}$$

In the case that the daily absorbed amount (a) is a function of time, this method makes the calculation much simpler than would be so if the differential equation were to be solved exactly.

If the daily absorbed amount (a) is approximated to be constant during yearly periods the equations will be as follows:

Year

$$1 \quad B_1 = a_1 \cdot \frac{1 - k^{365}}{1 - k}$$

$$2 \quad B_2 = a_1 \cdot \frac{1 - k^{365}}{1 - k} \cdot k^{365} + a_2 \cdot \frac{1 - k^{365}}{1 - k}$$

$$n \quad B_n = a_1 \cdot \frac{1 - k^{365}}{1 - k} \cdot (k^{365})^{n-1} + \ldots + a_n \cdot \frac{1 - k^{365}}{1 - k}$$

The equation for cadmium concentration in kidney cortex will be as follows:

$$C = 1.5 \cdot \frac{B}{3} \cdot \frac{1}{K} = \frac{B}{2K}$$

These equations, introduced by Kjellström et al., 1971, will be used in the following sections as a basis for discussion of the influence of various factors on the accumulation of cadmium in the renal cortex.

4.5.2 Basic Factors Influencing Cadmium Concentration in Renal Cortex

4.5.2.1 Variations of Kidney Weight with Age

In Figure 4:32 the values of kidney weight (Coppoletta and Wolbach, 1933, and Ivemark, unpublished data) used in our calculations are plotted in relation to age.

In the following calculations it will be assumed that kidney weight increases in proportion to body weight from ages 12 to 20 and thereafter remains constant at 300 g. In Figure 4:33 are shown calculated cadmium concentrations in kidney cortex with (Curve II) and without (Curve I) correction for the variation of kidney weight with age.

The effect of the correction for the variation in kidney weight will be a faster increase of the concentration in the age groups 0 to 12 years. Between 13 and 16 years there will be a decrease in the concentration. After that age the two curves are similar. This generally agrees with empirical data up to age 18 (Henke, Sachs, and Bohn, 1970) although the age trend after age 10 in their data will require further empirical information because of the large individual variations. Tsuchiya, Sugita, and Seki, 1972, used actual observations on kidney weight in their calculations. However, they presented only total amount of cadmium in kidneys and not the variation of kidney weight

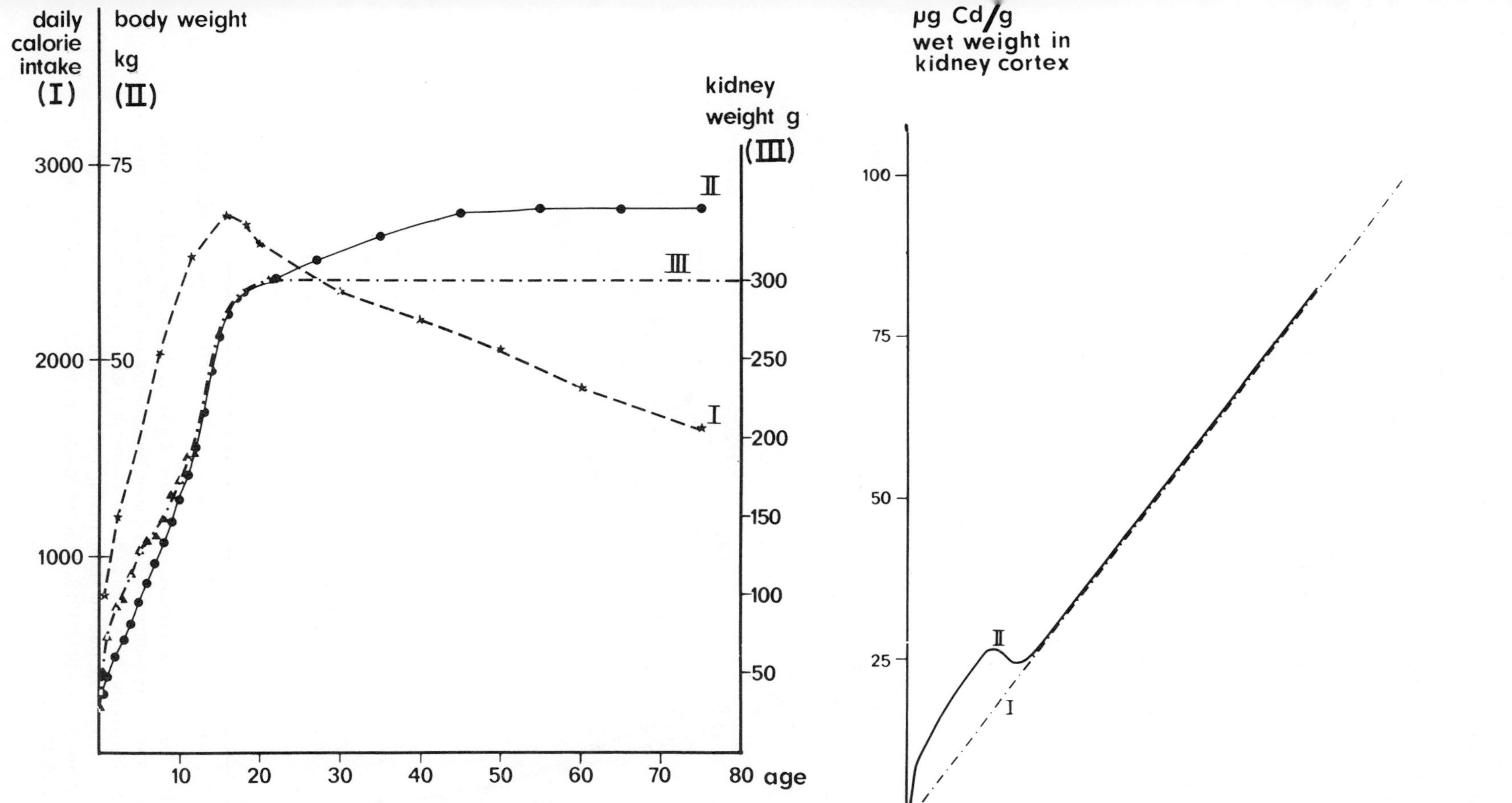

FIGURE 4:32. Estimation of daily calorie intake, body weight, and total kidney weight by age and averaged for both sexes. (Curve II for body weight from Society of Actuaries, 1959, and Nelson, 1964; Curve III for kidney weight from Coppoletta and Wolbach, 1933, and Ivemark, unpublished data).

FIGURE 4:33. Calculated cadmium concentration in kidney cortex in relation to age and kidney weight. Curve I: assuming constant kidney weight. Curve II: assuming variation in kidney weight according to Figure 4:32. Constant cadmium intake of 50 μg per day for all ages and 0% body burden excretion assumed for both curves.

with age; the latter figures would have been of interest for comparisons with the values used here.

4.5.2.2 Variations of Calorie Intake with Age

A large part of the cadmium intake in non-polluted areas comes via a relatively low background contamination of basic foodstuffs. One factor influencing daily cadmium intake would be variation of calorie intake with age. Although both dietary habits and geographical area of food production also can affect dietary cadmium intake, these factors have not been estimated in the present model.

Actual and recommended calorie intake in Sweden and Japan are depicted and referenced in Figure 4:34. The values to be used in our calculations (Figure 4:32) have been obtained from the Swedish data by taking the mean value

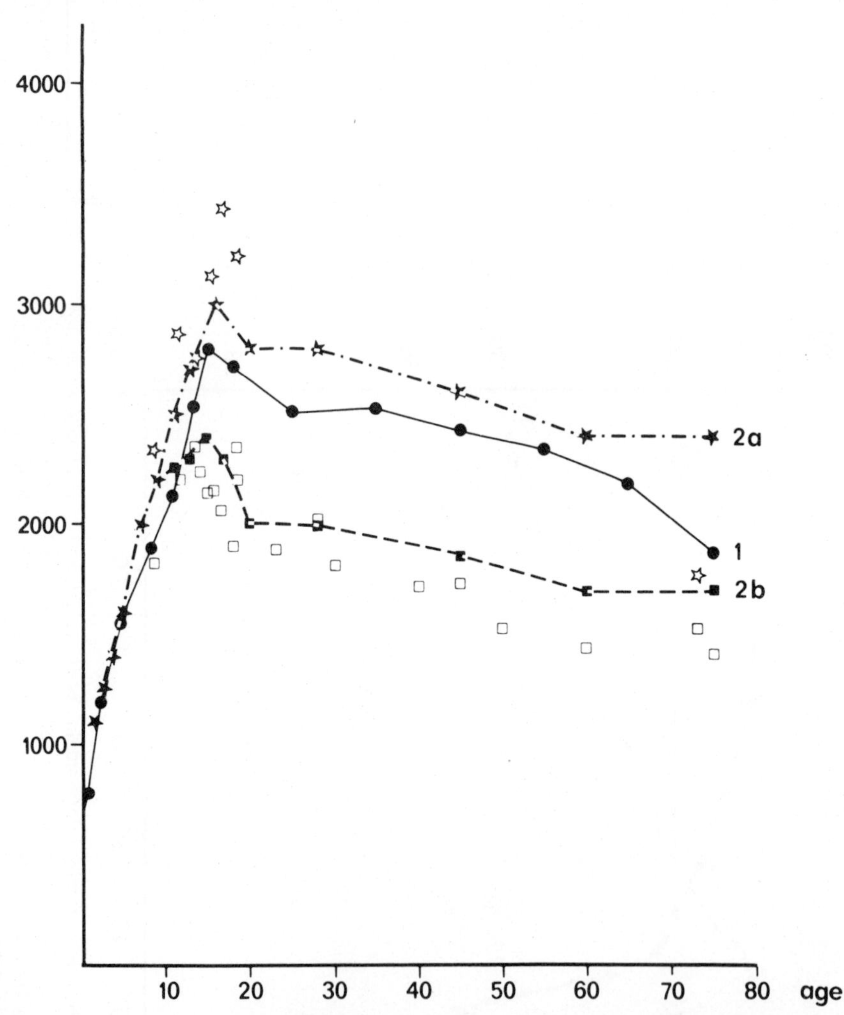

FIGURE 4:34. Daily calorie consumption by age. Curve 1: Japan, National Nutritional Survey, 1963. Curve 2a: recommended daily calorie consumption for Swedish men, National Institute of Public Health, 1969. Curve 2b: Recommended daily calorie consumption for Swedish women, National Institute of Public Health, 1969. ☆: actual male daily average calorie consumption in Sweden (Wretlind, 1968). □: actual female daily average calorie consumption in Sweden (Wretlind, 1968).

for both sexes and correcting for the relation between recommended and actual consumption. It is assumed that the average calorie intake in the United States is similar to the Swedish intake.

Figure 4:35 shows a comparison of renal cadmium accumulation both with and without the variation of daily calorie intake with age.

4.5.2.3 Different Excretion Rates of Cadmium

The approximate biological half-times of cadmium corresponding to different hypothetical excretion rates are seen in Table 4:12.

4.5.2.4 Calculated Cadmium Concentrations in Kidney Cortex Considering Excretion, Kidney Weight, and Calorie Intake Simultaneously

In Figure 4:36, curves for the renal accumulation of cadmium are calculated for different excretion rates. All curves pass through 50 μg/g at age 50, which is in accordance with empirical U.S. data by Schroeder and Balassa, 1961 (Section

TABLE 4:12

Daily Cadmium Excretion in Percentage of Body Burden and Corresponding Biological Half-time

Cadmium excretion per day (% of body burden)	Cadmium half-time (years)
0	∞
0.005	38
0.01	19
0.02	9.5
0.05	3.8

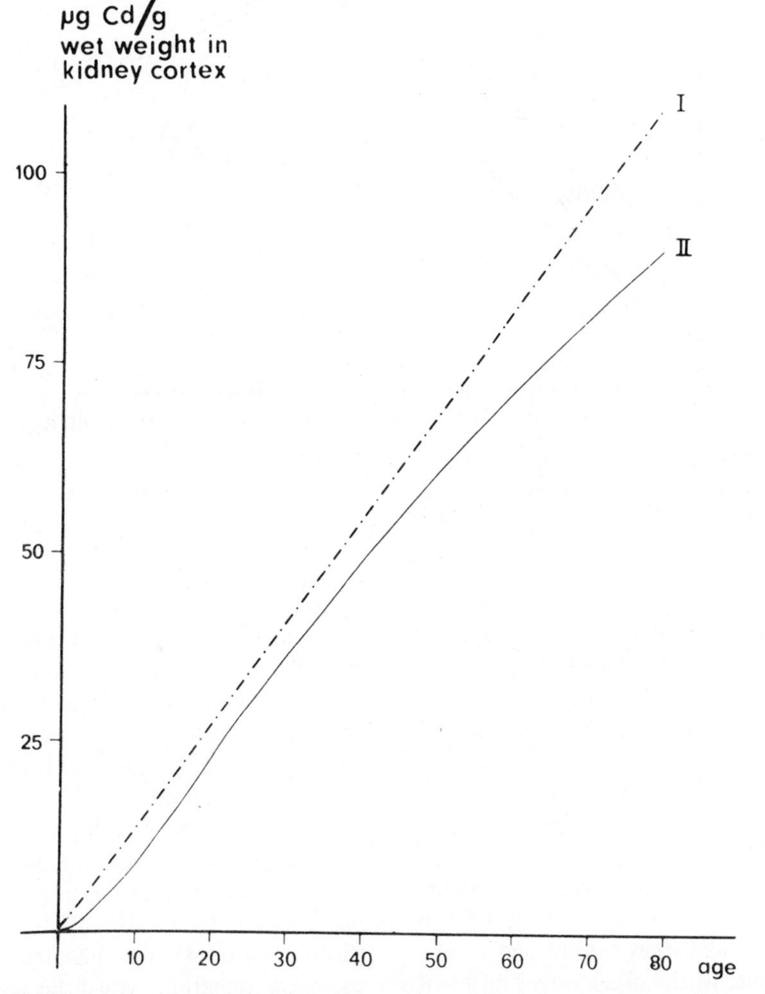

FIGURE 4:35. Calculated cadmium concentration in kidney cortex in relation to age and calorie intake. Curve I: assuming constant calorie intake. Curve II: adjusted for variation in calorie intake by age. Both curves assume 0% body burden excretion, 300 g kidney weight, constant.

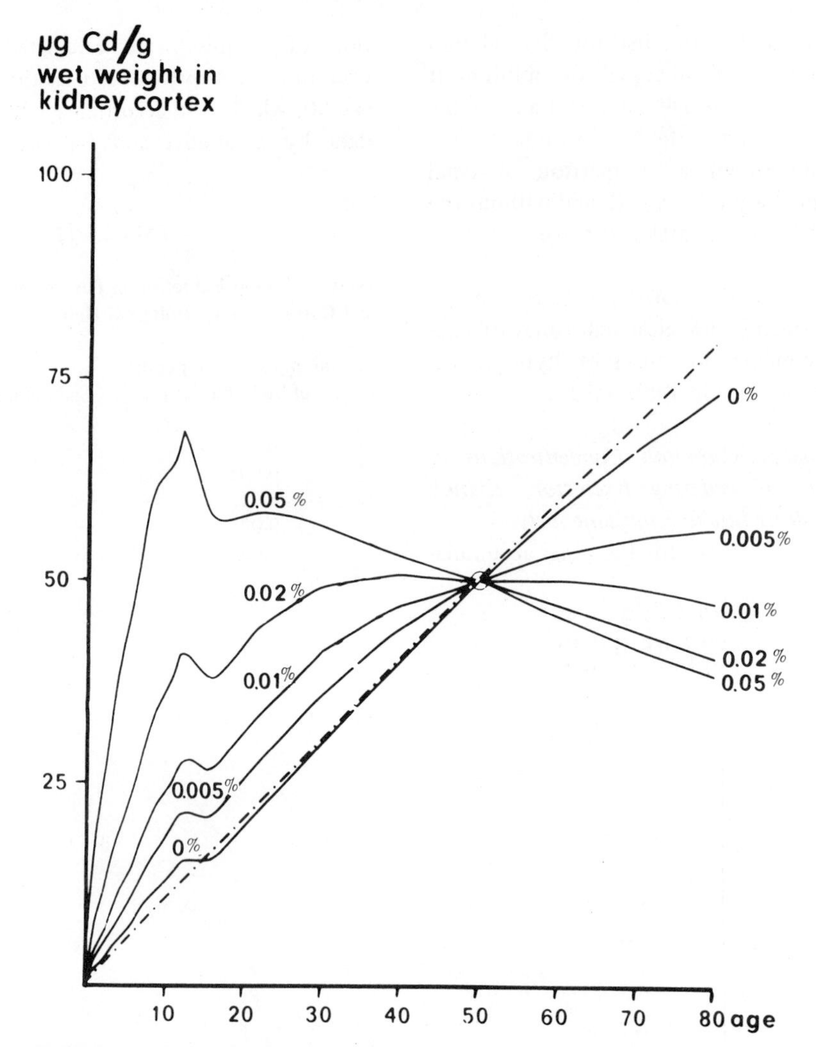

FIGURE 4:36. Calculated cadmium concentrations in kidney cortex by age for different body burden excretion alternatives. Adjusted for variation in calorie intake and kidney weight by age, except dotted line.

4.3.2.1, Figures 4:24 and 4:25), and only slightly higher than the figures by Gross, Cholak, and Pfitzer (unpublished data). Lower values have recently been reported by Hammer and Finklea, 1972, from U.S. samples, introducing the possibility that the data by Schroeder and Balassa do not necessarily represent an average for the U.S. as a whole.

The dotted straight line in Figure 4:36 represents the accumulation with age if body burden excretion is zero and kidney weight and calorie intake are constant. In the other curves different excretion alternatives are introduced and correction is made for kidney weight and calorie intake variation with age. Table 4:13 gives the necessary cadmium intake to reach 50 µg/g at age 50 under different assumptions (see also Chapter 9).

As was pointed out in Section 4.3.4, smoking may contribute to the total body burden. Cigarette consumption in the United States was on the increase until 1962, when the average consumption reached the maximum of about 4,000 cigarettes per year and adult (Beese, 1972). If each cigarette causes an inhalation of 0.1 to 0.2 µg (Section 3.1.5) and 25 to 50% is retained (Section 4.1.1.4) the maximum average daily cadmium retention would be about 1 µg. This figure should be compared with the 2.25 µg retained from 50 µg daily intake via food (Section 3.2) using the same assumptions as in the calcula-

TABLE 4:13

Daily Cadmium Excretion in Percentage of Body Burden and Corresponding Cadmium Intake

Cadmium excretion per day (% of body burden)	Necessary cadmium intake (μg per day; adults = 2,500 cal/day)
0	41
0.005	62
0.01	88
0.02	154
0.05	384

Assumptions:
(a) 4.5% Gastrointestinal absorption
(b) One third of body burden in kidney
(c) 50% Higher cadmium concentration in kidney cortex than the average kidney concentration
(d) Varying calorie intakes in different age groups (Figure 4:32)

tions above. Data on the changes of smoking habits over time within age groups have not been available to us but calculations based on the data at hand show that in a cross-sectional sample, smoking will contribute less than 40% to the body burden. Up to age 50 smoking will not distort the shape of the calculated accumulation curves to any large extent.

4.5.3 Some Possible Explanations for the Observed Decrease in Renal Cadmium Levels in Older Age Groups in Autopsy Studies

Studies in countries other than the United States have shown age curves for renal cadmium concentrations similar to U.S. data (Figure 4:25), and all of them show some decreaseafter middle age. This decrease does not fit in with the present simple metabolic model. There are several possible explanations for the consistently observed decrease in older age groups. The autopsy samples in older age groups may not have been comparable to those in younger age groups with respect to sex, diseases at death, and smoking habits. After 50 years of age, Swedish women have lower renal cadmium values than Swedish men (Piscator and Lind, 1971). Thus, unequal proportions of males and females in any specific age group could disturb the group average. Less smoking and lower calorie intake (Figure 4:34) might explain the lower values for women. Cadmium excretion in older age groups may increase due to renal morphological and functional changes. Less smoking among older people may also explain the shape of the curves.

The influence of changes in past exposure on the shape of the accumulation curves has been discussed by Kjellström and Friberg, 1972. Three types of past changes in exposure were considered. *Age-related* exposure variations include, for example, the effect of different dietary habits in different age groups apart from calorie intake variation. *Time-related* exposure variations could, for example, be caused by changes in cadmium concentrations with time in basic foodstuffs. *Cohort-related* exposure variations are exemplified by changes in smoking habits of successive cohorts.

In Figure 4:37 the effect on the shape of the curves of a certain type of age-related exposure variation is shown. Cadmium intake in relation to calorie intake increases linearly with age until age 40 and then decreases linearly.

Such an age-related cadmium intake may result from changes in dietary and smoking habits; a

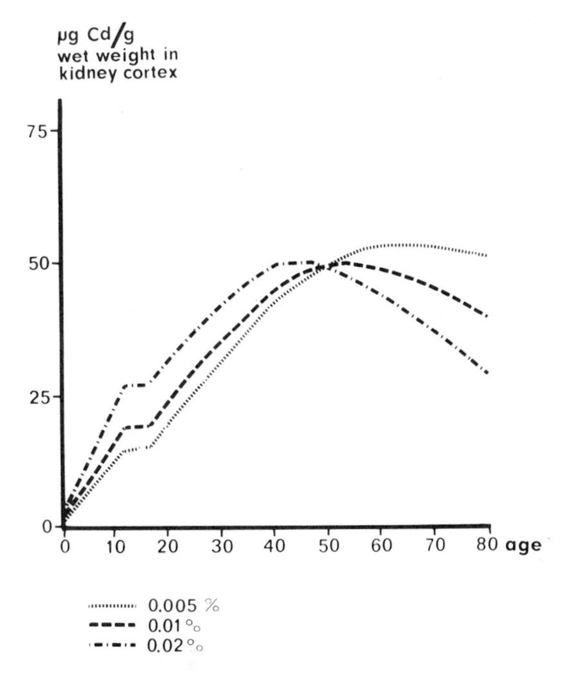

FIGURE 4:37. Calculated cadmium concentrations in kidney cortex by age for different body burden excretion alternatives. Daily cadmium intake proportional to an age-related factor (A) multiplied by daily calorie intake. Age 0 to 40, A increases linearly from 1 to 2; age 40 to 80, A decreases linearly from 2 to 1. (From Kjellström, T. and Friberg, L., Interpretation of Empirically Documented Body Burdens by Age of Metals with Long Biological Half-times with Special Reference to Past Changes in Exposure, Meeting on the Toxicology of Metals, Buenos Aires, September 1972, *Proc. XVII Int. Cong. Occup. Health,* 1974.)

large proportion of the calorie intake of infants and young children is made up of milk and milk products, which have a low cadmium content. Also, they do not smoke. With increasing age the consumption pattern changes, bringing an increase in the amount of ingested cadmium (e.g. in oysters, crabs and shellfish) and smoking is prevalent. In older age groups cadmium intake might decrease because the elderly usually eat more carbohydrates (e.g. buns) and smoke less.

Comparing Figures 4:36 and 4:37, the latter shows a straighter line of accumulation until age 50 for all excretion alternatives as well as a more distinct decrease in kidney cortex concentrations after age 50.

Various combinations of the three types of exposure variation may explain the empirical accumulation curves. Estimations about, for example, biological half-time based on such curves must therefore be regarded as very approximative if past changes in exposure cannot be accounted for.

4.6 BIOLOGICAL HALF-TIME

Information on biological half-times in man, as measured by observed decrease of retention of cadmium in an organ or in the whole body, is scarce.

In Section 4.1.2.2 an investigation by Rahola, Aaran, and Miettinen, 1971, is described, in which radioactive cadmium was given to male volunteers. The decrease in whole body retention of radioactive cadmium was very slow from 20 days to 2 months after exposure. It was not possible to give an exact value for the biological half-time for this slow component. The shortest estimate was approximately 130 days; the longest was infinity.

During 6 months, Sudo and Nomiyama, 1972, intermittently measured urinary cadmium excretion from a former cadmium worker who had proteinuria (Section 4.3.3.1). Calculating the biological half-time in urine, they arrived at a value of about 200 days. The authors did not account for the uncertainty in the value, which must have been considerable because of the reported marked variations in the daily urinary excretion and the short observation period compared to the biological half-time. These drawbacks, along with the facts that the person had proteinuria and that his fecal cadmium excretion was not taken into account, render it impossible to consider the value given by Sudo and Nomiyama as valid when calculating whole body biological half-time.

Biological half-time of cadmium in whole blood among workers after they have been removed from cadmium exposure can be roughly calculated to about 6 months from the data by Piscator (unpublished data) in Section 4.3.1. On the other hand Tsuchiya, 1970, showed a half-time in blood of about 1 year in two cadmium workers; in another worker there was no decrease subsequent to removal from exposure.

Another approach to the estimation of biological half-time is to compare directly excretion and body burden. As regards human beings, such data are only available on a group basis. Average excretion in a living population is compared with average body burden based on autopsy data. Tsuchiya, Seki, and Sugita, 1972a (Section 4.3.4) calculated that in Tokyo, Japan, body burden for a 50-year-old person is about 45 to 55 mg. Urine is considered the major route for body-burden-related excretion (Section 4.3.3.1), and in Tokyo urinary cadmium excretion is reported to be 1.5 to 2 μg/l (Figure 4:27). Assuming the urine volume to be 1.5 l/day these data taken together indicate that in Tokyo 0.004 to 0.006% of body burden is excreted daily. Similarly, in the United States body burden would be 15 to 20 mg (Section 4.3.4) and urinary excretion 1.6 μg/l (Table 4:7), which indicates a daily excretion of 0.012 to 0.015%. In Germany body burden can be calculated to be 8 to 12 mg (Section 4.3.2.1) and urinary excretion is reported to be about 1 μg (Table 4:7). These data indicate 0.008 to 0.013% body burden excretion. In summary, available data support 0.004 to 0.015% daily excretion of body burden, corresponding to a half-time of 13 to 47 years (Table 4:12).

Methods for calculating biological half-time using theoretical models of cadmium metabolism are discussed in Section 4.5. At present such models are the best means available to arrive at an estimate of the biological half-time for the body as a whole.

Intake via food is the major route for exposure to cadmium in a cross-sectional sample of the general population. Assuming that no past changes in exposure have occurred apart from age-related calorie intake variation, the following estimations can be made. The theoretically calculated curves in Figure 4:36 are compared with the empirically found curves in Figure 4:25; it is seen that the excretion rate seems to be 0.01 to 0.005%, which corresponds to a biological half-time of 19 to 38 years, as seen from Table 4:12. If the assumptions in the model concerning absorption rate (5%), kidney cortex level (50 μg/g at age 50), etc. are valid, Table 4:12 shows that an even longer half-time must be considered in light of U.S. dietary cadmium levels (Section 3.2).

All of these estimates are founded on the mentioned assumptions. If other assumptions are used, it is obvious that other results concerning biological half-times will be yielded. Hammer and Finklea, 1972, collected new autopsy data in the United States (Section 4.3.2): 25 μg/g Cd in renal cortex at age 50. They estimated the half-time in whole body to be 9 to 18 years, using the same model as used in section 4.5.1.

Tsuchiya, Sugita, and Seki, 1972, used a similar model, where, in addition to cadmium accumulation in kidney, cadmium accumulation in liver was included. Assuming no past changes in exposure, they calculated the half-time in kidney to be 17.6 years and in liver 6.2 years.

In conclusion, the question of the exact biological half-time of cadmium in the human body is still under discussion. Data favor a very long biological half-time, probably 10 to 30 years for the kidneys and whole body.

4.7 GENERAL CONCLUSIONS ON METABOLISM

4.7.1 In Animals

The fate of cadmium in experimental animals has been studied after exposure via the respiratory, gastrointestinal, or injection routes. Absorption is not known exactly with regard to inhaled or ingested cadmium, but reasonable estimates are about 10 to 40 and 2%, respectively. The absorption of ingested cadmium may increase considerably if the diet is deficient in calcium or protein or both. There are also considerable individual variations.

In blood, cadmium has been studied after single and repeated injections. After a single injection cadmium will initially be mainly in the plasma, but during the first 24 hr after injection, a rapid clearance from plasma takes place so that eventually the concentration in the cells will exceed that in the plasma.

After repeated exposure cadmium will mainly be found in the blood cells, bound to proteins such as metallothionein and hemoglobin. During continuous exposure to cadmium there will be a continuous increase in blood levels of cadmium, but at a certain level, a plateau will be reached and no further increase will be seen. When exposure has ceased, the concentration in blood will decrease.

Prior to the occurrence of renal damage, 50-75% of the total body burden of cadmium will be found in the liver and kidneys together. Immediately after single exposure the main part will be in the liver, but eventually renal levels will exceed liver levels. Repeated exposure to small amounts will also result in the renal concentrations surpassing the liver concentrations, but with increasing doses liver concentrations will exceed renal concentrations. The accumulation in the liver and kidneys seems to be mainly dependent on the storage of cadmium in the cadmium-binding protein, metallothionein.

In the kidney the highest concentrations of cadmium will be found in the cortex. It is possible that it is transported to the cortex with metallothionein. During exposure there will be a continuous accumulation of cadmium in the kidneys and urinary excretion of cadmium will be low. In mice the urinary excretion has been shown to constitute, depending upon the exposure level, 0.01 to 0.02% of the total body burden before renal damage has occurred. There is a considerable individual variation. When the cadmium level in the kidneys reaches about 200 to 300 μg/g wet weight (or about 400 μg/g wet weight in the cortex), there will be no further increase and urinary excretion of cadmium will increase considerably. As proteinuria occurs at this time, this change in metabolism is thought to be due to renal tubular dysfunction. Also, the excretion with feces will be low during the accumulation period. In mice the main part of the fecal cadmium excretion is related to recent exposure.

The biological half-time of cadmium after a single exposure has been estimated to be about 200 days in mice and rats, 400 days in dogs, and 1.5 years in squirrel monkeys. When daily doses of nonradioactive cadmium were given for 25 weeks after a single injection of radioactive cadmium in mice, the half-time of the radioactive dose was considerably longer, a couple of years.

The metabolism of cadmium is intimately connected with zinc metabolism. The low molecular weight protein, metallothionein, is able to bind both cadmium and zinc, and these two metals will thus be transported together. Zinc is an essential metal and many enzymes are zinc dependent. Cadmium seems to have the ability to exchange with zinc and thus cause changes in enzymatic activity. During exposure to cadmium organ levels of zinc will increase.

4.7.2 In Human Beings

Cadmium is absorbed via inhalation or ingestion. Observations on absorption from inhalation are scarce but data on smokers indicate that absorption of cadmium fumes is high and may well be 25 to 50%.

After ingestion an average absorption of about 6% (range: 4.7 to 7.0) has been observed in five persons given single doses of radioactive cadmium. An absorption of about 10%, or even higher, of ingested cadmium must be considered quite possible in individual cases under certain circumstances such as calcium and protein deficiency.

Estimations of retention starting from body burdens found at autopsies, together with estimated intake of cadmium in the past, point to an average via the oral route of somewhere between 3 and 8%.

The newborn contains less than 1 μg of cadmium, indicating that the placenta is an effective barrier to cadmium.

Normal concentrations of cadmium in blood are well below 10 ng/g blood. In exposed workers considerably higher values have been reported, usually between 10 and 100 ng/g. In exposed workers cadmium is found mainly in the blood cells. After exposure has ceased, concentrations of cadmium in blood will decrease slowly. After short-term exposure to high amounts of cadmium, the initial decrease may be fast.

In normal human beings the largest concentrations of cadmium will be found in the kidneys. This accumulation in the kidneys reaches its maximum when a person is around 50 years of age, when mean concentrations in the renal cortex have been shown to be 20 to 50 μg/g wet weight in persons living in the United Kingdom, the United States, and Sweden. In older persons lower values are seen. In three areas in Japan, mean normal levels of around 50, 90, and 125 μg/g, respectively, have been found in the cortex.

In exposed workers concentrations of around 300 μg/g have been found in the cortex, but values within the normal range have also been reported despite signs of cadmium intoxication. Losses of cadmium due to renal dysfunction are thought to be the cause of those low values.

In normal human beings, the kidneys and liver together contain about 50% of the total body burden of cadmium. The kidneys alone contain about one third of the total body burden. At high exposure, as seen sometimes in industry, the liver will contain a larger proportion of the total body burden. In some cases in industrial workers the liver concentration of cadmium has been shown to exceed the kidney level. The pancreas may contain high concentrations of cadmium.

As in animals cadmium metabolism in humans will be intimately related to zinc metabolism. When cadmium is accumulated in the kidneys, there will be a corresponding molar increase in renal concentrations of zinc. The cadmium and zinc-binding protein, metallothionein, has been found in normal human kidneys, and metallothionein probably plays an important role in the metabolism of cadmium in human beings.

In normal human beings, the excretion of cadmium via the urine is very low, around 2 μg/day or less. The excretion increases with age. On a group basis, for normal populations with moderate exposure to cadmium, this increase with age parallels the increase in the total body burden. A similar relationship has not been found in exposed workers without proteinuria. Furthermore, due to a wide scatter, and based on animal data, the predictive value of urinary cadmium as an index of total body burden or kidney burden on an individual basis must be considered low.

As judged from a study with radioactive cadmium, the excretion in feces is low if the very first postexposure period is disregarded. Small amounts of cadmium might also be excreted via hair, skin, sweat, breast milk, and saliva (in spitters) etc., but these routes of excretion with all probability will be of minor importance in relation to urinary and fecal excretion.

During occupational exposure to cadmium a worker's excretion of cadmium via urine might for some time be within normal limits. If and when tubular dysfunction occurs, as indicated by an increase in excretion of low molecular weight proteins, there will be an increase in the excretion of cadmium. This increase may be dramatic.

The biological half-time, as estimated based on mathematical models, may be between 10 to 30 years for total body. There are still considerable uncertainties and a proper accumulation model for cadmium in different organs, including the kidney, has yet to be found.

As cadmium concentrations in blood during exposure will vary independently of renal concentrations of cadmium, the determination of cadmium in blood will be of limited assistance for making an estimate of renal concentrations. On a group basis, it may be of value for estimating exposure.

APPENDIX 4:1

METALLOTHIONEIN: OCCURRENCE AND PROPERTIES

In 1957 Margoshes and Vallee reported that equine renal cortex possessed a cadmium-containing protein of low molecular weight. Further work by Kägi and Vallee, 1960 and 1961, resulted in the purification of a protein with a molecular weight around 10,000, which they named metallothionein because of its high content of metals, such as cadmium and zinc, and of sulfur.

This protein has many unique properties. In contrast to most other proteins, metallothionein absorbed UV-light of a wavelength of 250 nm, but not of 280 nm. The absence of aromatic amino acids, suggested by the lack of absorption at 280 nm, was verified by amino acid analysis. The absorption at 250 nm was shown to be dependent on the cadmium-mercaptide bond by the following procedure: the protein was dialyzed at a low pH or against EDTA whereby a metal free protein, thionein, was obtained. Thionein did not have any absorption at 250 nm because the cadmium mercaptide bond had disappeared. Addition of cadmium or zinc to thionein gave cadmiumthionein and zincthionein, respectively. By such addition the cadmium mercaptide bond was reestablished in the cadmiumthionein and the absorption at 250 nm reappeared.

The amino acid analysis showed that the high sulfur content, about 9%, was due to a very high percentage of cysteine in the protein. In metallothionein obtained from horse kidney, the cadmium content was 5.9% and the zinc content about 2.2%.

Proteins almost identical with metallothionein from horse kidney have been isolated and characterized from the liver of cadmium-exposed rabbits (Piscator, 1964, Kägi and Piscator, 1971, and Nordberg et al., 1972), from rat liver (Winge and Rajagopalan, 1972), and human liver (Bühler, 1973) as well as from human renal cortex (Pulido, Kägi, and Vallee, 1966). A protein has been isolated from horse liver that is identical with horse kidney metallothionein in amino acid composition and in other characteristics, but which has mainly zinc bound to it (Kägi, 1970). Zinc constituted more than 90% of the total metal content of this "hepatic metallothionein" which contained 11% sulfur, corresponding to 33% of thionein made up by cysteine. The ratio of cysteinyl residues to zinc was three, and Kägi suggested that a negatively charged complex had been formed: $Zn(Cys)_3$. The 3:1 ratio has been found in metallothionein from horse kidney and liver and from human kidney and liver, as well as from rabbit liver (Bühler, 1973, Kägi, unpublished data). The molecular weight was estimated from amino acid analysis to be about 6,600. The earlier mentioned higher values, around 10,000, had been determined by gel filtration. Data by Nordberg et al., 1972, and Bühler, 1973, confirm the results by Kägi, 1970, that the true molecular weight of thionein as shown by amino acid analysis is 6,000 to 7,000. Further results (Kägi, unpublished data) based on gel filtration in guanidine HCl, ultracentrifugation, and amino acid sequence analysis have also been consistent with the last mentioned value.

Similar proteins have been found in the liver and kidney as well as elsewhere in other animals, i.e., in the liver, kidney, and pancreas of cadmium-exposed rats and mice (Shaikh and Lucis, 1969, 1971, Jakubowski, Piotrowski, and Trojanowska, 1970, Nordberg, Piscator, and Lind, 1971, and Nordberg and Nordberg, 1973), in the spleen of the rat (Shaikh and Lucis, 1971), in the testis of the mouse (Nordberg, 1971a), and in the testis of the rat (Singh and Nath, 1972, Chen et al., 1972). Such proteins have been detected in the duodenum of the chicken (Starcher, 1969), the rat (Evans and Cornatzer, 1970), and the bovine (Evans, Majors, and Cornatzer, 1970).

In the blood of cadmium-exposed mice, cadmium is bound to a protein with a molecular weight similar to that of metallothionein (Nordberg, Piscator, and Nordberg, 1971, Nordberg, 1972a). A protein resembling metallothionein was also found in rat placenta, while in the lactating mammary glands of rats cadmium was found only in high molecular weight protein fractions (Lucis, Lucis, and Shaikh, 1972).

Some further data on the chemistry of metallothionein have been provided by Webb, 1972b, who suggested that in contrast to metallothionein from other species rat liver metallothionein needed only two SH-groups to bind cadmium. According to Webb these rat proteins are heat resistant, a property that can be used in their preparation. Webb, 1972b, 1972c, has also thoroughly discussed the evidence behind the claim that metallothioneins are mainly synthesized in response to exposure to cadmium and zinc. Webb stimulated synthesis of zinc-containing, cadmium-binding low molecular weight proteins by injecting zinc, which is in contrast to earlier reported results of Shaikh and Lucis, 1970, who stated in a very short communication that they could not induce synthesis of cadmium-binding proteins by giving zinc. The purity of the zinc compounds given was not stated in either case. Since zinc salts may contain cadmium as an impurity, an explanation may lie in a possible difference in cadmium content in the zinc salts used by the respective authors. The amount of contaminating cadmium might have been enough to stimulate synthesis of metallothionein in Webb's experiments.

Webb, 1972a, when separating liver and kidney supernatants by gel filtration, also found some slight differences between hepatic and renal low molecular weight cadmium-binding proteins. For instance, they differed in electrophoretic mobilities.

That there are different forms of metallothionein has been shown by Nordberg et al., 1972, who obtained two major forms of rabbit liver metallothionein with different metal contents and different isoelectric points by using isoelectric focusing. Amino acid analysis revealed that the amino acid composition of both forms was similar and also similar to metallothionein from horse kidney, earlier characterized by amino acid analysis by Kägi, 1970. Some differences between the rabbit liver and horse kidney proteins were notable, e.g., the rabbit protein lacked the arginine, valine, and leucine reported in the horse protein.

The two forms of metallothionein purified by Nordberg et al., 1972, differed as to their contents of lysine, proline, and glutamate. One of these forms contained cadmium and zinc in a molar ratio of 50:1 and the other one in a ratio of 1:1.2. The total molar amount of metal was about the same. The differences in electrophoretic mobilities and in isoelectric points may be due to both different charges and sizes, dependent on metal content and type of metal, but also on differences in amino acids. A third form, containing only zinc, was also observed, but in smaller amounts, which did not allow further characterization.

More data on horse metallothionein have been furnished by Roosemont, 1972, who found two metalloproteins in horse renal cortex. These differed in metal content and in ratios between cadmium and zinc.

Regarding metallothionein in human beings, our knowledge still comes mainly from the studies of Pulido et al. mentioned earlier. Some additional information has recently been obtainable from reports by Wisniewska-Knypl, Jablonska, and Myslak, 1971, and Bühler, 1973. Wisniewska-Knypl, Jablonska, and Myslak, 1971, investigated a case of acute cadmium poisoning in a man who swallowed large amounts of cadmium iodide (25 mg Cd/kg body weight) in a suicidal attempt. He suffered anuria, among other things. Treatment with a calcium EDTA compound was tried but did not improve his condition and he died after 7 days. Cadmium accumulated mainly in the liver and renal cortex, and it was possible from these two organs to identify low molecular weight cadmium protein complexes by separations on Sephadex G-75. In renal medulla, testis, and heart, the main cadmium fraction was also found in a low molecular weight fraction.

Bühler, 1973, prepared two main forms of metallothionein from human liver, both of which contained zinc for the most part. Like the rabbit protein, both forms of human liver metallothionein lacked arginine and one of them (MT-2) was also devoid of leucine. The molecular weight for both forms was between 6,000 to 7,000, as mentioned above.

Metals other than cadmium and zinc can be bound by metallothionein. Pulido, Kägi, and

Vallee, 1966, found that renal metallothionein from human beings treated with mercurial diuretics contained mercury. By experiments with equine metallothionein, Kägi and Vallee, 1961, showed that the bond between metallothionein and mercury was stronger than the one to cadmium, which in turn is about 3,000 times stronger than the one to zinc. Jakubowski, Piotrowski, and Trojanowska, 1970, found that part of the renal mercury from rats injected with $^{203}HgCl_2$ was in a low molecular weight protein, which had characteristics in common with metallothionein.

Wisniewska-Knypl et al., 1972, and Piotrowski et al., 1972, have recently been able to isolate a mercury-binding metallothionein from rat liver. Determination with gel filtration showed a molecular weight around 10,000 and an absorption curve typical for metallothionein. Injections of mercury did not induce synthesis of metallothionein to the same extent as cadmium, and it was also necessary to inject cadmium to stimulate the production of metallothionein. It has not been shown that exposure to mercury alone causes increases in metallothionein in the liver, but such a course of events has been observed in the kidney (Piotrowski et al., 1972).

Bryan and Hayes, 1972, claim that they found an increase in low molecular weight metal-binding proteins in liver of mice after oral exposure to mercury chloride for several weeks. They separated soluble liver components on a short Sephadex G-75 column (void volume 4.2 ml) at pH 7 and with an eluant of very low ionic strength, and found increases in the 250/280-nm ratio in the low molecular weight fractions, which they estimated to have a molecular weight of 11,000. This estimation cannot be correct. It is difficult to see how reliable molecular weight determinations could be obtained on such a small column without a valid marker procedure. Moreover, a careful examination of the gel filtration patterns in the report reveals that the fraction that is called low molecular weight protein is most probably composed of compounds in the molecular weight range of amino acids.

A protein similar to metallothionein but containing mainly copper was found in duodena of different species (Starcher, 1969, and Evans, Majors, and Cornatzer, 1970). Both zinc and cadmium could displace the copper bound to this protein.

It is not known to what extent the above mentioned proteins which bind the different metals are identical, but all contained a large amount of metal and were of low molecular weight. These two properties make it highly probable that they play an important role for the absorption, transport, and excretion of metals, especially cadmium.

In a report by Nordberg, 1971a, a pharmacological approach was chosen to study the effects of administration of metallothionein to mice. Graded doses of metallothionein were injected together with cadmium and zinc. Although the total dose of cadmium was the same in all groups (1.1 mg/kg body weight), animals that were given a certain amount of metallothionein were protected from the testicular damage evident in the group not given any metallothionein. In the metallothionein-injected group histological evidence of tubular kidney damage was observed. Further studies on mice have been performed by Nordberg, Goyer, and Nordberg, 1973. These studies revealed that the toxicity of metallothionein-bound cadmium was greater than that of cadmium chloride. The greater toxicity was probably due to the histologically demonstrated kidney damage induced by metallothionein-bound cadmium. It was also found that more cadmium was in the kidneys and less in the livers of the animals given metallothionein-bound cadmium as compared with the animals given the same dose of cadmium chloride. This is in accordance with earlier theories that cadmium bound to metallothionein will be filtered through the glomeruli and subsequently be reabsorbed and stored in the tubules.

The importance of metallothionein for absorption, transport, excretion, and toxicity of cadmium has been discussed elsewhere in this book (Section 4.2.8) and will not be dealt with further here.

Chapter 5

RESPIRATORY EFFECTS AND DOSE-RESPONSE RELATIONSHIPS

Much evidence has been collected and entered in the literature showing the severe and extensive respiratory effects in human beings exposed in industries and in animals exposed in laboratories to various cadmium compounds. These symptoms can be noted in cases of acute exposure to high concentrations as well as in cases of chronic exposure to low concentrations of cadmium. In this review special attention will be given to the effects of chronic exposure. As the acute effects serve to show the potential hazards of cadmium, a brief review is given also of the nature of such effects as well as the concentration ranges associated with them.

5.1 ACUTE EFFECTS AND DOSE-RESPONSE RELATIONSHIPS

In 1932, Prodan reported acute effects in animals which had inhaled cadmium compounds. Since that time several reports verifying the deleterious effects on the lungs of both animals and human beings have been published. Some data have been recorded regarding dose-response relationships for different species.

Paterson, 1947, described in detail the pathology in rats acutely poisoned with cadmium through inhalation of finely dispersed cadmium oxide or cadmium chloride aerosols. The findings, similar for both exposure types, were confined to the lungs. The three clearly demarcated stages were (1) acute pulmonary edema, developing within 24 hr of exposure, (2) proliferative interstitial pneumonitis, observed from the 3rd to 10th days after exposure, and (3) permanent lung damage in the form of perivascular and peribronchial fibrosis. Similar findings, including long-term effects of an emphysematous nature, have been reported by Thurlbeck and Foley, 1963.

The first two of the demarcated stages described in animal studies could be confirmed clinically or at autopsy on human beings exposed during, for example, welding or cutting materials containing cadmium (Paterson, 1947, Huck, 1947, Reinl, 1961, Lamy et al., 1963, Kleinfeld, 1965, Blejer, Caplan, and Alcocer, 1966, Beton et al., 1966, and Townshend, 1968).

The question of whether a nonfatal acute exposure can produce long-term effects in human beings does not seem to have been properly studied on a large number of cases. Townshend, 1968, reported one case of pulmonary edema caused by acute cadmium poisoning which was observed over a period of 4 years. Tests showed a gradual improvement of the lung function during the first 6 months. The CO diffusing capacity was normal 4 years later but the forced vital capacity was less than 80% of the predicted value. Similar studies on larger groups exposed acutely to cadmium are greatly needed.

Human data regarding dose-response relationships are quite scarce. From animal experiments evidence is fairly well documented concerning the acute mortality rate after short-time exposure to cadmium oxide fumes, the type of exposure occurring in connection with most of the reported human cases.

Earlier mentioned studies by Harrison et al., 1947 (Section 4.1.1.2) showed an LD_{90} in dogs of 320 mg Cd/m^3 over a 30-min exposure to a cadmium chloride aerosol (9,600 min · mg/m^3). Barrett, Irwin, and Semmons, 1947, gave data from exposure of a substantial number of rats, mice, guinea pigs, rabbits, dogs, and monkeys to cadmium oxide fumes. The exposure periods lasted from 10 to 30 min. The LD_{50} (cumulative mortality up to 7 to 28 days) varied between 500 and 15,000 min · mg CdO/m^3 depending on animal species (Table 5:1).

Based on the amount of cadmium found in the lungs and the lung ventilation for the different species, the authors calculated the retention of cadmium oxide to about 11%. This did not differ markedly among the species. They also calculated the dose for two fatal human cases described by Bulmer, Rothwell, and Frankish, 1938, and Paterson, 1947, to 2,500 min · mg/m^3. Starting points were observed values of cadmium in the lungs corresponding to 18 and 17 μg cadmium oxide per gram of dry tissue, respectively, and an assumption that the percentage of retention of cadmium oxide fumes for man was the same as for animals.

An independent check on this calculated dose was made by Barrett and Card, 1947. They tried to reproduce the actual exposure conditions and

TABLE 5:1

LD_{50} by Inhalation of Arced Cadmium Oxide Fume for Various Species. Time Exposed: 10 to 30 Min

Species	LD_{50} min · mg/m³	Remarks
Rats	500	160 animals used in groups of 10–25
Mice	Less than 700. Probably about the same as for rats.	Approximate only
Rabbits	2,500	Approximate only
Guinea pigs	3,500	Approximate only
Dogs	4,000	Approximate only
Monkeys	15,000	Approximate only

From Barrett, Irwin, and Semmons, 1947.

measured the cadmium concentrations formed. They concluded that a lethal concentration of thermally generated cadmium oxide fume for a man doing light work is not over 2,900 min · mg/m³ and would possibly be as little as half of this value for arc-produced fume. Beton et al., 1966, applied to a fatal case of cadmium poisoning the type of calculations used by Barrett, Irwin, and Semmons, 1947. With an observed concentration of cadmium oxide in the lungs of 2.5 μg per gram wet weight, they reached the conclusion that the concentration times exposure time should have been around 2,600 min · mg/m³. The exposure time in this case was 5 hr and the authors calculated an average lethal concentration for this exposure time to be around 8.6 mg/m³ for cadmium oxide fumes. For 8 hr exposure a lethal concentration would then be around 5 mg/m³.

The authors pointed out the uncertainty in regard to some of the assumptions necessary for the calculations. There could be no doubt that such uncertainties existed. Furthermore, the used "retention value" of 11% was with all probability considerably lower than the "retention value" immediately after the exposure. As a value of 11% was used in both the animal studies and in the calculations for human beings, any error tends to be cancelled out, however. There are reasons to accept a figure of about 5 mg cadmium oxide fumes/m³ over an 8 hr period as a probable lethal concentration for man.

By no means must 5 mg/m³ be the lowest that can give rise to fatal poisoning. It is known from animal experiments (Paterson, 1947) that exposure to concentrations of only about one fourth the LD_{50} can give rise to acute symptoms and to a significant degree of permanent lung damage. Even without applying a safety factor when extrapolating from animal data, there is thus reason to regard a concentration of around 1 mg/m³ over an 8 hr period as immediately dangerous for human beings with reference to cadmium oxide fumes.

For cadmium dust the data necessary to perform similar estimations are not available. Friberg, 1950, however, determined LD_{50} for seven groups of eight rabbits for cadmium-iron oxide dust (proportion Cd to Fe: 3:1, observation period: 14 days). On an average, about 95% of the particles were smaller than 5 μm and about 55% smaller than 1 μm (coniometer method). The rabbits were exposed for 4 hr, and LD_{50} calculated as the product of concentration and exposure time was about 11,000 min · mg/m³. Taking only the cadmium oxide part of the dust into consideration, the LD_{50} for cadmium oxide dust would be about 8,000 min · mg/m³, i.e., about three to four times the values for cadmium oxide fumes (for rabbits: 2,500). For an 8-hr exposure, this gives a concentration of about 17 mg/m³ of cadmium oxide dust for rabbits.

5.2 CHRONIC EFFECTS AND DOSE-RESPONSE RELATIONSHIPS

5.2.1 In Human Beings

5.2.1.1 Industrial Exposure

In the earlier literature (Stephens, 1920) there are scanty data suggesting a chronic cadmium poisoning in human beings. With the knowledge available today, it can further be said that some early reports of plumbism might instead have been cadmium poisoning. When Seiffert, 1897, described 65 cases of chronic lead poisoning among workers in a zinc smeltery, he thus reported emphysema among 83% and proteinuria among 82% of these workers. No doubt there must have been a considerable exposure also to cadmium, and the symptoms described fit chronic cadmium poisoning.

5.2.1.1.1 Respiratory Effects Reported – Exposure to Cadmium Dust

The first reports of lung emphysema in chronic cadmium poisoning were given by Friberg, 1948a and b, and in more detail by Friberg, 1950, in studies among male workers exposed to cadmium oxide dust in an alkaline accumulator factory in Sweden. Complaints of shortness of breath were common. An impaired lung function was demonstrated as an increased residual capacity in relation to total lung capacity. It was shown that the impairment of lung function was closely associated with a poor physical working capacity evaluated with a standardized working test on a bicycle ergometer (Figure 5:1).

The Swedish reports referred to above covered 43 male workers with an average time of employment of 20 years (range: 9 to 34 years). Another group of 15 workers, similarly exposed, among whom lung function was normal, had been employed for only 1 to 4 years. All workers had been exposed to a mixture of cadmium iron oxide dust and nickel graphite oxide dust. Quantitative data concerning the exposure were incomplete as air analyses were made only upon one occasion and only at five places (over about 30 min at each place) in the working rooms. The amount of cadmium in the air varied between 3 to 15 mg/m^3. Time-weighted 8-hr average values were not calculated but were probably lower. Respiratory masks were provided but stated to be used rarely.

The Swedish findings were confirmed at a

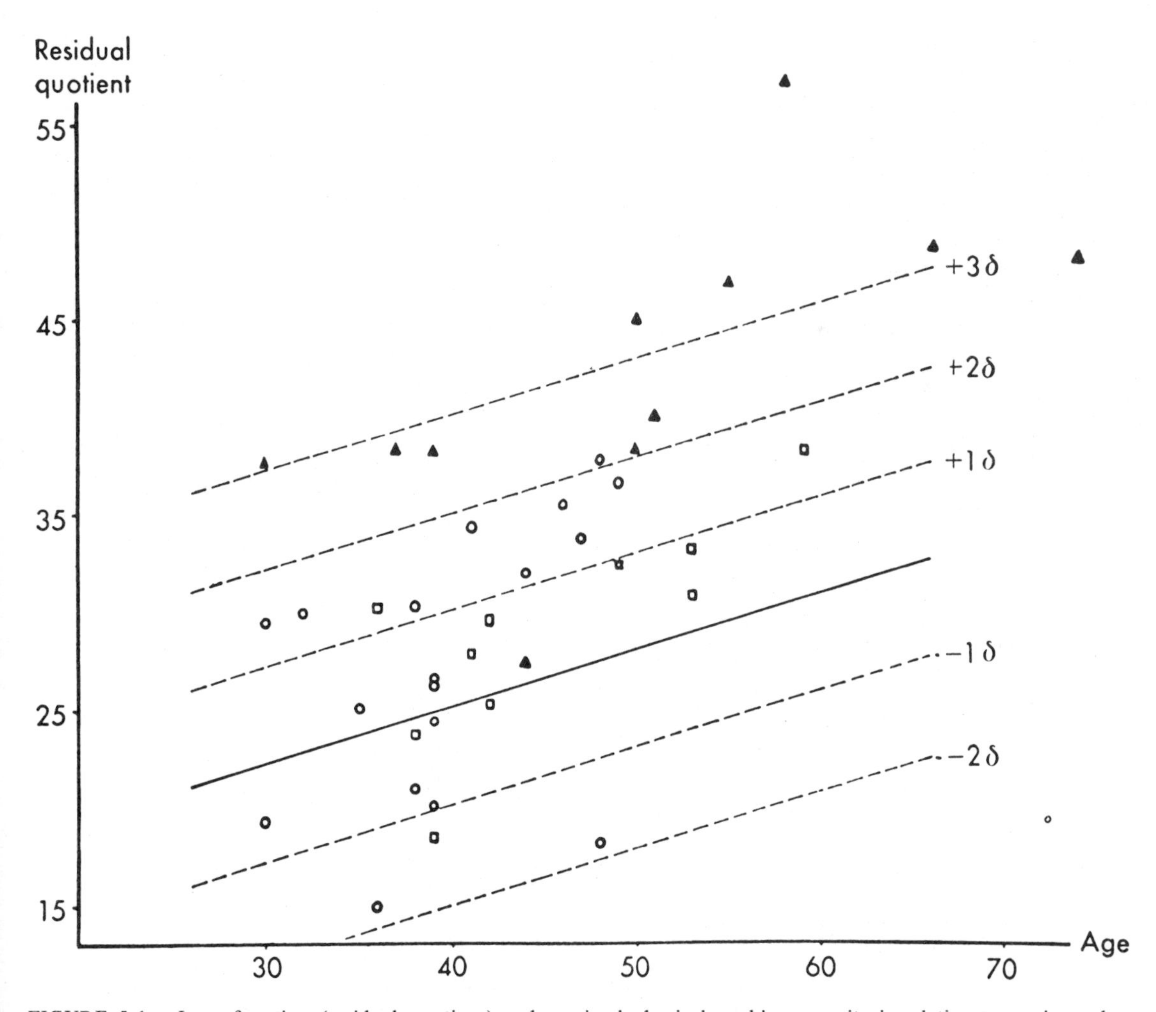

FIGURE 5:1. Lung function (residual quotient) and maximal physical working capacity in relation to age in workers exposed to cadmium iron oxide dust. Maximal physical working capacity with working test on bicycle ergometer: ▲∩≤ 600 kg/min; o = 900 kg/min; □∩≥ 1,200 kg/min. The regression line represents the mean residual quotient for 200 nonexposed workers. The individual symbols represent exposed workers. (From Friberg, 1950.)

similar German factory by Baader, 1951 and 1952. Out of eight workers exposed 8 to 19 years, six had emphysema as judged by clinical and X-ray examinations. Complaints of coughing and shortness of breath were common. No quantitative analysis about the exposure was given.

Several other reports of lung damage have since been made. Vorobjeva, 1957a, reported "diffuse pulmonary sclerosis" among female workers in the production of alkaline accumulators. From the same type of factory in Britain, Potts, 1965, reported bronchitis in six men and associated emphysema in four. Diagnostic methods were not discussed. In a later report from the same factory, Adams, Harrison, and Scott, 1969, found a decrease in forced expiratory volume (FEV_1) in male workers (Figure 5:2); 27 male workers were covered. Their exposure times were not obvious from the report but seem to have varied from 15 to 40 years. Cadmium in air estimations were reported beginning from 1957; before that time, regular analyses were not carried out. In one working area in the factory most values varied between 0.5 to 5 mg Cd/m^3, in a second between 0.1 to 0.5, and in a third, mostly around 0.1 mg Cd/m^3. It is not possible from the report to evaluate the exposure before 1957. The extent to which respirators were worn was not stated.

Kazantzis et al., 1963, reported emphysema in cadmium pigment workers. Thirteen men had been exposed to a variety of cadmium compounds of which cadmium sulfide based pigments were by far the most abundant. In addition to cadmium sulfide, the compounds included cadmium selenosulfide, cadmium zinc sulfide, cadmium carbonate, cadmium hydroxide, and cadmium oxide dust and fume. Out of six men engaged in the manufacture of cadmium pigments for 25 years or more, three had mild respiratory symptoms and showed slight but definite impairment of ventilatory function with a low FEV_1 (forced expiratory volume – 1 sec) percentage (56 to 61) and a high time constant (1.08 to 1.38 sec). A fourth man had died at the age of 46 from respiratory insufficiency and right-sided heart failure due to emphysema. It was concluded that the emphysema was caused by exposure to cadmium. Concerning the exposure, the report contains no data on air concentrations of cadmium. The exposure times were as follows: 25 to 31 years for six of the workers, 12 to 14 years for four of the workers, and ½ to 2 years for three of them.

5.2.1.1.2 Respiratory Effects Reported – Exposure to Cadmium Oxide Fumes

British researchers have published several reports on chronic cadmium poisoning in men exposed to cadmium oxide fumes in casting copper cadmium alloys. Emphysema has been prominent among the symptoms. Lane and Campbell, 1954, thus reported two fatal cases

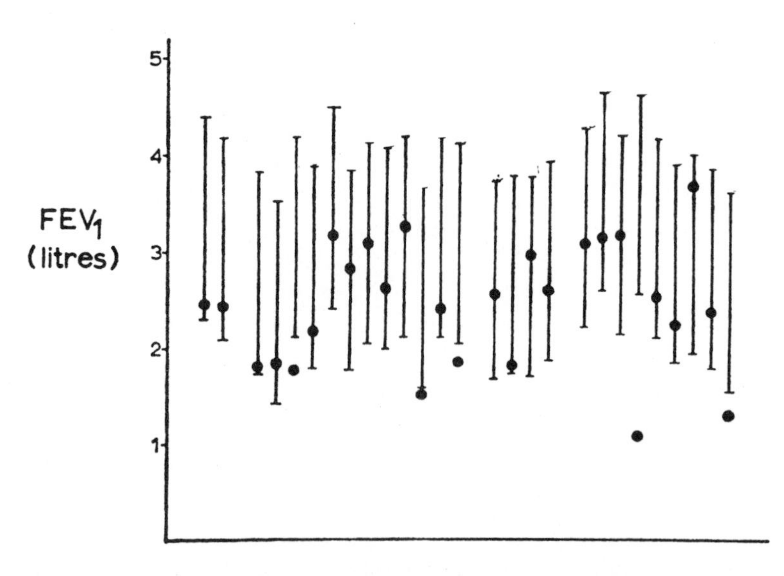

FIGURE 5:2. FEV_1 among male cadmium workers in the production of alkaline accumulators. The normal range for men of the same age and height (from Kory et al., 1961) is shown in each case. (From Adams, Harrison, and Scott, 1969.)

among men making a cadmium copper alloy for less than 2 years. The general exposure was given as 0.1 to 0.4 mg Cd/m^3. For very short periods, however, the men had been exposed to high concentrations of cadmium metal fume (Actual concentrations not given). Bonnell, 1955, examined 100 exposed men and 104 controls from two factories. The respiratory function was tested by measuring the vital capacity, the maximum ventilatory capacity at controlled rates of breathing, and the expiratory fast vital capacity. Swept fractions at given respiratory rates were calculated. Among the exposed workers 11 were diagnosed as having emphysema. There is no mention of emphysema in the controls. On a group basis significant differences were found between exposed and controls. In Factory A the vital capacities and the maximum ventilatory capacities were similar in the two groups, but the mean swept fractions at 30, 50, and 70 respirations per minute were significantly lower in the exposed group. A significant difference was also found between the groups for the mean time constant of the expiratory fast vital capacity curve. In Factory B the mean time constant for the expiratory fast vital capacity curve differed significantly between the exposed group and the control group. The other test results were similar in the two groups. The results of these tests are discussed in more detail by Kazantzis, 1956.

Some further studies were made on 37 exposed men from Factory B (Buxton, 1956). The men with more than 10 years exposure showed a significant increase in the mean value of the residual air expressed as a percentage of the total lung volume (mean: 43.9%, S.D.: 10.3%) compared with a control group (mean: 34.6%, S.D.: 8.3%) and workers exposed less than 10 years (mean: 36.6%, S.D.: 10.9%). The workers in the "cadmium groups" had been exposed from 1½ to 28 years, and those with a recognized emphysema from 7 to 27 years. King, 1955, has made a careful study to evaluate the exposure conditions at the time of Buxton's medical investigations. It is obvious from King's data that the concentrations of cadmium were considerably lower than in the accumulator industries referred to. At one working area he found average values for 8-hr working shifts over a 5-day period of from 13 to 89 μg Cd/m^3, at another place (two different positions) over a 9-day period, of from 4 to 132 μg/m^3, and at a third working place (four different positions) over an 8-day period of from 1 to 270 μg/m^3. About 90% by weight of the particles had a size of less than 0.5 μm. It is obvious from the report that the mean exposure for the majority of the workers must have been considerably less than 0.1 mg Cd/m^3. Unfortunately the study did not show the earlier exposure conditions. King pointed out that during recent years the working conditions in the industry had improved.

5.2.1.1.3 Respiratory Effects Not Reported

The studies discussed above are those in which pulmonary effects of chronic cadmium intoxication have been observed. There are also reports with negative findings.

Suzuki, Suzuki, and Ashizawa, 1965, examined workers exposed to cadmium stearate dust and lead in a vinyl chloride film plant. The study comprised 27 male workers in 1963 and 19 male workers (mean age: 22.8 years, S.D.: 5.5 years, time of employment: 3.3 years, S.D.: 1.9 years) as well as 24 controls in 1964. They found at the examination in 1964 no increased occurrence of respiratory symptoms in the exposed group compared to the controls. Furthermore, there was no difference between the two groups in the outcome of lung function tests. The concentrations of cadmium in 1963 (standard impingers) at one occasion (sampling time not given) were at four different operations 0.03, 0.26, 0.55, and 0.69 mg Cd/m^3. The size of the particles varied from 0.4 to 20 μm (cascade impactor dark-field microscopy). The exposure took place three or four times a day for 20 min each time. Nothing was mentioned about the use of respirators. Apart from the above mentioned study in which the time of employment was short, no reports are known to us in which a careful pulmonary examination, including lung function tests, has revealed negative findings. There are some reports, however, where the use of more insensitive methods has shown normal or only very slight changes. These reports are referred to below.

Princi, 1947, examined 20 workers in a cadmium smeltery. In the clinical examination he found no subjective symptoms that could be connected with exposure to cadmium (no controls were examined). Pulmonary X-rays showed normal conditions. Lung function studies were not carried out. The average period of employment for the workers examined was 8 years (range: 0.5 to 22 years, seven workers with more than 10 years). All of them were exposed to cadmium in the form of dust or fumes (CdO and/or CdS). Princi carried out air analyses on three different occasions at 11 different areas of work. In this connection he found values which, calculated as metallic cadmium, varied as a rule from 0.04 to 1.44 mg/m^3. In a few of the working areas there were considerably higher concentrations of dust. Thus, in two places in which, to judge from the description of the working conditions, the men were probably exposed to cadmium sulfide, values of 19 and 31 mg Cd/m^3 were measured. In another area of work, where the workers were probably exposed to cadmium oxide, a content of 17 mg Cd/m^3 was measured. Since many of the men were stationed sometimes at the one and sometimes at the other area of work, an accurate estimate of the individual exposure was not possible. Respirators were provided but worn infrequently.

Hardy and Skinner, 1947, left open the question of the possibility of chronic cadmium poisoning. They examined five workers exposed for periods from 4 to 8 years in the production of cadmium-faced bearings. The workers complained of unspecific symptoms, including respiratory symptoms on damp days. Chest X-rays were normal; no lung function tests were carried out. The authors gave the average exposure to cadmium as approximately 0.1 mg Cd/m^3 air.

L'Epée et al., 1968, examined 22 workers in an alkaline accumulator industry, 14 of whom had been employed for less than 5 years and 8 for more than 5 years. Unspecific respiratory symptoms were found among five workers. No pulmonary studies were made. The report included no quantitative data on cadmium exposure.

Tsuchiya, 1967, examined 13 workers (age range: 19 to 32 years) and 13 controls. The workers had been exposed to cadmium fumes while smelting alloys of silver and cadmium. No lung function tests were reported. Chest X-ray examinations did not reveal any abnormalities. No persistent subjective symptoms could be detected. The cadmium concentration in the air was reported as time-weighted averages (electrostatic precipitator) at nose level of the workers for 5 days, and values varying from 68 to 241 $\mu g/m^3$ were found.

5.2.1.2 Nonindustrial Exposure

Lewis, Lyle, and Miller, 1969, studied the cadmium, copper, and iron concentrations in water-soluble protein fractions of liver extracts in connection with autopsies of patients with chronic bronchitis and emphysema. The samples were analyzed by atomic absorption spectrophotometry. They found a higher concentration of cadmium compared with a control group but no differences for iron and copper. The groups consisted of adult patients, but no age distribution was stated. For 20 patients with chronic respiratory problems, the cadmium concentrations had a mean of 40.4 (S.E.: 6.0) μg cadmium per gram protein. The corresponding value for 38 controls was 10.6 (S.E.: 17) μg cadmium per gram protein. Morgan, 1971, in a similar study also found an increased hepatic concentration of cadmium in deceased patients with a diagnosis of emphysema. For a discussion of this study, see Section 7.2.

Peacock (unpublished data, 1970) reported a study of 127 residents aged 50 to 69 of Birmingham, Alabama. They were given a detailed questionnaire on their health status supplemented by a medical examination including determination of serum cadmium levels (atomic absorption, analyses carried out by Morgan). The 127 participants were part of a total sample of 464 individuals who in turn comprised 78% of those requested to attend. Peacock compared the prevalence of an extensive number of variables such as symptoms, food, drinking, and smoking habits, and social characteristics in groups of high and low cadmium values. The only statistically significant ($p < .05$) association he found (with vertigo) was no more than would have been expected by chance. He also observed a relation ($p < .01$) between place of residence and cadmium levels, with most of the high serum cadmium levels among groups who might have been exposed to cadmium as an air pollutant presumably derived from the industrial steel complex in Birmingham. The study may serve as a basis for further investigations with specified hypotheses.

In summary, there are a few data showing an increased cadmium storage in persons with a diagnosis of emphysema and/or chronic bronchitis at autopsies. In view of the well-known association between smoking and chronic bronchitis as well as emphysema (*The Health Consequences of Smoking,* 1971) and the considerable exposure to cadmium via smoking (see Section 3.1.5), there is reason to believe that smoking may be of importance.

5.2.1.3 Prognosis

Friberg, 1950, showed that workers with emphysema had a poor physical working capacity. He also reported one fatal case in which the cause of death was a pronounced emphysema, as well as another case in which pronounced pulmonary symptoms were present and hypertrophy of the right ventricular chamber was found at post-mortem. Baader, 1951, reported one fatal case due to severe emphysema. From England reports of several cases in which the cause of death was emphysema have been published (Lane and Campbell, 1954, Bonnell, 1955, Smith, Kench, and Smith, 1957, Smith, Smith, and McCall, 1960, and Kazantzis et al., 1963). The British researchers have pointed out that they did not find evidence either clinically or pathologically of chronic bronchitis.

The prognosis of chronic cadmium poisoning is further elucidated by data from two 5-year follow-ups by Friberg and Nyström, 1952, and by Bonnell, Kazantzis, and King, 1959, of the two earlier mentioned studies by Friberg, 1950, and Bonnell, 1955. When given respiratory function tests, exposed groups showed a greater deterioration with increase in age than the control group in the British study. The results in individual cases showed a deterioration in the men who had emphysema at the time of the original survey despite the fact that the majority with chronic cadmium poisoning had not been exposed to cadmium since the original examination. In the Swedish follow-up study subjective symptoms increased in several of the cases. Though a general tendency toward poor lung function remained, further impairment had not taken place. Most of the workers had not been exposed to significant concentrations of cadmium during the period of the follow-up study.

In a follow-up of the cadmium workers studied by Friberg in the 1950's (Section 5.2.1.1.1), a significant overmortality was found (Friberg and Kjellström, unpublished data). The observed mortality of these workers was compared with the sum of the individual probable mortalities as calculated from Swedish life tables for men (= expected mortality) (Statistical Yearbook 1971). For the 43 persons with the longest exposure times, the observed mortality in 1952 was 6 and in 1956 was 7, which should be compared to the respective expected mortalities of 1.75 and 3.38. Assuming that the number of deceased workers is a Poisson-distributed function, it can be calculated that for 1952 the difference between observed and expected is statistically significant ($p < 0.01$) while for 1956 it is not significant. Corresponding figures were calculated for 19 cadmium workers with very high exposure who were referred to the hospital for detailed clinical investigations. In Figure 5:3 the observed and expected deaths for different years are shown. In 1955 the expected number of deaths was 1.1 and the observed 7. This difference is statistically significant ($p < 0.001$).

5.2.2 In Animals

Pulmonary changes have been found in animal experiments. Friberg, 1950, exposed 25 rabbits for 3 hours a day for 20 days per month for 8 months to about 8 mg cadmium iron oxide dust/m^3 (taken from the alkaline accumulator factory where the clinical studies were carried out; see Section 5.2.1.1.1). This corresponded to approximately 5 mg cadmium/m^3, or about 900 min · mg/m^3 per exposure day. All rabbits showed signs of emphysema in addition to inflammatory changes. As the workers in the accumulator factory were exposed also to nickel graphite dust, similar animal experiments were carried out with such dust. Emphysema and inflammatory changes were seen but to a much lesser extent, despite the fact that the actual dust concentrations were 10 to 20 times higher than in the cadmium experiments.

Vorobjeva, 1957a, installed intratracheally in rats cadmium oxide dust (2.5 mg/kg body weight), cadmium iron dust (3.5 mg/kg body weight) as well as nickel graphite dust (20 mg/kg body weight). After 4 to 7 months the animals were killed. In the cadmium and cadmium iron groups there were signs of interstitial pneumonia, sclerosis, and emphysema. Similar changes were not seen among the animals in the nickel graphite group. In a brief report from the U.S.S.R., Shabalina, 1968, mentioned that cadmium stearate did affect the lung tissue 2 months after an intratracheal installation. The studies were performed to evaluate symptoms of cadmium intoxication in human beings in the form of subatrophic cathars of the upper respiratory tract and hyposmia.

Against these findings stands a study by Princi and Geever, 1950. They exposed dogs to cadmium oxide dust (ten dogs, 6 hr a day, 5 days a week for 35 weeks) and cadmium sulfide dust (ten dogs, 6

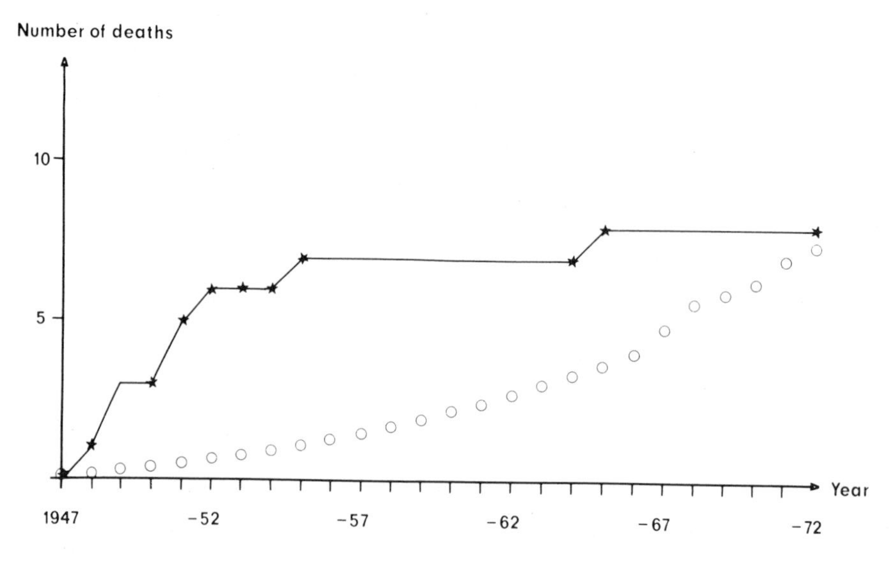

FIGURE 5:3. Observed (★) and expected (○) mortality of 19 Swedish cadmium workers with very high exposure at different times after the first clinical investigation in 1947. (From Friberg and Kjellström, unpublished data.)

hr a day, 5 days a week for about 30 weeks) without finding any respiratory changes at postmortem compared with a control group. The concentrations of cadmium in the air varied from 3 to 7 mg/m^3 with an average concentration of 4 mg/m^3. Of the particles, 98% were less than 3 μg in diameter. It should be mentioned that several of the animals had to be killed because of severe injuries received while fighting among themselves. Dogs that died of injuries had bronchopneumonia.

5.3 CONCLUSIONS

Brief inhalations of high concentrations of cadmium compounds can give rise to severe, often fatal, pulmonary changes (pulmonary edema). This has been shown in human beings with regard to cadmium oxide fumes and in animals with regard to cadmium oxide fumes, cadmium chloride aerosols, and cadmium oxide dust.

For human beings a fatal exposure to cadmium oxide fumes and cadmium chloride aerosols is not higher, probably lower, than about 2,500 min · mg/m^3 (corresponding to about 5 mg/m^3 for 8 hr). An exposure to about 500 min · mg/m^3 (corresponding to about 1 mg/m^3 over 8 hr) is considered immediately dangerous.

Dose-response relationships for cadmium oxide dust are scarce and uncertain. In rabbits, LD_{50} was about 8,000 min · mg/m^3, which would correspond to about 15 mg cadmium oxide dust/m^3 for an 8-hr exposure.

Occupational exposure for longer periods of time to lower concentrations of cadmium compounds can give rise to chronic pulmonary disorders, characterized as emphysematous changes. These changes as a rule have taken several years to develop in human beings but have been observed already after a few years' exposure. Lung damage has been observed after exposure to cadmium oxide fumes as well as to cadmium oxide dust and cadmium pigment dust.

Dose-response relationships are very uncertain because time-weighted averages are not available or only available for short periods. Furthermore, it is not known what reduction of exposure was brought about due to the use of respiratory masks. There is reason to believe that cadmium oxide fumes are more dangerous than cadmium oxide dust. For cadmium oxide fumes, a prolonged industrial exposure to well below 0.1 mg/m^3 might well be considered hazardous with reference to emphysema.

There exists evidence that nonoccupationally exposed people with chronic bronchitis and emphysema have on an average a higher body burden of cadmium than "controls." Smoking may be one cause of the increased body burden.

Chapter 6

SYSTEMIC EFFECTS AND DOSE-RESPONSE RELATIONSHIPS

6.1 RENAL EFFECTS AND DOSE-RESPONSE RELATIONSHIPS

It is well documented that exposure to cadmium will cause impairment of the kidneys. Investigations have covered workers exposed to different cadmium compounds in industry and general population groups exposed to cadmium-contaminated food, as well as animal experiments. Almost all of the effects referred to have resulted from prolonged exposure to cadmium. A brief mention of the acute effects will be made, particularly against the background of possible enhancement of the effects due to simultaneous exposure to chelating agents.

6.1.1 Acute Effects and Dose-response Relationships

6.1.1.1 In Human Beings

Exposure to high concentrations of cadmium oxide fumes has caused severe acute lung damage (see Section 5.1). In two fatal cases described by Bulmer, Rothwell, and Frankish, 1938, "cloudy swelling" was found in the kidneys at microscopic examination. The authors ascribed this to a general toxemia. Beton et al., 1966, reported on a fatal case in which bilateral cortical necrosis of the kidneys was found at autopsy. They thought that this was mainly due to vascular changes associated with massive pulmonary changes. The concentration of cadmium in the kidneys was given as 5.7 μg/g wet weight, which is below the average "normal" level for people of the patient's age (53 at death). In two other patients, who recovered after exposure, transient proteinuria was noted. Electrophoretic examination of the urine from one patient showed that mainly albumin was excreted.

Wisniewska-Knypl, Jablonska, and Myslak, 1971, reported on a case of acute cadmium poisoning. A 23-year-old man drank 5 g CdI_2 dissolved in water in order to commit suicide (about 25 mg Cd/kg body weight). The patient developed transient anuria during the first day. The authors stated that the laboratory findings showed pronounced signs of damage to liver and kidneys, hypoproteinemia with hypoalbuminemia, and metabolic acidosis.

6.1.1.2 In Animals

Acute effects of *inhalation* of cadmium have been studied on cats by Prodan, 1932. Two cats were exposed to high concentrations (precise amount not stated) of cadmium oxide fumes for 30 min. One of them showed fatty degeneration, especially in the convoluted tubules. In another experiment three cats were exposed for 24 hr to cadmium oxide fumes. The concentration of cadmium in air at the beginning of the exposure, 18 mg/m^3, was gradually reduced to 4 mg/m^3 at the end of the experiment. The cats were killed 5 and 9 days after exposure. Histological examinations disclosed in all three a moderate amount of fat in the tubular epithelium. It was calculated that the cats had inhaled between 6 and 12 mg of cadmium.

Regarding the dose-response relationship after respiratory exposure, no certain conclusions can be drawn as in these experiments the lung lesions were dominant and might have caused a general toxemia. As severe, probably lethal, lung damage will occur at cadmium concentrations that cannot cause renal damage, the kidney is not a critical organ after acute exposure.

Experiments involving a *subcutaneous injection* have revealed that the kidneys were not damaged when rats were given soluble cadmium salts (2.2 mg Cd/kg), a dose which did cause testicular necrosis (Parizek and Zahor, 1956, Parizek, 1957, 1960, and Kennedy, 1968). Favino and Nazari, 1967, observed tubular lesions of the nephrotic type in rats given a single subcutaneous injection of cadmium chloride (10 mg Cd/kg).

In the rabbit, Foster and Cameron, 1963, produced renal lesions with two subcutaneous injections of cadmium chloride (9 mg Cd/kg). The most striking changes were found in the proximal tubules.

6.1.1.3 Influence of Chelating Agents

It was earlier said that the treatment of cadmium poisoning will not be reviewed. However, some chelating agents (e.g., EDTA and NTA) have been introduced as components of detergents. A possible role of these agents in changing the toxicity of cadmium in man's environment has been discussed. As most of the work on chelating agents is concerned with more or less acute effects of cadmium, a brief mention of a few chelating agents will be made in this section.

In 1946, Gilman et al. documented a decrease in the acute mortality normally incurred after a single injection of cadmium when rabbits were treated with 2,3-dimercaptopropanol (BAL). However, several of the rabbits succumbed *later* from kidney damage. In a follow-up of these studies, Tepperman, 1947, showed that BAL increased the uptake of cadmium in the kidneys, an observation confirmed by Niemeier, 1967. Also, EDTA has been shown to decrease the acute lethality from a single injection of cadmium (Friberg, 1956, and Eybl, Sýkora, and Mertl, 1966a) whereas the agent increased the nephrotoxicity of cadmium after repeated exposure to cadmium (Friberg, 1956). Dalhamn and Friberg, 1955, have presented evidence of a similar unfavorable effect of BAL upon cadmium toxicity during prolonged exposure. Tobias et al., 1946, showed that *prophylactic* treatment with BAL had a deleterious effect upon mice exposed to cadmium chloride dust by inhalation. However, when BAL was given promptly *after* exposure to the cadmium dust, in an optimal course of repeated injections, it could reduce the mortality considerably. Also in these experiments BAL increased the cadmium content of the kidneys, but this seems not to have been of importance for mortality or pathological changes as no renal pathology was observed following BAL treatment of mice poisoned by inhaling cadmium. The experiments by Tobias et al. were performed with an aerosol of $CdCl_2$, an exposure situation less common in practice than exposure to CdO aerosol (cadmium fume). This latter exposure type was used in experiments by MacFarland, 1960, who showed that BAL treatment after the exposure could considerably reduce the mortality.

All of the experiments performed by Tobias et al. and MacFarland consisted of a short, single exposure. When the data by Gilman et al., Tepperman, and Dalhamn and Friberg are considered together with what is known about renal lesions as a main feature of chronic cadmium poisoning, it is highly probable that BAL also increases the toxicity of inhaled cadmium during long-term exposure.

Chelating agents such as hydroxyethylene diaminetriacetic acid (HEDTA), diethylenetriamine pentaacetic acid (DTPA), and others also had a preventive effect upon acute cadmium toxicity. As mentioned in Section 4.2.7, both EDTA and DTPA (Eybl, Sykora, and Mertl, 1966a) treatments decrease the uptake of cadmium in the kidneys. However, to what extent this decrease in uptake was responsible for the decrease in toxicity is not known.

In some contrast to the above mentioned diminutions in acute toxicities of injected cadmium, Chernoff and Courtney, 1970, reported an increased acute toxicity and increased fetal mortality when pregnant rats were given a single injection of a chelate of cadmium with nitrilotriacetic acid (NTA). This observation has been partly confirmed by Nordberg, 1972c, 1973, who found an increased mortality in certain dose intervals and at certain post injection survival times in mice given injections of cadmium chloride together with NTA. However, a similar effect was noted when STPP (sodium tripolyphosphate) was injected along with cadmium. The mentioned experiments on effect on acute toxicity have led to studies on chronic oral exposure to NTA. As taken up in Section 4.2.7, NTA has so far not been shown to change the distribution or the toxicity of cadmium in the body. If anything, large doses of NTA seem to accelerate the fecal elimination of cadmium.

6.1.2 Chronic Effects and Dose-response Relationships

6.1.2.1 In Human Beings

6.1.2.1.1 Proteinuria and Renal Function

In 1950, Friberg, having investigated a large group of workers exposed to cadmium oxide dust in an accumulator factory (for further details see Section 5.2.1.1.1), showed that prolonged exposure to cadmium gave rise to renal damage. He studied two groups, one of 43 workers with a mean exposure time of 20 years (range: 9 to 34 years) and one of 15 workers with a mean exposure time of 2 years (range: 1 to 4 years). In the former group proteinuria could be demonstrated with the nitric acid test in 65% of the workers and with the trichloracetic acid test in 81%. However, only 11% of the nitric acid positive urines reacted positively to the picric acid test. In the other group, with the mean exposure time of 2 years, no positive reactions were obtained with any of the tests. Tests of renal function disclosed that in the first group, 21% (9 out of 42) had decreased ability to concentrate the urine (maximum specific gravity $\leqslant 1.019$). Out of those nine people, eight had a constant proteinuria. In 33 workers with a higher concentration capacity (maximum specific gravity $\geqslant 1.020$), only 11 (33%) had a constant proteinuria.

As the proteinuria differed from ordinary albuminuria concerning precipitation reactions, a special investigation was made (Olhagen, 1950). Using electrophoretic and ultracentrifugal analysis, Olhagen found that in the urinary proteins the amount of albumin was relatively small and that the dominating proteins migrated as α-globulins. The molecular weight for the main component was between 20,000 and 30,000.

Friberg, 1950, concluded that prolonged exposure through inhalation of cadmium oxide dust gave rise to, as a rule, relatively mild kidney damage. He suggested that the kidney injury was secondary to the excretion of low molecular weight proteins.

As late as 1947 Princi had claimed that prolonged exposure to cadmium did not constitute any serious health hazard. Having investigated 20 smeltery workers who had been exposed to different cadmium compounds during a range of 1 to 20 years, he found no proteinuria. As he had used the boiling test (personal communication), proteinuria would not have been detected. Friberg in 1950 and Piscator in 1962b showed that urinary proteins from cadmium workers were not precipitated with this test. The studies by Princi thus did not prove that renal damage had not occurred.

In 1950, Friberg described the common findings of proteinuria in workers in alkaline accumulator factories also in Germany, England, and France. Further evidence for the high prevalence of proteinuria in cadmium workers has been provided by Baader in 1951, Smith, Kench, and Lane in 1955, Bonnell in 1955, Smith and Kench in 1957, Bonnell, Kazantzis, and King in 1959, Kazantzis et al. in 1963, Potts in 1965, Suzuki, Suzuki, and Ashizawa in 1965, Tsuchiya in 1967, Adams, Harrison, and Scott in 1969, and Harada in 1973. From these investigations it is seen that not only cadmium oxide dust, but also cadmium oxide fume, cadmium sulfide, and cadmium stearate may give rise to proteinuria, provided exposure has been prolonged. That proteinuria could appear after the workers had been removed from exposure has been indicated by the findings of Friberg and Nyström in 1952. They reexamined 38 workers and found proteinuria in 4 who had earlier had negative test results.

In many of the above mentioned investigations only qualitative tests had been used for detecting proteinuria. When quantitative determinations were performed by Piscator in 1962a and b, it was first possible to study protein excretion in more detail in relation to exposure times and removal from work. In 1962a Piscator studied a group of 40 workers of whom 39 had been included in Friberg's investigations in the 1950 report. By using a quantitative method for the determination of the total urine protein, Piscator, 1962a, has shown that in these workers, who had not been exposed to cadmium for the last 10 years, there was a relationship between the exposure times, i.e., total dose, and protein excretion, as shown in Figure 6:1. Half a year later, Piscator, 1962b, examined 14 urine samples. Essentially the same protein excretion values were obtained, indicating that this proteinuria was constant. In additional follow-ups it was found that in 18 workers reexamined in 1963 (Piscator, 1966a) and in 27 workers reexamined in 1969 (Piscator, 1971) from the above mentioned group of 40 workers, the protein excretion was in most cases virtually unchanged 3 and 10 years after the first quantitative determinations in 1959. As Friberg had not used a quantitative method for the determination of protein, a similar comparison could not be made with his original results. However, a comparison was made between the results of nitric acid and trichloracetic acid tests from the two investigations. No marked changes had occurred in the group (Piscator, 1962a).

Proteinuria has thus been a common finding in cadmium workers. Further evidence of a disturbed renal function has been given by the findings of glucosuria documented by Bonnell, Kazantzis, and King, 1959, Smith, Wells, and Kench, 1961, Kazantzis et al., 1963, Suzuki, Suzuki, and Ashizawa, 1965, Piscator, 1966a, and Adams, Harrison, and Scott, 1969. Aminoaciduria has been observed by Clarkson and Kench, 1956, Kazantzis et al., 1963, Piscator, 1966a, Adams, Harrison, and Scott, 1969, and Goyer et al., 1972. The last mentioned authors could not find specific increases in the excretion of certain amino acids as reported by Clarkson and Kench, 1956. Toyoshima, Seino, and Tsuchiya, 1973, found increased excretion of most amino acids in urine of cadmium workers. Especially great increases

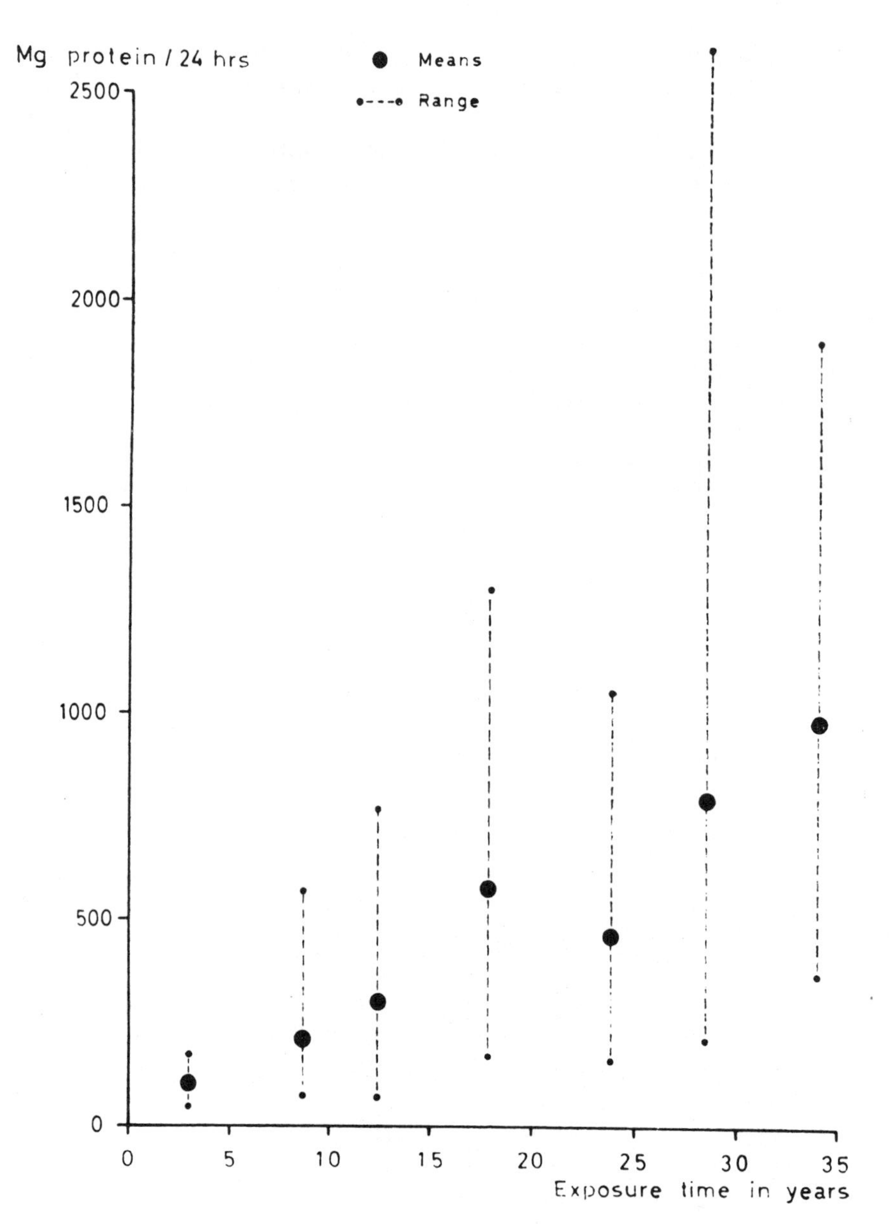

FIGURE 6:1. Relation between exposure time and protein excretion for 40 cadmium workers divided into 5-year exposure groups. The values were obtained about 10 years after the last exposure. (From Piscator, 1966c.)

occurred for citrulline and arginine. A comparison with the Itai-itai disease patients (Section 8.2.2.2.2) showed similar excretion patterns, although the high excretion of proline and hydroxyproline among patients was absent among workers.

A decreased concentrating capacity was detected by Friberg, 1950, and by Ahlmark et al., 1961. Kazantzis et al., 1963, and Adams, Harrison, and Scott, 1969, found changes also in renal handling of uric acid, calcium, and phorphorus. Proteinuria can appear alone without the other above mentioned changes, indicating that proteinuria is often an early sign of cadmium intoxication.

By improved methods for examination of urinary proteins, it has recently been shown that slight increases in certain low molecular weight urinary proteins may exist while the total excretion of urinary proteins may still be within the normal range (Piscator, 1972, Piscator, Evrin, and Vesterberg, to be published). These findings furnish proof that an increase in the excretion of certain urinary proteins is the first sign of cadmium effects on the kidneys.

Though glomerular function has not been extensively investigated, the studies of Friberg, 1950, Ahlmark et al., 1961, and Adams, Harrison, and Scott, 1969, have shown that decreases in glomerular filtration rates will appear in cadmium workers. This symptom is rather late, appearing after the other above mentioned changes which are more indicative of tubular dysfunction. Kazantzis et al., 1963, found profound changes in tubular function but no definite evidence of glomerular malfunction.

The above mentioned findings in workers are similar to those in groups in the general population in Japan who have been exposed to high amounts of cadmium via rice for prolonged periods. These findings will be discussed in more detail in Chapter 8.

6.1.2.1.2 Nature of Proteinuria

In 1958 Butler and Flynn described the so-called tubular proteinuria which they found in cases with tubular dysfunction, as in the Fanconi syndrome, galactosemia, etc. This proteinuria was characterized by a relatively small albumin fraction and dominance of proteins with mobility as α,- β- and γ-globulins in paper electrophoresis. By comparing urinary proteins from cadmium workers with urinary proteins from persons with other tubular disorders, Butler and Flynn, 1961, Butler et al., 1962, Piscator, 1962a, and Kazantzis et al., 1963, were able to show that the proteinuria in chronic cadmium poisoning was similar to the tubular proteinuria described by Butler and Flynn in 1958. Typical paper electrophoretic patterns of urinary proteins in chronic cadmium poisoning are shown in Figures 6:2 and 8:6. Comparative patterns resulting from disc electrophoresis are shown in Figures 8:7 and 8:8. Further investigations by Piscator, 1966a and b, showed that the proteins excreted by cadmium workers were composed mainly of low molecular weight proteins, chiefly derived from the serum. Among these were β_2-microglobulin, enzymes such as muramidase and ribonuclease, and immune globulin chains, the latter constituting nearly half of the low molecular weight fraction. It was not possible to show the existence of any proteins specific for cadmium poisoning.

Most researchers have accepted that the early proteinuria in chronic cadmium poisoning is of the tubular type and caused by decreased reabsorption of proteins (Harrison et al., 1968, Davis, Flynn, and Platt, 1968, Berggard and Bearn, 1968, Peterson, Evrin, and Berggard, 1969, Rennie, 1971, and Evrin, 1973. However, Kench, Gain, and Sutherland, 1965, found in urine protein from two cases of chronic cadmium poisoning a low molecular weight albumin, immunologically indistinguishable from normal serum albumin but with a considerably lower molecular weight. They introduced the word "minialbumin" for this protein which they considered the main feature of chronic cadmium poisoning. Since their conclusions were founded upon investigations of only two samples, lyophilized and stored for 8 years at room temperature before examination, further data are needed (see also Section 6.1.2.2, however). Using the same methods, Piscator, 1971, could detect a small quantity of albumin in low molecular weight fractions obtained by gel filtration from the urine of both normal men and workers with chronic cadmium poisoning. Piscator considered this due to the "tailing" of ordinary albumin on the Sephadex® column.

Vigliani, Pernis, and Luisa, 1966, using specific antisera, demonstrated that some 30% of the protein in the urine from three Swedish cadmium workers consisted of L chains of γ-globulin. These findings do not contradict the above mentioned findings by Piscator despite the fact that Vigliani interpreted his results as unfavorable to cadmium

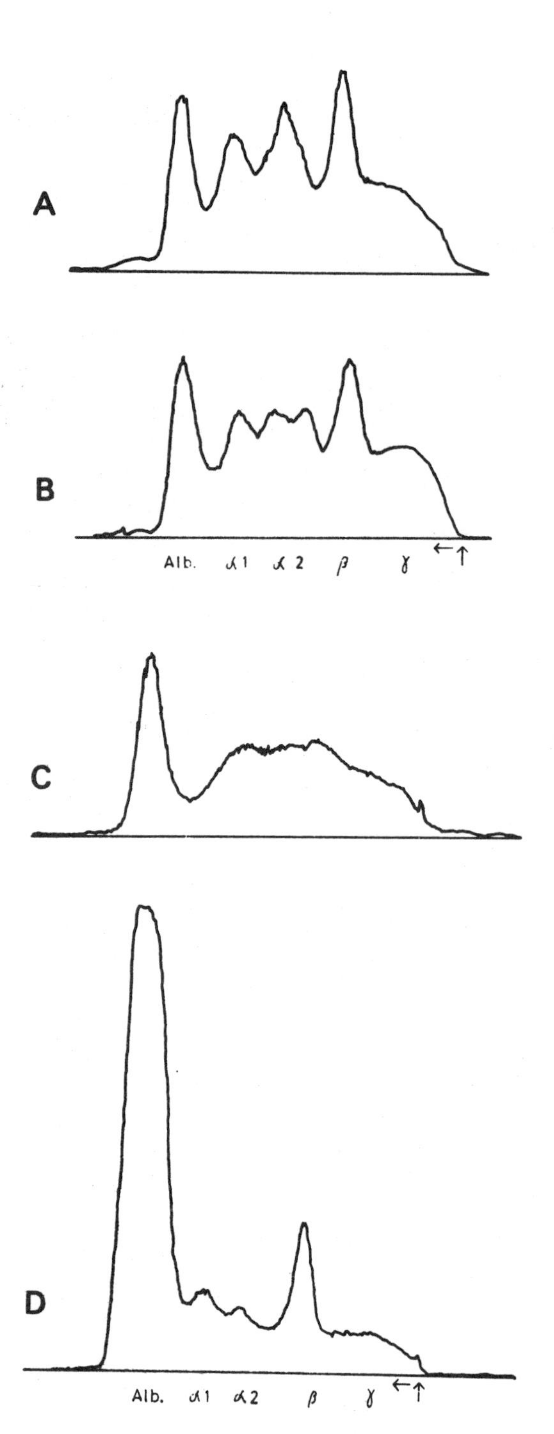

FIGURE 6:2. Scanned paper electrophoretic patterns for urinary proteins. A and B. Cadmium workers. C. Normal man. D. Person with chronic nephritis. (From Piscator, 1966c.)

proteinuria's being simply an increased normal proteinuria caused by impaired tubular reabsorption. In their first report, Vigliani, Pernis, and Luisa, 1966, expressed the opinion that cadmium proteinuria is mainly caused by a defect in the assembling of polypeptide chains of immunoglobulins, resulting in the excretion of low molecular weight chains. In 1969, however, Vigliani abandoned this theory in favor of the hypothesis that cadmium has an effect upon the catabolism of immunoglobulins and other proteins in the kidneys.

The nature of proteinuria among Japanese Itai-itai disease patients in comparison with that of cadmium workers is discussed in Section 8.2.2.2.2.

6.1.2.1.3 Renal Stones

Especially in workers in Sweden, the formation of renal stones has been a common feature (Friberg, 1950, and Ahlmark et al., 1961). Ahlmark et al., 1961, found that 44% of a group of workers exposed to cadmium dust for more than 15 years had a history of renal stones. In 1963, Axelsson reported that the stones were mainly composed of basic calcium phosphate. He stated that the formation of such stones was evidence of a disturbed metabolism of calcium and phosphorus.

During later years kidney stones have also been found in British accumulator factory workers by Adams, Harrison, and Scott, 1969. It is noteworthy that both this later research group and Axelsson in 1963 found a higher prevalence of renal stones in non proteinuric men, indicating that more developed tubular dysfunction means less risk of developing kidney stones. This finding also indicates that changes in reabsorption of calcium might occur early. Adams, Harrison, and Scott, 1969, stated that some of their workers had hypercalcuria without proteinuria. However, the methods employed to detect proteinuria were not sensitive enough to show minor increases in urinary protein excretion.

In a follow-up examination of the group of 12 workers previously referred to (Kazantzis et al., 1963), Kazantzis, 1970, described recurrent renal colic with renal stone formation in two workers and nephrocalcinosis in a third worker. These three men all had more than 25 years' exposure to cadmium when they were first seen. Kazantzis (to be published) considers that the abnormality in the handling of calcium by the kidney is likely to be related to the presence of an acquired distal renal tubular acidosis which is additional to the proximal renal tubular defects making up the acquired Fanconi syndrome more commonly seen.

6.1.2.1.4 Autopsy and Biopsy Findings

Information about morphological changes and cadmium levels in the kidneys is available from a total of 11 autopsies (six persons exposed to cadmium oxide dust and five to cadmium oxide fume) and 9 biopsies (all exposed to cadmium oxide dust). In Table 6:1 data are given from six autopsies and nine biopsies on workers exposed to cadmium oxide dust. In four of the autopsy reports (S.W.H., K.J., H.B., and A.B.) the morphological changes in the kidneys were mainly confined to the proximal tubules, whereas the glomeruli were less affected. Similar but less pronounced changes were seen in two biopsies. In the other cases with morphological changes, the exact nature of these has not been specified.

The data in the table indicate that cadmium levels were lower in cases with morphological changes than in cases without detectable morphological alterations, or with only minor changes. As exposure times were longer in the former cases and the concentrations of cadmium dust conceivably higher (after 1950 improvements in working conditions were made in the Swedish factory), the lower levels in these cases cannot be explained by less exposure. There is some difference between the two groups in regard to the time that had elapsed between the end of exposure and autopsy or biopsy. This fact does not seem to have been of significance, though, as can be seen if patients examined different times after the end of exposure are compared. The combination of heavy exposure to cadmium, severe renal damage, and relatively low levels of cadmium in the kidneys indicates instead a considerable excretion of cadmium. In workers with no impairment or only minor dysfunction (latter group), the renal losses must have been smaller and thus the renal levels of cadmium were higher.

The above mentioned findings in workers exposed to cadmium oxide *dust* are further supported by data from autopsies on workers exposed to cadmium oxide *fumes.* Smith, Smith, and McCall, 1960, reported on autopsies on four workers exposed for less than 10 years. The autopsies were performed 5 to 18 years after the last exposure. These workers had been removed from exposure because of respiratory symptoms. Proteinuria had been the only sign of renal impairment in three of them, and histological examination did not reveal any renal alterations. Cadmium levels in the cortex were between 150 and 395 μg/g wet weight, i.e., in the same range as in workers exposed to dust and without major renal changes. Bonnell, 1955, on the other hand, reported on a worker exposed for 32 years to fumes and autopsied only 2 years after the last exposure. There was severe renal damage with both tubular and glomerular involvement. The cadmium concentration in cortex was relatively low, 62 μg/g wet weight.

There are also some data from Japan on renal levels of cadmium in relation to morphological changes in persons exposed to cadmium in the general environment. In three inhabitants of the Annaka district, Nomiyama, 1971b, reported liver concentrations of cadmium to be about the same, ranging from 29 to 35 μg/g wet weight. In two of these persons, who had morphological kidney changes, the concentrations of cadmium in renal cortex were 29 and 134 μg/g wet weight, respectively. The other person, showing no morphological changes, had a concentration of 264 μg/g wet weight in renal cortex.

In animal experiments (Section 6.1.2.2) it has been shown that morphological changes became evident and renal levels of cadmium ceased to rise after a certain period of exposure. Further exposure then resulted in a lowering of the cadmium concentration. These findings agree with the findings in human beings.

In conclusion, when a person suffers severe renal damage through the toxic action of cadmium, his kidney concentration of cadmium will decrease considerably. Thus, he will have lower levels in kidneys than a person with only slight renal disturbances.

6.1.2.2 In Animals

Prodan, 1932, stated that renal lesions consisting of fatty degenerations in the tubules developed in cats given daily oral doses of 2, 10, and 100 mg of cadmium for 1 to 2 months. However, controls were not included in the experiment and no certain conclusions can be drawn from these results. Wilson, deEds, and Cox, 1941, added cadmium chloride to the diets of rats in concentrations of 31, 63, 125, 250, and 500 μg/g. Exposure time was 100 days, after which the rats were

TABLE 6:1

Concentrations of Cadmium in Kidney Cortex (μg/g Wet Weight) in Workers Exposed to Cadmium Oxide Dust in Relation to Morphological Kidney Changes Seen at Autopsy or Biopsy

Worker	Morphological changes**	Proteinuria††	Cadmium in cortex* (μg/g wet weight)	Age	Years exposed	Years since last exposure	Years of autopsy(A) or biopsy(B)	References
S.W.H.	++	+	83	46	28	1	1960(A)	Kazantzis et al., 1963.
K.J.	++	+	75	49	22	9	1951(A)	Friberg, 1950, 1952, and 1957.
K.N.	†	+	20	57	18	6	1952(A)	Friberg, 1950 and 1957.
H.B.	++	+	33	60	26	3	1949(A)	Friberg, 1950 and 1957.
A.B.	++	+	174	39	16	4	1950(A)	Baader, 1951; Hörstebrock, 1951
O.J.	++	+	63	62	20	1	1967(A)	Piscator, 1971
G.J.	(+)	+	321	44	11	12	1959(B)	Axelsson and Piscator, 1971
G.K.	(+)	+	152	46	15	0	1959(B)	Axelsson and Piscator, 1971
A.L.	–	(+)	220	36	6	0	1959(B)	Axelsson and Piscator, 1971
E.Y.	–	–	446	39	7	0	1959(B)	Axelsson and Piscator, 1971
E.H.	–	+	320	40	15	10	1959(B)	Axelsson and Piscator, 1971
J.P.	–	+	330	43	20	6	1959(B)	Axelsson and Piscator, 1971
N.U.	–	–	180	44	12	0	1959(B)	Axelsson and Piscator, 1971
H.N.	–	+	21	45	13	0	1959(B)	Axelsson and Piscator, 1971
K.N.	–	–	190	50	15	2	1959(B)	Axelsson and Piscator, 1971

*Underlined figures are based on cadmium concentrations in whole kidney, assuming that the cadmium concentration in cortex is 1.5 times the average kidney concentration (Section 4.3.2.1).

†Results from histological examinations not reported, but at examination in 1946 (Friberg, 1950) this worker had the lowest kidney function tests of all (Inulin clearance: 42 ml/min; concentration capacity: 1.016; NPN: 44 mg%).

**– = No morphological changes
(+) = Slight morphological changes
++ = Profound morphological changes

††– = Negative results on repeated testing with trichloracetic acid
(+) = Varying results
+ = Positive results on repeated testing

killed. Histological examination showed slight tubular changes and no glomerular alterations in the animals exposed to 63 μg/g cadmium in the diet and no changes in animals exposed to 31 μg/g cadmium. Cadmium determinations were not performed on the kidneys.

In 1950, Princi and Geever reported that they could not find evidence of morphological changes in the kidneys of dogs subjected to prolonged exposure to cadmium oxide dust or cadmium sulfide (average concentration: 4 mg Cd/m^3). They made routine analyses of urine but used only the boiling test and did not detect protein. In 1950, Friberg made a long-term experiment (for details, see Section 5.2.2) in which he exposed rabbits for 3 hr a day for 28 days per month for 8 months to about 8 mg cadmium iron oxide $dust/m^3$. After 4 months of exposure, moderate proteinuria could be demonstrated with the trichloracetic acid test. When the animals were killed after 7 to 9 months of exposure, histological examination did not disclose any structural changes. Cadmium concentrations in the kidneys were generally between 300 and 700 μg/g wet weight. When a group of rabbits was given subcutaneous injection of cadmium sulfate, 0.65 mg Cd/kg, 6 days a week, proteinuria was demonstrated after 2 months. Analyses of cadmium in the kidneys were not performed. Electrophoretic analysis of urine proteins showed that the main component was not the same as the one found in rabbits with kidney damage due to uranium salts. In further experiments, Dalhamn and Friberg, 1957, found that mainly tubular changes took place in rabbits injected with cadmium, whereas the glomeruli were not affected.

Bonnell, Ross, and King, 1960, gave rats of both sexes an intraperitoneal dose of cadmium chloride (0.75 mg Cd/kg) thrice weekly (for details see Section 4.2.2.2). After 5 to 6 months injections were discontinued for 1 to 2 months. Thereafter, some animals were given a reduced dose while others were given no further injections. The total time for the experiment was 1 year. Every month animals were killed and histological examinations performed. After 4 months of exposure, tubular damage was evident. From that time despite continued exposure there was no further increase in kidney levels of cadmium; in fact, some decrease in kidney concentration could be demonstrated (see Figure 4:11). After 4 months the renal concentration of cadmium was about 275 μg/g wet weight and after 5 months, about 225 μg/g. The corresponding concentrations in cortex would be about 400 and 325 μg/g, respectively.

These data are in accord with the findings by Friberg, 1952. He showed that rabbits given daily subcutaneous injections of radioactive cadmium sulfate had very low urinary excretion of cadmium during the first 2 months while after 2 months of exposure, the cadmium excretion increased 50- to 100-fold (see Figure 4:12). After 1 month of exposure the renal concentration of cadmium was on an average 126 μg/g wet weight whereas after 2½ months of exposure the average concentration was 213 μg/g. Friberg reported also that the concentration in cortex was about five times higher than in medulla, which means that in cortex the concentrations would have been about 200 and 350 μg/g, respectively, under the assumption that the mass of cortex is equal to or slightly larger than that of the medulla. In later experiments, Friberg, 1955, alternating radioactive and non-radioactive cadmium, demonstrated that a large part of the excreted cadmium must have come from deposits in the kidneys.

Axelsson and Piscator, 1966a, gave rabbits subcutaneous injections of cadmium chloride, 0.25 mg Cd/kg 5 days a week. After 17 weeks, alkaline phosphatase activity of the renal cortex decreased. After 23 weeks there was a decrease in the capacity to reabsorb glucose, together with considerable proteinuria and excretion of cadmium. Histological examination (Axelsson, Dahlgren, and Piscator, 1968) revealed that already after 11 weeks mild alterations took place in the proximal tubules. At that time cadmium concentration in the renal cortex was about 250 μg/g wet weight (range: 194 to 315). After 23 weeks pronounced changes were found not only in the proximal tubules, but also in other parts of the nephron. The collecting tubules were not detectably affected. Though the glomeruli showed some slight alterations in some cases, most cases were without demonstrable alterations.

In another group of rabbits exposure conditions were similar to those stated above, but exposure was discontinued after 24 weeks, after which the animals were followed for another 30 weeks. Urine investigations showed that protein excretion was significantly higher than in controls after 12 weeks of exposure, with a maximum at about 1 month after exposure had ceased. Then came a reduction in protein excretion. At the end of the observation time there was no significant difference between exposed animals and controls. Electrophoretic examination of urinary proteins showed a predominance of proteins with a mobility as α- and β-proteins. It was concluded that this corresponded to the tubular proteinuria found in human beings. Before these animals were killed, kidney function was investigated. After death, alkaline phosphatase activity and cadmium in renal cortex were determined as described by Piscator and Axelsson, 1970. Glucose reabsorption and protein excretion did not differ significantly between exposed and control animals. However, a highly significant decrease in the activity of alkaline phosphatase was found. The renal cortex in the exposed animals contained large amounts of cadmium, about 250 μg/g wet weight, despite the fact that the rabbits had been without exposure to cadmium for 30 weeks (see also Section 4.2.2.2). Morphologically, there was an almost complete regeneration of the tubular epithelium. It was concluded that in the rabbit, exposure to cadmium would give rise to renal tubular damage, partly reversible, as judged by functional and morphological investigations.

Castano and Vigliani, 1972, found that intravenously injected horseradish peroxidase, a low molecular weight protein, was reabsorbed in a normal way in rats given a dosage of 0.51 mg Cd/kg body weight, three times a week for 1 to 4 weeks intraperitoneally as cadmium chloride. Electron microscopy revealed normal pinocytosis. The authors claimed this as proof that cadmium did not affect reabsorption processes and as support for the hypothesis that catabolic changes in the kidneys constitute the main cause of proteinuria (Vigliani, 1969). No determinations of cadmium in the kidneys were performed. An estimate of the concentration of cadmium in renal cortex can be obtained by comparison with an experiment by Bonnell, Ross, and King, 1960, who gave rats intraperitoneal injections three times a week in doses of 0.75 mg/kg. This dose resulted in an average renal concentration of about 60 μg/g (cortex levels not above 100 μg/g wet weight) after 1 month. Since the dose given by Castano and Vigliani was 2/3 of that given by Bonnell, Ross, and King, the renal cortex level could be expected to be below 100 μg/g. It is hence unlikely that significant changes in reabsorption would have been found at this level.

It was mentioned earlier (Section 6.1.2.1.2) that Kench and co-workers have reached the opinion that the excretion of a low molecular weight albumin is the main feature of the proteinuria in chronic cadmium poisoning in man. In experiments on rabbits (Kench, Wells, and Smith, 1962), on monkeys (Kench, Gain, and Sutherland, 1965, and Kench and Sutherland, 1966), and on rats (Kench and Sutherland, 1966), they were able to isolate an albumin of low molecular weight (10,000) from urine. In these experiments, however, animals were given very large doses of cadmium by intravenous injections (2 mg/kg twice a week) or a single intraperitoneal injection (15 mg/kg). Thus, the experimental conditions did not reproduce the slow chronic cadmium poisoning in man. At present these experiments are indicative that large doses of cadmium may cause profound changes in protein metabolism. More evidence is needed before these experimental results can be applied to chronic cadmium poisoning in man.

In some reports, such as those by Princi and Geever in 1950 and by Anwar et al. in 1961, no effects on the kidneys were reported despite prolonged exposure to cadmium. Princi and Geever stated, "In no case was tubular damage found in the kidneys (such as might be expected from severe metal poisoning)." Milder tubular lesion, often reported in chronic cadmium poisoning, might not have been registered. Princi and Geever found relatively large amounts of cadmium in the blood (380 to about 1,200 ng/g whole blood) and urine (32 to about 380 ng/g) of dogs exposed to cadmium oxide dust and, to some extent, in dogs exposed to cadmium sulfide. However, kidney levels were low, in the same range as those for normal human beings. On the other hand, relatively high concentrations of cadmium were found in the liver (12 to 47 μg/g in dogs exposed to cadmium oxide), bile, and feces. The authors do not mention whether the values are calculated as μg/g wet weight or dry weight. We have assumed that they are wet weight values.

Anwar et al. gave a total of eight dogs cadmium in drinking water for 4 years in concentrations ranging from 0.5 to 10 μg/g. One additional dog was used as control. They stated in their summary that pathological changes were not produced in dogs receiving cadmium. However, in their textual discussion they described the findings of larger amounts of fat in the proximal convoluted tubules and in some areas, atrophied tubules, in one dog from each of the two groups receiving the higher concentrations (two dogs: 5 μg/g, 2 dogs: 10 μg/g). As the authors thought that some animals also were infected with Leptospira, they concluded that some of the changes might be due to leptospirosis. Because of the small number of dogs used, the data are of limited value and in any case cannot be taken as evidence that long-term exposure to orally ingested cadmium will not give rise to kidney damage.

Nordberg and Piscator, 1972, found a decrease in the excretion of total protein in male mice given subcutaneous injections of cadmium chloride (0.25 or 0.5 mg Cd/kg body weight 5 days per week for 6 months). The main urinary protein in male mice is a low molecular weight protein, synthesized in the liver, and this synthesis is stimulated by testosterone. A possible cadmium influence on testosterone synthesis could explain a decrease in the synthesis rate of that protein, and in turn, a decrease in total urinary protein excretion. After 21 weeks of exposure, cadmium excretion increased in mice given 0.5 mg/kg. At that time there were also changes in the urinary protein electrophoretic pattern, which earlier had been dominated by the above mentioned testosterone-dependent protein. Other proteins appeared which had a pattern indicating renal damage. Further examinations (Nordberg, unpublished data) at the end of the experiment, i.e., after 24 weeks of exposure, showed average concentrations of 170 μg/g of cadmium in whole kidney (Table 4:4) in both exposure groups, which corresponds to renal cortex levels of about 170 to 250 μg/g (Section 4.2.2.2).

Piscator and Larsson, 1972, gave cadmium as the chloride in drinking water (0 to 7.5 μg/g) to groups of female rats on normal and low calcium intake, respectively. The animals were killed after 1 year. In rats on a calcium-deficient diet and drinking water with 7.5 μg/g cadmium, the mean cadmium level in renal cortex was about 90 μg/g. The total protein excretion was not significantly increased in these rats compared to control rats on a calcium-deficient diet but without exposure to cadmium. There was a highly significant increase in ribonuclease, which may be in accord with findings in humans with chronic cadmium poisoning. It may well be that the excretion of ribonuclease in the rats was a very early sign of renal tubular dysfunction.

Nishizumi, 1972, gave male rats cadmium in drinking water at concentrations of 10, 50, and 300 μg/g. Animals were killed after 6, 12, 24, and 40 weeks and the kidneys examined by electron microscopy. In rats given water containing 10 μg/g cadmium, minor changes were noticed already after 6 to 12 weeks. After 40 weeks slight but obvious changes were seen such as swollen mitochondria and vacuoles containing cell debris. At high dose levels obvious changes were noted already after 6 to 12 weeks and pronounced mitochondrial changes after 24 to 40 weeks. Cadmium concentrations in organs were not determined.

Berry, 1972, gave male rats subcutaneous injections of 1.4 mg of cadmium sulfate three times a week for 5 months. At this high dose level (2 to 3 mg Cd/kg body weight) proteinuria was said to have appeared after 1 month. The method was not stated. Electron microscopy revealed glomerular and tubular lesions after 1 month. No determinations of the cadmium content of the cortex were performed. At the doses given, very severe renal lesions would be expected.

Ogawa et al., 1972b, reported on an experiment in which mice were given water containing 146 μg/g of cadmium. After 3 months a considerable reduction in carbonic anhydrase activity in blood, liver, and kidney was observed.

Stowe, Wilson, and Goyer, 1972, gave ten rabbits cadmium in drinking water (160 μg/g) for 6 months. The mean exposure was 15.5 mg Cd/kg body weight per day. Assuming that 1 to 2% had been absorbed, this dose would result in a mean daily absorption of 0.16 to 0.32 mg/kg. Cadmium was determined in whole kidney and histopathological examinations were carried out by light and electron microscopy. The mean renal concentration was 170 (S.D. 14) μg/g wet weight. Calculating with a ratio of 1.5 between cortex:whole kidney (Section 4.2.3.2) would give about 250 μg/g wet weight in the cortex. Extensive interstitial fibrosis was seen by light microscopy in the kidneys of exposed animals. Pronounced changes were seen in the proximal tubules, including pyknosis, karyorrhexis, and epithelial sloughing. Electron microscopy also showed seemingly unaffected mitochondria. Collagen was deposited in the glomeruli. Amyloid deposits could not be demonstrated.

Nomiyama, Sato and Yamamoto, 1973, on the basis of experiments on rabbits given daily s.c. injections of $CdCl_2$ (1.5 mg/kg) for 45 days, stated that increased urinary excretion of alkaline and acid phosphatase was an early sign of renal injury. It should be pointed out, however, that of seven animals in the study, four died between days 15 to 36. The rabbits also displayed decreased inulin clearance and decreased tubular mass para-aminohippurate, as well as increased urinary protein, amino acids and sugar.

Nomiyama, 1973, and Nomiyama (unpublished data) performed studies on rabbits given repeated daily injections of nonradioactive $CdCl_2$ (1.5 mg/kg) for various times up to 35 days, after which he studied inulin clearance as well as clearance of ^{115m}Cd using single intracardiac infusions. The results showed an increasing urinary excretion of radioactive cadmium with increasing kidney cortex concentrations of nonradioactive cadmium. Nomiyama interpreted his findings as showing that the increase in urinary cadmium excretion is unrelated to tubular dysfunction. This conclusion is hard to accept. The rabbits must have suffered from considerable tubular dysfunction as substantiated by his own data (Tm_{PAH} = 5.1 and 12.7 in two rabbits as compared with his own normal value of 37.1) as well as experience from several other reports as documented in this chapter. In some rabbits treated with uranyl in the same experiment there was neither an increased quotient between clearance of ^{115m}Cd and inulin clearance nor the same extent of tubular dysfunction as in "cadmium rabbits" (Tm_{PAH} = 20.2 and 19.3 in two rabbits). This may be in agreement with the findings reported by Friberg, 1950, in which protein in "cadmium rabbits" had a distinctly different electrophoretic mobility than that in "uranyl rabbits."

6.1.2.3 Mechanisms for the Development of the Renal Injury

It has been shown that cadmium will accumulate in the kidneys. During the period of accumulation only small amounts of cadmium will be excreted in the urine. This excretion has been shown on a group basis to be related to the body burden of cadmium (Figure 4:27). When a certain concentration has been reached in the kidneys, cadmium excretion will increase considerably and the accumulation rate of cadmium in the kidney will decrease.

A hypothesis on the development of renal injury as advanced at our institute will now be set forth. In Section 4.2.8 it was said that cadmium is probably transported to the tubules bound to the low molecular weight protein, metallothionein. During normal conditions this protein will be reabsorbed in the tubules just as other proteins, and cadmium will accumulate in the renal tissue. Cadmium excretion in "normal" people and in workers with short periods of exposure to low air concentrations of cadmium thus is low because proteins are almost completely reabsorbed. With increasing exposure more cadmium than can be bound by metallothionein will eventually be accumulated in the kidneys. Cadmium will then exchange with zinc in enzymes necessary for reabsorption and catabolism of proteins. Chiappino, Repetto, and Pernis, 1968, found inhibition of the leucine-aminopeptidase activity in the renal cortex of cadmium-exposed rats and in rabbits exposed under similar conditions as in the experiments by Axelsson and Piscator in 1966a. Leucine-aminopeptidase is a zinc-dependent enzyme thought to play a role in the renal handling of proteins. As a result of these anti-enzymatic actions less protein will be catabolized or reabsorbed, causing tubular proteinuria. Cadmium excretion will increase also as less metallothionein will be reabsorbed. At this stage the accumulation rate of cadmium will become slower, but cadmium will still be reabsorbed and cadmium levels in the tissue may get still higher. The reabsorption defect will be greater and eventually renal cadmium will cease to increase. Tubular cells will be damaged by cadmium, and it is conceivable that cadmium will be excreted together with desquamated tubular cells, resulting in a decrease in renal levels of cadmium. If glomerular function is impaired, there will also be less filtration of metallothionein.

What has been described above might be true in continuous low level exposure to cadmium via air or food. As pointed out in Section 4.3.3.1, short-term exposure to high amounts of cadmium may cause a considerable excretion of cadmium independent of renal function. If exposure is discontinued when there is slight or no proteinuria, there is a possibility that cadmium levels in kidneys will increase for some time, probably depending upon liver levels. Proteinuria thus may appear some time after exposure has ceased, as supported by findings in workers removed from exposure (Friberg and Nyström, 1952) and in animals (Axelsson and Piscator, 1966a). The mechanisms mentioned above might also explain the poor correlations between proteinuria and cadmium excretion found earlier in workers (Smith and Kench, 1957). Piscator, 1972, showed that cadmium excretion generally is low in workers exposed for prolonged periods of time to relatively low concentrations of cadmium oxide dust and who do not have proteinuria as tested by specific determination of low molecular proteins and electrophoretic examination. In contrast, in exposed workers with a slight increase in urinary protein excretion, cadmium excretion is considerably higher.

The findings in autopsy cases of chronic cadmium poisoning and in biopsies from exposed workers presented a paradox; i.e., high levels of renal cadmium were combined with histologically normal kidneys while relatively low cadmium levels were combined with long-standing proteinuria and gross structural changes in the kidneys. The above mentioned mechanisms will explain this paradox, i.e., the more severe and long lasting the renal lesion, the more cadmium will be excreted.

6.1.2.4 Dose-response Relationships

The kidneys are considered to be the critical organs at long-term low-level exposure to cadmium. Accordingly, effects on the kidney set the dimensions for cadmium levels in the environment necessary to prevent any such effect. Because of the importance of a discussion of dose-response relationships for chronic effects on the kidneys, a separate chapter (Chapter 9) is devoted to dose-response relationships with regard to environmental cadmium and daily intake. In this section mainly relationships between effects and cadmium concentrations in the kidney, blood, and urine will be discussed.

6.1.2.4.1 Evaluations Starting from Data on Cadmium Concentrations in Blood and Urine

As can be realized from the detailed discussion in Section 4.4, cadmium levels in blood are of very limited value for evaluating dose-response relationships because they might vary considerably during exposure while the kidneys will accumulate cadmium continuously up to a certain level (when damage will occur). Though increased blood levels show that exposure has taken place, it will be impossible to differentiate whether they reflect accumulated cadmium or recent exposure.

Cadmium concentrations in urine seem to be useful on a group basis as indicators of body burdens in normal populations with long-term low-level exposure to cadmium via food as discussed in Section 4.3.3.1. Available data from industrially exposed workers, however, do not yet allow an estimate of the amount of cadmium in urine that reflects a critical renal level. In Section 4.3.3.1 it was mentioned that in workers without increased excretion of low molecular weight proteins cadmium excretion generally was below 10 μg/g of creatinine. The workers involved had long-term exposure to relatively low concentrations of cadmium oxide dust.

Since investigations in Japan (Tsuchiya, Seki, and Sugita, 1972a) have shown that the increase in urinary excretion of cadmium in a normal population seems to be parallel to the increase in body burden, some estimates can be obtained from these data, keeping in mind that the urinary data were obtained on healthy persons whereas autopsy data had been obtained from persons with varying degrees of diseases (Figures 4:24 and 4:25).

At a mean renal cortex level of about 100 μg/g wet weight, urinary excretion was about 1.8 μg/l (S.D. 1 μg/l). Assuming daily urinary volume of 1.5 l, this would correspond to a daily excretion of 2.7 μg of cadmium. Provided that there is a similar relationship between urinary cadmium excretion and kidney cortex levels also at somewhat higher concentrations, but before renal tubular dysfunction has occurred, a group mean urinary excretion of about 5 μg/day would correspond to about 200 μg/g in kidney cortex.

6.1.2.4.2 Evaluations Starting from Data on Cadmium Concentrations in the Kidneys

In Table 6:1 data are given on workers industrially exposed for long periods to considerable amounts of cadmium. As can be seen, the concentrations of cadmium in the renal cortex have varied within wide ranges. In Section 6.1.2.1.4 it was concluded that the low values found in persons with morphologically pronounced kidney damage resulted not from low exposure but from high excretion of cadmium. In Table 6:1 there also can be seen a group of workers with long exposure to cadmium but with no or only slight morphologically detectable kidney damage. Only some of them had proteinuria. It can be assumed that the kidney levels of cadmium in such a group would be fairly representative for concentrations at the time at which the first signs of renal dysfunction appeared. Reported levels in such cases did, with one exception, vary between 150 to 450 μg/g wet weight in renal cortex.

In the experiments on rabbits, rats and mice (Section 6.1.2.2) functional and morphological changes appeared when the cadmium concentration in the renal cortex reached about 200 to 400 μg/g wet weight. Renal levels of about 130 μg/g wet weight, corresponding to about 200 μg/g wet weight in renal cortex (only one month observation time), were found in rabbits with neither proteinuria nor excretion of cadmium. In one group of mice exposed for 6 months to subcutaneous injections of 0.25 mg/kg and without obvious signs of renal dysfunction, renal cortex levels were 170 to 250 μg/g wet weight, whereas in another group exposed for 6 months to 0.5 mg/kg and with tubular proteinuria, renal cadmium concentrations were approximately the same. The animal experiments thus suggest that about 200 μg/g wet weight is a critical concentration in the cortex.

Even if animal data are not immediately valid for man, it is striking that they are within the range for human beings exposed to cadmium for several years and with slight or no impairment. It is considered justified to start out from a value of 200 μg/g cadmium in renal cortex (wet weight) when discussing which exposures can bring about cadmium-induced detectable renal dysfunction in man. This of course does not mean that 200 μg/g would give rise to renal tubular dysfunction in *all* persons exposed. As always in biological experience, at a certain low concentration only a fraction of an exposed population will show signs of effects. On the other hand it should be emphasized that with renal dysfunction here is meant a dysfunction that has been detected with fairly crude methods. With all probability cellular effects will appear at considerably lower levels. In calcium-deficient rats an increase in the excretion of ribonuclease was found at a mean renal cortex level of 90 μg/g, indicating that such effects may be possible.

6.2 EFFECTS ON THE HEMATOPOIETIC SYSTEM AND DOSE-RESPONSE RELATIONSHIPS

6.2.1 Acute Effects and Dose-response Relationships

Effects on the blood after respiratory exposure to high concentrations of cadmium have been noted both in human beings (Beton et al., 1966) and in animals (Prodan, 1932). Beton et al. found elevated levels of hemoglobin in some subjects. This elevation was probably connected with the hemoconcentration caused by the edema of the lungs. A hemolytic effect after a single injection of cadmium sulfate into a dog (about 25 mg Cd/kg) was noted by Athanasiu and Langlois in 1896, but it is not known if acute respiratory exposure in human beings has ever given rise to such an effect.

It is not possible to establish any dose-response relationships with regard to effects on the hematopoietic system in man or animals after acute exposure. There is a lack of data and the edema of the lungs and the general toxemia will probably overshadow any other effects.

6.2.2 Chronic Effects and Dose-response Relationships

6.2.2.1 In Human Beings

Anemia has been described in workers exposed to cadmium oxide dust (Nicaud, Lafitte, and Gros, 1942, and Friberg, 1950) and cadmium oxide fumes (Tsuchiya, 1967) but is usually moderate. In a group of 16 workers exposed to cadmium for 5 to 30 years, a significant correlation was found between high cadmium levels in blood and low hemoglobin levels (Piscator, 1971). In this group of workers the mean haptoglobin level was below normal. Moreover, in several cases the haptoglobin concentration was at the lower normal limit, indicating that slight hemolysis might have occurred. In a group of workers who had not been exposed to cadmium for at least 10 years, haptoglobin concentrations were generally normal or elevated and did not differ from levels in a group of workers never exposed to cadmium. The number of white cells is generally normal, but there is an increase in the number of eosinophil cells in exposed workers (Nicaud, Lafitte, and Gros, 1942, Princi, 1947, and Friberg, 1950). No explanation has yet been given.

The bone marrow was examined microscopically by Friberg, 1950, in 19 cadmium-exposed workers but no pathological changes were found.

6.2.2.2 In Animals

Anemia has been a common finding. Wilson, deEds, and Cox, 1941, gave rats food containing 31 to 500 μg/g cadmium and found that the lowest dose gave animals anemia after a couple of months. After exposure had ceased, there was a tendency to normalization in hemoglobin levels. In a group given 125 μg/g cadmium for up to 7 months they noted also increases in reticulocytes and eosinophils. Examination of the bone marrow showed that it was hyperplastic.

Decker et al., 1958, noted that hemoglobin levels had already decreased considerably after 2 weeks in rats given 50 μg/g cadmium in drinking water. On the other hand, rats given 10 μg/g in water for 1 year did not become anemic. Friberg, 1950, found that rabbits exposed to cadmium oxide dust developed a slight anemia. A prominent finding was eosinophilia, as 25% of the white cells in exposed rabbits were eosinophils compared with 3% in a control group. In rabbits injected with cadmium sulfate (0.65 mg Cd/kg), 6 days a week, pronounced anemia was evident after 2 months of exposure.

In further studies (Friberg, 1955; for exposure conditions, see Section 4.2.1.2) it was found that simultaneous administration of iron had a beneficial effect on the anemia, indicating that the anemia was partly due to iron deficiency. By giving ^{59}Fe to cadmium-exposed rabbits (1 mg Cd/kg, 6 days a week, subcutaneous injection), one group receiving the isotope before exposure, another group receiving it after 2 months' exposure, and comparing with controls, Berlin and Friberg, 1960, could show that there was an increased destruction of erythrocytes. There was no certain difference between controls and exposed animals with regard to utilization of iron for hemoglobin synthesis. The beneficial effect of parenterally given iron also indicated that cadmium did not block hemoglobin synthesis, but that there could be a decreased uptake of iron from the intestines. Berlin and Friberg did not observe any decreased osmotic resistance in the erythrocytes as described by Swensson, 1957. Berlin, Fredricsson, and Linge, 1961, found that in rabbits given cadmium there were increased deposits of iron in the bone marrow but no decrease in erythropoietic activity. They found some other changes in the bone marrow, but as the rabbits were rather severely intoxicated after having received subcutaneous injections of 1 mg Cd/kg daily for several months, these changes were regarded as nonspecific and produced by general toxic effects. In a group of rabbits followed after exposure (24 weeks, 0.25 mg Cd/kg, 5 days a week) ceased, hemoglobin levels rose again but were still significantly lower than those in a control group 30 weeks later (Piscator and Axelsson, 1970).

Fox and Fry, 1970, and Fox et al., 1971, studied Japanese quail fed a diet containing 75 μg/g of cadmium. In addition to iron, ascorbic acid prevented anemia. According to the authors ascorbic acid was a good prophylactic because it brought about increased absorption of iron from the intestinal tract. Pond and Walker, 1972, showed that a single prophylactic injection of iron can prevent the occurrence of cadmium anemia in growing rats exposed for 4 weeks to 100 μg/g of cadmium in the diet. Addition of iron to the food had the same preventive effect.

Berlin and Piscator, 1961, studied blood and plasma volume in cadmium-poisoned rabbits and concluded that the anemia could partly be explained as caused by increases in plasma volume. Axelsson and Piscator, 1966b, performed long-term experiments on rabbits (for details, see Section 4.2.2.2) and found that after 11 weeks of exposure, there was ahaptoglobinemia, indicating hemolysis. The hemolytic anemia persisted in some animals for up to 6 months after cessation of exposure. These animals also continued to excrete large amounts of cadmium (Piscator and Axelsson, 1970). Nordberg, Piscator, and Nordberg, 1971, found that cadmium in erythrocytes from exposed mice is stored mainly in a low molecular weight protein, probably metallothionein. The high cadmium excretion in rabbits with hemolytic anemia can thus partly be explained as being due to the release of cadmium-containing low molecular weight proteins from the erythrocytes.

6.2.3 Dose-response Relationships

The investigations on animals given injections of cadmium have indicated that during exposure anemia will be related to the accumulated dose (Friberg, 1955, Berlin and Friberg, 1960, Berlin and Piscator, 1961, and Axelsson and Piscator, 1966a), but that as soon as exposure has ceased, hemoglobin levels will tend to rise again. When given cadmium in drinking water in concentrations between 0.5 and 10 μg/g for 1 year, groups of rats did not differ from one another with regard to hemoglobin levels (Decker et al., 1958). At these exposure levels there was thus no dose-response relationship. Hemoglobin levels dropped rapidly, however, in rats given 50 μg/g in the water. The accumulated amount of cadmium could not be related to the anemia since levels in kidneys and liver were higher in rats exposed for 1 year to 10 μg/g in water than in rats given 50 μg/g for 3 months.

With regard to human exposure, it is impossible at present to give any statement on dose-response relationships.

6.3 EFFECTS ON THE CARDIOVASCULAR SYSTEM AND DOSE-RESPONSE RELATIONSHIPS

6.3.1 Acute Effects and Dose-response Relationships

Dalhamn and Friberg, 1954, found that an intravenous injection of cadmium sulfate (0.3 to 0.5 mg Cd/kg) caused a rapid fall in blood pressure within 15 sec in the cat and the rabbit. Perry and Yunice, 1965, found that an intraarterial injection of 0.1 to 0.4 mg/kg caused increased diastolic blood pressure in rats, whereas larger doses (0.8 to 3 mg/kg) caused decreased pressure. Schroeder et al., 1966, found an increase in the systolic pressure after an intraperitoneal injection of 1 to 2 mg cadmium as cadmium citrate in rats. Hypertension was produced in rats by intravenous injections of cadmium in doses of 0.02 to 2 mg/kg (Perry et al., 1970) and by intraperitoneal injections of 0.2 to 2.4 mg/kg (Perry and Erlanger, 1971). The last mentioned authors also found considerable increases in renin activity in blood of rats given an intraperitoneal dose of about 1 mg Cd/kg. The maximum activity was observed 8 hr after injection but a significant increase took place up to 1 month after injection (Perry and Erlanger, 1973).

6.3.2 Chronic Effects and Dose-response Relationships

6.3.2.1 In Human Beings

6.3.2.1.1 In Exposed Workers

Friberg, 1950, examined 43 workers with a mean exposure time to cadmium oxide dust of 20 years and 15 workers with a mean exposure time of 2 years. The study included physical and roentgenological examination of the heart, electrocardiographic examination at rest and after exercise, and the measurement of blood pressure. No increased prevalence of cardiac disease was found. Electrocardiographic changes occurred to the same extent as in a group of sawmill workers without exposure to cadmium. The majority of the workers had completely normal blood pressures. Nothing supported an assumption of higher frequency of hypertension in the cadmium workers than that normally expected in workers in relevant age groups. As Friberg did not examine a control group with regard to the prevalence of hypertension, it is impossible to use his data for a more precise evaluation.

Chest examinations and blood pressure measurements have also been made in other studies (Bonnell, 1955, Bonnell, Kazantzis, and King, 1959, Kazantzis et al., 1963, and Holden, 1969), but none of the resulting reports contained findings of cardiac disease or hypertension caused by cadmium exposure. Hammer et al., 1972, could not find a relationship between exposure to cadmium and blood pressure in superphosphate workers.

6.3.2.1.2 In the General Population

Cadmium has been incriminated as one of the factors that may cause hypertension in human beings (Schroeder, 1964, 1965, and 1967). Data by Tipton and Cook, 1963, show that the cadmium to zinc ratio was higher in people whose deaths were related to hypertension than in other groups. Subjects dying of malignant hypertension with renal failure had low values of both cadmium and zinc in the renal tissues, presumably due to loss of renal tissue. In earlier studies, it had been shown that patients with hypertension excreted more cadmium than did controls (Perry and Schroeder, 1955). These findings, supported by findings in cadmium-exposed animals (see next section), initiated epidemiological research.

Carroll, 1966, found a correlation between cadmium concentrations in the air of 28 American cities and death rates from hypertension and arteriosclerotic heart disease. Hickey, Schoff, and Clelland, 1967, made a similar study covering 26 cities and found that cadmium together with vanadium was correlated with mortality in heart disease. Hunt et al., 1971, upon reanalysis of Carroll's data, reported a higher correlation between population density and death rates than between cadmium concentrations in air and death rates.

Hunt et al., 1971, made a study of 77 cities in the Middle West of the United States with populations varying between 100,000 and 1,000,000. Cadmium, lead, and zinc were measured in the dustfall in residential, commercial, and industrial areas from September to December, 1968. Cadmium in milk also was determined in the selected areas. The age-adjusted cardiovascular death rate for 1959 to 1961 was studied against ten independent variables: population density, dustfall, cadmium, lead and zinc fallout, total precipitation, average maximum and minimum temperatures, and cadmium and lead concentrations in milk. The authors did not find a significant relationship between cadmium fallout and mortality from cardiovascular disease in residential, commercial, or industrial areas. There was, however, a significant relationship between mortality and each of the following independent variables: dustfall, lead fallout, maximum and minimum temperatures, and residential areas. Minimum temperature was the only variable significantly related to cardiovascular mortality in the commerical and industrial areas.

Pinkerton et al., 1972b, found a significant correlation between cadmium levels in milk and cardiovascular mortality in 58 cities in the United States. Population density was shown to have a significant effect and the association between cadmium in milk and cardiovascular disease was shown to be secondary.

Morgan, 1969, determined renal cadmium in autopsy cases and did not find differences between a group with hypertension and a control group. Lewis et al., 1972b, reported that cadmium concentrations in autopsy samples of kidney, liver, and lung were correlated to smoking habits and that this association was not affected by patient characteristics such as arterial hypertension (Section 4.3.4). However, in Czechoslovakia, Lener and Bibr, 1971, found that in a group of 12 people (mean age: 62) with histories of hypertension, cadmium concentrations in kidney were 36 μg/g wet weight, significantly higher than in a control group of 10 people (mean age: 59) with a concentration of 27 μg/g wet weight. Finally, it must be mentioned that according to reports by Nogawa and Kawano, 1969, and Tsuchiya, 1971, hypertension has not been found in patients with Itai-itai disease, nor in people living near the endemic district where the disease is found (Section 8.2.2.1), nor in people living in other cadmium-contaminated areas of Japan.

Thind, 1972, claimed that hypertensive subjects have higher plasma levels of cadmium than normal persons. Since the data were not presented and the difficulties in determining plasma cadmium are well known, this work cannot be taken as support of a relationship between cadmium and hypertension. By neutron activation, Wester, 1971, determined cadmium levels in coronary arteries from autopsied persons with and without atherosclerosis. Cadmium concentration tended to be lower in the atherosclerotic arteries, but since the groups differed highly in age distribution, no conclusions can be drawn.

6.3.2.2 In Animals

Hypertension has been produced in rats of the Long Evans strain, especially in female rats, by giving them cadmium in drinking water for long periods of time (Schroeder and Vinton, 1962, Schroeder, 1964, and Kanisawa and Schroeder, 1969b). Control rats were given a special cadmium-free diet and the animals to be exposed were given 5 μg/g of cadmium (as chloride) in double deionized water together with essential elements. In rats receiving cadmium in drinking water, hypertension usually was manifest after 1 year. When cadmium was given in hard water (Schroeder, Nason, and Balassa, 1967) there was less hypertension. Analysis of cadmium in tissues revealed that cadmium concentrations in the kidneys and liver were of the same magnitude as in American adults, around 40 and 6 μg/g wet weight, on an average, respectively. (In an earlier investigation by Schroeder and Vinton, 1962, with the same exposure conditions, values of only about one tenth of these were reported.) When the cadmium to zinc ratio was above 0.8, the animals were always hypertensive. Histological examination (Kanisawa and Schroeder, 1969b) disclosed renal, arterial, and arteriolar lesions. According to these authors, the changes were indistinguishable from those accompanying benign hypertension from other causes. Hypertension has also been induced by parenteral injection of cadmium salts. Schroeder et al., 1966, and Schroeder and Buckman, 1967, found that when one or two intraperitoneal injections of cadmium citrate (1 to 2 mg Cd/kg) were given, hypertension developed and, over a month's period, became equal to the hypertension appearing in rats on which partial constriction of one renal artery had been performed. Similar results are reported by Chiappino and Baroni, 1969, in Spraque-Dawley rats. They also found a hyperplasia of the juxtaglomerular apparatus and of the glomerular zone of the adrenal cortex. They concluded that the renin-aldosterone system probably was stimulated.

Thind et al., 1970, were able to induce hypertension in rabbits given weekly intraperitoneal injections of cadmium acetate (2 mg/kg) for 7 weeks. During that period mortality was 25%. A decrease in vascular responsiveness to angiotensin was suggested as playing a role in the pathogenesis of cadmium hypertension.

Schroeder, Nason, and Mitchener, 1968, discontinued exposure to cadmium after 515 days when rats were hypertensive. They found a lowering of the blood pressure, and after 715 days blood pressure was normal.

The hypertension induced by cadmium could be reversed by giving the chelating agent cyclohexane-1,2,-diamine-NNN'N'-tetraacetic acid (CDTA), which binds cadmium more firmly than zinc (Schroeder and Buckman, 1967). When the disodium zinc chelate was given by intraperitoneal injection to nine rats previously given cadmium in drinking water, blood pressure was normal in all after 2 weeks. However, in rats given cadmium by injections, repeated treatment with the chelate was sometimes necessary to achieve normal blood pressure. After treatment there was a decrease in cadmium levels in the kidneys, and the authors concluded that the lowering of the cadmium to zinc ratio could be the cause of the decrease in blood pressure. For a discussion of the possible toxic effects of cadmium chelates see Section 6.1.1.3.

There are also negative findings in regard to cadmium and hypertension. Lener, 1968, and Lener and Bibr, 1970, reported that hypertension did not appear when Wistar female rats were given cadmium in drinking water (5 μg/g) and observed for 16 months. When Long Evans rats were put on 2% NaCl for 12 days and then given three intraperitoneal doses of cadmium citrate (1, 1, and 2 mg/kg) with intervals of 1 and 2 weeks, repeated determinations of blood pressure up to 4 weeks after the last injection did not show hypertension. Castenfors and Piscator (unpublished data), using female Sprague-Dawley rats, could not induce hypertension by giving them injections of cadmium chloride (0.5 mg Cd/kg) 3 days a week for 6 months. In another experiment the rats were given cadmium in drinking water (5 μg/g) for 1 year, but monthly determinations of blood pressure did not show any difference compared with controls. At the end of this experiment the intraarterial blood pressure had a mean value in 13 exposed rats of 153 mm Hg (S.D.: 16) compared with 151 (S.D.: 22) in nine controls. There are a few reports that indicate that cadmium may act on the renal handling of sodium. Vander, 1962, showed that a single intravenous dose of a cadmium-cysteine complex enhanced sodium reabsorption in the renal tubules. Perry, Perry, and Purifoy, 1971, produced permanent sodium retention in female rats with four injections of cadmium (1 mg/kg) at 1-month intervals. Lener and Musil, 1970, exposed rats for 16 months to 5 μg/g of cadmium as acetate in drinking water. The animals were also exposed to an increased amount of sodium chloride in their drinking water. At the end of the exposure period, a reduced excretion of sodium was found after a sodium chloride load. This was interpreted as being the result of cadmium enhancing proximal tubular reabsorption of sodium. The renal concentrations of cadmium were on an average 4.79 and 1.29 μg/g wet weight in the exposed and control groups, respectively. Taking into account the relatively high concentrations of cadmium in the drinking water and the long exposure time, the values reported for the exposed group are surprisingly low. In the experiments on rats by Decker et al., 1958, and Piscator and Larsson, 1972, much higher concentrations were found after 1 year at the same exposure level. There is a need for further studies on the relationship between cadmium and sodium. If cadmium can cause retention of sodium in susceptible individuals this could be of great impact.

Retention of sodium may increase the susceptibility to cadmium as indicated by the findings of Lener and Bibr, 1973. When a group of Long Evans rats was put on 2% NaCl for 50 days before receiving two doses of cadmium citrate (2 mg Cd/kg) with an interval of 84 days, there was a significant increase in systolic blood pressure in the cadmium-exposed animals on high sodium intake, whereas cadmium alone did not cause hypertension. Thus, both strain differences and sodium intake might be determining factors in the development of hypertension in experimental animals.

6.3.3 Discussion and Conclusions

Since the possibility that cadmium produces cardiovascular disease, especially hypertension, has attracted much interest during recent years, a special discussion on this subject is motivated.

The fact that hypertension has been associated with high kidney levels of cadmium in certain studies cannot be used for firm conclusions concerning causality. The data have been obtained by analyses of kidneys from people who died from vascular disease. Nothing has been stated about the previous history or the course of the disease, smoking habits, etc., and the interpretation of the data is subject to the usual difficulties met in this type of epidemiological analysis.

The results from epidemiological investigations are hitherto ambiguous. They have been obtained by associating cardiovascular disease with dustfall data or cadmium concentrations in air. Other more important sources of cadmium exposure have not been considered. The average cadmium concentration in a "standard American man" seems to be around 30 mg in the age range of 40 to 50 years (Section 4.3.4). Even with such a high average cadmium concentration in the ambient air as 0.01 $\mu g/m^3$, the total amount of cadmium absorbed over a 50-year period would amount to only a small fraction of the total body burden. An absorption rate of inhaled cadmium of 25% and a lung ventilation of 20 m^3/day thus would give a total uptake of less than 1 mg of cadmium over a 50-year period. The cadmium fallout, of course, could influence the cadmium levels of various foodstuffs. It does not seem probable, however, that the average cadmium intake via food by city dwellers in the U.S. is very much dependent upon the local ambient air concentration of cadmium. It was pointed out by Hunt et al., 1971, that population density is an important factor. Furthermore, smoking must be taken into account.

In certain areas of Japan where there has been a considerable exposure to cadmium for decades, hypertension has not been associated with cadmium. In addition, there are no reports showing that workers exposed to cadmium have a higher prevalence of hypertension than other groups.

The results from investigations on animals are also ambiguous. It is true that studies show that intraperitoneal injections of cadmium will induce hypertension in rats and rabbits. Long-term exposure via the oral route with relatively low concentrations of cadmium has induced hypertension in rats. There are also negative findings and strain differences as well as other factors that might influence the response in rats when cadmium is administered.

Even if thus far the available data do not support a hypothesis that cadmium is causally associated with cardiovascular disease in man, there are reasons to study the question further. The findings, particularly in the animal studies, but even in the epidemiological studies in human beings, merit further attention. Epidemiological studies should be carried out on a longitudinal basis with due consideration given to routes of exposure other than air. Industrial workers with known exposure to cadmium should be included in such studies. More animal experiments should be developed, with particular emphasis upon the mechanisms of cadmium-induced hypertension, as the reasons for the differing results reported by various investigators are not clear.

6.4 EFFECTS ON BONE AND DOSE-RESPONSE RELATIONSHIPS

Cadmium does not accumulate to any considerable extent in osseous tissue, as discussed in Section 4.3.2.2. A direct effect of cadmium upon such tissue is therefore unlikely. However, there is evidence that the regulation of the calcium and phosphorus balance by the kidneys can be disturbed by cadmium exposure (see Section 6.1.2.1). A disturbed calcium and phosphorus balance can in turn give rise to bone changes but even though the effect upon the calcium balance becomes grave within a very short time, the symptoms and changes in the skeleton require considerable time to develop. Acute effects on bone are not reported and the following account of effects on bone is concerned only with long-term effects.

6.4.1 In Human Beings

Nicaud, Lafitte, and Gros, 1942, described a group of workers exposed to cadmium oxide dust in an accumulator factory. In two men (ages: 41 and 60; exposure time: 11 years), three married women (ages: 48, 51, and 60; exposure times: 8, 16, and 14 years), and one unmarried woman (age: 37; exposure time: 10 years), there were symptoms of pain in the back and extremities and difficulties in walking. Roentgenological examination disclosed in all six workers lines of pseudofractures, especially in the shoulder blades, pelvis, femur, and tibia. Proteinuria was not found, but only the boiling test was used. The possibility of tubular proteinuria cannot be excluded. Later studies using nitric and trichloracetic acid on other workers in that factory (Friberg, 1950) revealed the presence of proteinuria. Electrophoretic examination of urine from one worker revealed that it was the same type of proteinuria as was seen in the Swedish workers.

Friberg made a roentgenological study of the bones of 11 workers from a group of 43 exposed to cadmium oxide dust for 9 to 34 years (for further details, see Section 5.2.1.1) but changes were not found. Bonnell, 1955, found at autopsy severe decalcification in a cadmium worker exposed to cadmium oxide fume for 32 years. He died from chronic renal failure. Before his death proteinuria had been found at repeated examinations. Electrophoretic analysis revealed that this proteinuria was of the type common in chronic cadmium poisoning.

Gervais and Delpech, 1963, found in eight male workers (ages: 49 to 63; exposure times: 8 to 30 years) exposed to cadmium oxide fume (seven workers) and cadmium oxide dust (one worker) the same lines of pseudofractures as described by Nicaud, Lafitte, and Gros, 1942. These workers had not been exposed for 10 to 20 years and the symptoms were diagnosed after exposure had ceased. In most cases there was proteinuria which the authors, probably erroneously, blamed upon lead (Section 6.8). Pujol et al., 1970, have shown that in a worker with bone changes in the same factory, tubular proteinuria and other signs of tubular impairment were present. Adams, Harrison, and Scott, 1969, found one case of osteomalacia in a group of workers exposed to cadmium oxide dust. These authors expressed little doubt that this osteomalacia was caused by renal disease since the worker had multiple tubular defects.

Chronic cadmium poisoning in combination with certain dietary deficiencies, i.e., lack of calcium and vitamin D, seems to be the causative agent for the Itai-itai disease, a bone disease in Japan (see Chapter 8). Of special interest is that the workers in the Swedish factory had a heavy exposure to cadmium oxide dust which gave rise to severe pulmonary and renal changes but not to bone changes (Friberg, 1950). Also, the British investigations have not disclosed a high incidence of bone disease, except the two above mentioned cases, even though disturbances in calcium and phosphorus metabolism have been common findings (Kazantzis et al., 1963, Adams, Harrison, and Scott, 1969). Different nutritional habits will probably play a role. The intake of fat-soluble vitamins such as Vitamin D and of calcium is much higher in Sweden and the United Kingdom than in Japan (see Chapter 8, Table 8:8). In France the intake of milk is lower than in Sweden or the United Kingdom. Furthermore, the poor nutritional conditions during the Second World War must have played an important role. It should be pointed out that some French patients and almost all of the Japanese patients have been women and thus more susceptible to calcium and Vitamin D deficiencies. As for the Japanese women, it is further known that most of them had more than six pregnancies.

6.4.2 In Animals

Animal experiments have also indicated that exposure to cadmium can cause bone changes. Ceresa, 1945, injected 1 mg $CdSO_4$ daily into rabbits. After 2 months he observed decreased calcium levels in serum and decreased mineral content in bone. Maehara, 1968, gave groups of male rats, female rats, and ovariectomized rats cadmium in a diet low in calcium (50 μg/g Cd for 45 weeks followed by 25 μg/g for 5 weeks). He found a decalcification and concluded that this was probably caused by a disturbance of calcium absorption in the digestive tract. He did not report on levels of cadmium in the organs or on proteinuria, so it cannot be said to what extent the bone changes were caused by renal dysfunction. Larsson and Piscator, 1971, gave female rats 25 μg/g cadmium as the chloride in drinking water. Exposure was for 1 and 2 months and diets with normal and low contents, respectively, of calcium were given. For data after 2 months, see Table 4:1. A significant decrease in the mineral content of the bones was found in groups exposed to cadmium and fed a low calcium diet for 2 months. Cadmium exposure alone and a low calcium diet alone did not cause such a decrease. Furthermore, cadmium levels in liver and kidneys from animals exposed to cadmium and on a low calcium diet were 50% higher than in animals exposed to cadmium but on a normal intake of calcium, indicating that a low intake of calcium will increase the absorption of cadmium. Studies with injected ^{45}Ca showed that there was the same calcium accretion in the tibia in animals on low calcium and in animals on this diet combined with cadmium exposure. These findings suggested that there was not a direct action of cadmium on bone tissue but that the osteoporosis in cadmium-exposed rats on a low calcium diet was caused by an increased bone reabsorption for maintenance of blood calcium. Further results from this study have been recently reported by Piscator and Larsson, 1972. Calcium deficiency alone gave rise to a compensatory increase in the parathyroid volume. In animals also exposed to cadmium, such as increase in parathyroid volume did not occur. New experiments (Piscator and Larsson, 1972) revealed an increase in excretion of ribonuclease in female rats after 1 year of exposure to 7.5 μg/g of cadmium in drinking water and a diet low in calcium. This finding indicates renal tubular damage. There was also a significant reduction in the mineral content of bone. Whether the bone changes were due to the changes in renal function or due to an effect of cadmium on intestinal absorption of calcium is not known.

More evidence that exposure to cadmium may cause bone changes in animals comes from Hirota, 1971. He exposed rabbits for about 1 year to cadmium in water corresponding to daily doses of 5 and 20 mg Cd per rabbit. The rabbits exposed to 20 mg Cd/day showed histological signs of renal tubular changes at the end of the exposure. An increase in serum calcium and decrease in serum phosphorus as well as an increase in the urinary excretion of calcium and phosphorus were also seen. Proteinuria was noted in some animals. Itokawa, Abe, and Tanaka, 1973, studied weight curves and bone changes after 30 days of treatment with cadmium in male rats. Regardless of the amount of calcium or protein in food, 200 μg/g cadmium in the diet gave a strong depression of weight development per gram of food intake. A group of five rats given the same cadmium dose but along with low protein and calcium intake had an abnormal curvature of the spine, diminished density of bones, and cortical thinning of the femur. Taking food intake into consideration, the rats in this group had been exposed to about 1.4 mg Cd/day or about 8 mg Cd/kg body weight. Unfortunately no data on cadmium concentrations in kidney or liver are available.

Zinc has been shown to be important for bone formation. In calves given cadmium orally (350 μg/g in the diet for 7 days) the uptake of zinc was increased in most organs but not in bone, in which there was instead a decrease (Powell, Miller, and Blackmon, 1967). Lease, 1968, found in the chick that cadmium caused a decrease in the zinc uptake in the tibia.

6.4.3. Discussion and Conclusions

A direct action of cadmium upon bone is unlikely, as stated in the introduction to this section. No data have been brought forward to support such a possibility so it will not be further discussed.

An effect of cadmium upon zinc metabolism, essential for bone tissue, cannot be ruled out. Available data, however, do not permit any conclusions as to whether this possible action is critical for bone effects of cadmium.

The effects on bone with all probability are secondary to effects on calcium-phosphorus metabolism. Theoretically, cadmium exposure could influence metabolism of bone minerals in several ways: by affecting the absorption of such minerals in the gastrointestinal tract directly or by influence on vitamin D activity, by changing the parathyroid activity, or by directly affecting the renal regulation of the Ca/P balance.

None of these possibilities has been ruled out entirely by available data. It was shown in Section 6.1 that tubular kidney damage and tubular proteinuria are typical features of chronic cadmium poisoning. Increased excretion of bone minerals is not an uncommon cause of kidney stones. An impairment of the renal tubular regulation of Ca/P balance, therefore, is with all probability the most important factor for elicitation of bone changes in workers with proteinuria. The bone changes reported in the French cadmium workers have features typical of osteomalacia, and it is well known that other forms of tubular dysfunction, hereditary forms (Dent and Harris, 1956), and acquired forms (deSeze et al., 1964) can give rise to osteomalacia. Further discussion on the mechanism for elicitation of the so-called Itai-itai disease can be found in Chapter 8.

Concerning the other possibilities for influence on the Ca/P metabolism, the following could be said: the parathyroids regulate blood calcium levels mainly by influencing bone resorption and renal excretion of phosphate and therefore will be involved invariably when disturbances in bone salt metabolism occur. However, at present there is no reason to believe that cadmium influences primarily the parathyroids. The rats in the experiment by Piscator and Larsson, 1972, were exposed to relatively large amounts of cadmium, and the inhibition of compensatory increase in parathyroid volume may well have been due to effects of cadmium upon other endocrine organs.

An action on the uptake of calcium from the gastrointestinal tract is most likely of some importance, especially when cadmium exposure is by the oral route. The evidence from the animal experiments by Larsson and Piscator, 1971, and Piscator and Larsson, 1972, speaks in favor of such an action. Kidney levels were below those likely to give rise to severe renal impairment.

The importance of dietary calcium supply for the elicitation of bone changes makes precise estimations regarding dose-response relationships impossible. The evidence by Larsson and Piscator, however, shows that even a relatively short exposure to moderate concentrations of cadmium can influence bone mineralization when the calcium content of the diet is low.

6.5 EFFECTS ON THE LIVER AND DOSE-RESPONSE RELATIONSHIPS

6.5.1 Acute Effects and Dose-response Relationships

An influence on liver function has been recorded in several workers with acute cadmium poisoning after exposure to cadmium oxide fumes (see Section 6.1.1). In lethal cases microscopic changes have been noted in the liver and attributed to the general toxemia. These changes have been slight compared to the changes in the lungs.

Prodan, 1932, exposed cats to high concentrations of cadmium oxide fumes for short periods of time and found fatty infiltration in the livers. Similar changes were found in cats exposed to cadmium oxide dust.

Andreuzzi and Odescalchi, 1958, gave rabbits single intravenous injections of doses varying from 1.25 to 3 mg Cd/kg and determined the GOT (glutamic-oxaloacetic-transaminase) activity in the serum at different times during 72 hr. In the group given 3 mg/kg, the GOT activity increased considerably after 17 hr, but 60% of the animals were dead within 24 hr. When 2.5 mg/kg was given, the GOT activity increased more than tenfold after 24 hr, but 40% of the rabbits died within 48 hr. Rabbits given 2 and 1.25 mg/kg, respectively, survived more than 72 hr. After 24 hr the increase in GOT activity was about fourfold, but after 72 hr the activity was within the normal range again. These results indicate that a dose near the LD_{50} is necessary in order to induce a severe liver lesion and that the effect on GOT activity produced by lower doses is reversible.

Kapoor, Agarwala, and Kar, 1961, gave rats single subcutaneous injections of 10 mg Cd/kg as chloride, a dose later shown to produce morphologic alterations in rat liver (Favino and Nazari, 1967). The subcellular distribution of cadmium was studied at times varying from 6 to 168 hr after injection. Initially the cadmium concentration was greatest in the nuclear fraction, but later, more was found in the supernatant. Nordberg, Lind, and Piscator, 1971, found a considerable decrease in liver weights within the first 6 days in mice given a single subcutaneous injection of 3 mg Cd/kg but no difference in liver weights compared to controls after 18 days.

That acute exposure will give rise to morphological and functional changes in the liver is clear from the foregoing experimental results. In Chapter 4 it was shown that after acute exposure, both via the respiratory tract and injection, cadmium was accumulated in the liver. It is not known if changes found at autopsy in livers from human beings exposed to high concentrations of cadmium oxide fumes are due to a direct action of cadmium or to the general changes produced by the edema of the lungs. It is conceivable that cadmium will have a direct action on certain liver functions in experimental animals, especially because there is no binding of cadmium to metallothionein during the first hours after injection (see section 4.2.8). However, the possibility that changes in the cardiovascular system also influence the function of the liver cannot be excluded.

6.5.2 Chronic Effects and Dose-response Relationships

6.5.2.1 In Human Beings

Friberg, 1950, found that the Takata reaction was positive in 2 and the thymol test in 3 of 19 workers exposed to cadmium oxide dust with a mean exposure time of 20 years. Increases in serum gamma globulin levels were found in several workers. In most other investigations liver function has been little studied, but it will be apparent from the reports of Bonnell, 1955, Kazantzis et al., 1963, and Adams, Harrison, and Scott, 1969, that compared to the pronounced changes in renal function, gross changes in liver function are unusual findings in cadmium-exposed workers.

6.5.2.2 In Animals

Prodan, 1932, found changes in the livers of

cats exposed to cadmium via the respiratory or the oral route. Wilson, deEds, and Cox, 1941, made similar observations in rats given cadmium orally, 250 to 500 μg/g in the diet for several months. Friberg, 1950, injected cadmium sulfate (0.65 mg Cd/kg) into rabbits 6 days a week for 2 to 4 months and found cirrhotic changes. Axelsson and Piscator, 1966a, determined the GOT activity in serum from rabbits given injections of cadmium chloride (0.25 mg Cd/kg) 5 days a week for 11 to 29 weeks. They noted that compared with controls, there was no difference after 11 weeks of exposure, whereas there were significant increases in activity after 17 weeks. At that time the concentration of cadmium was around 450 μg/g wet weight. Piscator and Axelsson, 1970, made similar determinations in a group of rabbits exposed in the same way for 24 weeks and thereafter followed for another 30 weeks. At that time the GOT activity was the same in serum from the exposed rabbits and controls. The concentration of cadmium in the liver was about 180 μg/g. There was no difference in alkaline phosphatase activity of serum. These findings suggest that the liver damage was reversible. Kimura, 1971, has reported increased levels of serum GOT and GPT in rabbits exposed to 2 mg Cd/kg body weight subcutaneously 6 days a week for 2 weeks. He also reported a slower clearance of galactose from blood in these rabbits compared with controls. All of these changes are suggestive of hepatic involvement; the author could demonstrate morphological changes in the livers of these rabbits.

Long-term studies on effects of cadmium on activities of certain liver enzymes have been performed by Sporn, Dinu, and Stoenescu, 1970. In one experiment, rats were given cadmium chloride (1 μg/g cadmium) in drinking water for 335 days. An increase in the activity of phosphorylase a and a decrease in the aldolase activity were found, indicating that cadmium may interfere with carbohydrate metabolism in the liver. When rats were given larger amounts of cadmium (10 μg/g Cd in food) for shorter periods, an influence of cadmium on the oxidative phosphorylation in the liver mitochondria not seen in the above mentioned long-term experiment was noted.

When zinc was administered simultaneously via the oral route to rats given 10 μg/g cadmium in food for 60 days, it was found that whereas zinc prevented the action of cadmium on the oxidative phosphorylation in the mitochondria, it did not prevent changes in the activity of phosphorylase a or aldolase (Sporn et al., 1969).

Cadmium was not determined in organs, so the effect on the enzymes cannot be directly related to liver concentrations. However, Sporn, Dinu, and Stoenescu, 1970, estimated that the mean total intake of cadmium in the rats receiving 1 μg/g in drinking water for 335 days was 4 mg.

In an experiment by Decker et al., 1958, it was found that animals receiving 0.5 μg/g cadmium in drinking water for 1 year had ingested a total amount of 5.8 mg. In these animals the mean cadmium concentration in the liver was 1.1 μg/g. It is conceivable that in the above mentioned experiment by Sporn, Dinu, and Stoenescu, 1970, cadmium levels in livers were not much different from the ones reported by Decker et al., 1958. This indicates that cadmium might act upon certain enzyme activities at a liver concentration of the same magnitude as can be found in normal human adults (see Figure 4:22).

A recent investigation on rabbits by Stowe, Wilson, and Goyer, 1972, gives additional information. The experimental conditions have been described in Sections 4.2.2.2 and 6.1.2.2. After 6 months of oral exposure to cadmium, mean liver concentrations were 188 μg/g wet weight. Light microscopy revealed that, in contrast to controls, the cadmium-exposed rabbits showed depletion of glycogen and deposits of collagen. Inflammatory cell infiltrates were frequent in the portal regions and biliary hyperplasia was often present. Electron microscopy revealed that the most striking changes took place in the endoplasmic reticulum, which was increased in exposed animals. Liver function tests such as determination of activity of GOT, GPT, alkaline phosphatase, and LDH isoenzymes in serum, BSP test, and blood coagulation tests did not show any difference between controls and exposed. These results indicate that negative liver function tests do not exclude morphological liver changes and that some earlier conclusions regarding liver damage might not be valid.

6.5.2.3 Dose-response Relationships

As the liver will accumulate large amounts of cadmium during exposure, functional disturbances could be expected. Liver dysfunction has not been a common finding in exposed workers. The results by Sporn, Dinu, and Stoenescu, 1970, indicate that in rats certain liver enzymes may be influenced by cadmium concentrations of the same magnitude as found in normal human adults, but it is not known at present to what extent these findings can be applied to the human liver. These results and some recent findings on cadmium-exposed rabbits also suggest that commonly used clinical tests for liver function may not be suitable for evaluation of cadmium effects on the liver.

6.6 EFFECTS ON TESTICLES AND DOSE-RESPONSE RELATIONSHIPS

Little attention had been drawn to effects of cadmium on testicles prior to the mid-1950's, when Parizek described the destructive effect on testicular tissue (Parizek and Zahor, 1956, and Parizek, 1957). As early as 1919, however, a brief mention was made of a "bluish discoloration of the testicles" of experimental animals in a report on the pharmacology of cadmium (Alsberg and Schwartze, 1919, and Schwartze and Alsberg, 1923), but this information seems to have escaped attention at that time. During the last 17 years, interest in the different aspects of the destructive action of cadmium on this organ has been revived. A vast literature on the subject has accumulated and has been reviewed and discussed by Gunn and Gould, 1970. The brief account that follows does not cover all published articles but intends to give the more important aspects.

6.6.1 Acute Effects in Animals

6.6.1.1 Events in Cadmium-induced Testicular Necrosis

According to Parizek's description in 1957, a subcutaneous injection of 0.02 mmol of cadmium chloride or lactate per kilogram (2.24 mg Cd/kg) caused macroscopic changes in the testicles within the first few hours after injection. The organs first became swollen and dark red or purple. The weight then decreased rapidly and the testicles became small, hard, and yellowish. At the same time the weight of the seminal vesicles and the prostate decreased as a result of a decreased endocrine activity of the testicles. Histologically a capillary stasis and edema of the interstitium were observed 2 to 4 hr after injection, followed by extensive hemorrhages, according to Parizek and Zahor, 1956. Regressive changes of the seminiferous epithelium were seen 4 to 6 hr after injection and progressed to a total necrosis within 24 to 48 hr.

At a certain time (about 1 month in the rat) after the acute necrosis, a revascularization of testicles (Niemi and Kormano, 1965, and Kormano, 1970) and a regeneration of Leydig cells occurred (Allanson and Deanesly, 1962). Simultaneously, the endocrine activity of the testes returned (Parizek, 1957, Allanson and Deanesly, 1962). When the dose injected was large enough, the germinal epithelium did not regenerate and the testicles thus functioned as endocrine organs only (Kar and Das, 1960, Allanson and Deanesly, 1962, and Gunn, Gould, and Anderson, 1963a). About 1 year after injection interstitial cell tumors developed (Gunn, Gould, and Anderson, 1963a, and Roe et al., 1964; see also Chapter 7).

Simultaneously with the alterations in the testicles, morphological changes in the spermatozoa of the ductus deferens and proximal parts of the epididymis occur, but spermatozoa in distal parts of the epididymis are sometimes unaltered. Animals in some cases become permanently sterile as early as 24 hr after injection (Kar and Das, 1962a) but if androgens are administered, fertility can be maintained as long as 9 days after cadmium injection (Gunn, Gould, and Anderson, 1970). Cadmium is also extremely toxic for sperm cells in vitro (White, 1955). The sterilizing effect has been studied in a number of animal species both by systemic (see below) and intratesticular (Kar, 1961, 1962, Setty and Kar, 1964, Chatterjee and Kar, 1968, Kar and Kamboj, 1963, and Kar and Das, 1962b) injections.

6.6.1.2 Sensitivity of Different Animals to Testicular Necrosis and Doses Applied

The observations of testicular necrosis by Parizek and Zahor, 1956, and Parizek, 1957, concerned mice and rats. Extensive later studies have confirmed these observations (Meek, 1959, Kar and Das, 1960, Gunn, Gould, and Anderson, 1961, 1963a and b, and Allanson and Deanesly, 1962). Similar changes have been shown to occur after systemic administration of cadmium salts to other experimental animals such as rabbits (Parizek, 1960, Cameron and Foster, 1963), monkeys (Girod, 1964a and b), guinea pigs (Parizek, 1960, Johnson, 1969), and golden hamsters (Parizek, 1960), as well as domestic animals, calves (Pate, Johnson, and Miller, 1970). However, some animal species such as frog, pigeon, rooster, armadillo, and opossum (Chiquoine, 1964), and domestic fowl (Erickson and Pincus, 1964, and Johnson, Gomes, and VanDemark, 1970) did not develop testicular necrosis, in spite of such doses as 10 and 20 mg $CdCl_2$/kg (corresponding to 6.2 and 12.4 mg Cd/kg) given to the species tested by Chiquoine, 1964. On the basis of these findings, Chiquoine, 1964, suggested that cadmium necrosis is common to species possessing scrotal testes and absent from those possessing abdominal testes. Opossum was an exception to that generalization.

Some strains of mice were also insensitive to subcutaneous injections of 0.02 and 0.04 mmol $CdCl_2$/kg (2.2 to 4.4 mg Cd/kg) (Chiquoine and Suntzeff, 1965). Even doses in the lethal range did not produce any testicular effects in some strains of mice, according to experiments performed by Gunn, Gould, and Anderson, 1965. They concluded that the amount of cadmium needed to produce minimal testicular damage varied within the susceptible strains. It is of interest in this connection that Lucis and Lucis, 1969, have shown that strains of mice susceptible to cadmium-induced testicular necrosis exhibit a greater uptake of cadmium in the testes than strains that are resistant to this action. The lowest subcutaneous dose that has been reported constantly to give rise to testicular necrosis in mice is probably 0.012 mmol/kg (1.34 mg Cd/kg), stated by Gunn, Gould, and Anderson, 1968a, to be effective for the Cd-1 strain.

Gunn, Gould, and Anderson, 1965, also tested different doses in different strains of rats. They concluded that the dose of cadmium needed to produce minimal testicular damage differed among various strains and even within the same strains of rats derived from different commercial sources. Usually 0.02 to 0.05 mmol $CdCl_2$/kg (2.2 to 5.6 mg Cd/kg) has been used in rats (Parizek, 1957, Kar and Das, 1960, and Gunn, Gould, and Anderson, 1961, 1963a and b). In calves, intravenous administration of doses as low as 0.20 mg $CdCl_2$ has been reported to give rise to testicular necrosis (Pate, Johnson, and Miller, 1970).

6.6.1.3 Mode of Administration

All the studies mentioned are concerned with acute effects and sequelae of a single injection of cadmium. Subcutaneous, intraperitoneal, and intravenous injections of cadmium salts are effective with regard to testicular necrosis in various animal species. Bouissou and Fabre, 1965, tested several routes of administration in rats. Large doses (0.25 to 0.57 mmol/kg, i.e., 28 and 64 mg Cd/kg) administered by the oral route were also reported to give rise to testicular changes similar to those produced by the injection routes. In some large animals such as calves (Pate, Johnson, and Miller, 1970), only the intravenous injection was effective. Kar and Kamboj, 1963, reported that "scrotal inunction" (i.e., painting of $CdCl_2$ solution on the scrotal skin) gave rise to testicular atrophy and sterility in several animal species.

6.6.1.4 Mechanisms for Testicular Necrosis

The mechanism for development of acute necrosis of the testicles after cadmium injection has been discussed in a number of papers. Parizek, 1957 and 1960, advanced two possibilities for the primary action of cadmium: (1) circulatory failure and (2) action on spermiogenic epithelium. Parizek, 1957, considered the latter possibility more probable. As zinc is essential for the maintenance of germinal epithelium (Elcoate et al., 1955), Parizek considered his own observation (Parizek, 1956) of an antagonistic action of zinc against the necrotizing effect of cadmium supportive of this last mentioned theory. Later Dimow and Knorre, 1967, found changes in the enzymes of the germinal epithelium before histological changes were evident in the testicles. They considered their results to be in favor of a direct action of cadmium on germinal epithelium. Hodgen, Gomes, and VanDemark, 1970, have detected an isoenzyme of carbonic anhydrase in rat testis not found in rat kidneys or erythrocytes. They suggested that this organ-specific carbonic anhydrase was the primary site of action of cadmium in the testicle. Hodgen and his colleagues (Hodgen, Butler, and Gomes, 1969, and Hodgen, Gomes, and VanDemark, 1970) also showed that the activity of this isoenzyme decreased already 30 min after cadmium injection and later ceased entirely. Hodgen, Gomes, and VanDemark, 1969, showed that the testicular isoenzyme was found in both rat testicular artery and testicular parenchyma. Thus, these studies did not reveal whether the action of cadmium took place in the interstitium or in the tubules of the testicle.

Whatever may be the molecular basis for the cadmium action, it has been well established that the vascular bed and the blood flow of the testicles are affected very early after injection of cadmium. This has been shown with histological (Kar and Das, 1960, and Gunn, Gould, and Anderson, 1963b), electron microscopical (Chiquoine, 1964, and Clegg and Carr, 1967), angiographic (Niemi and Kormano, 1965), and functional (Waites and Setchell, 1966, Clegg and Carr, 1967, Carr and Niemi, 1966, Johnson, 1969, and Setchell and Waites, 1970) techniques. Because the injected cadmium is concentrated in the interstitial tissue (Berlin and Ullberg, 1963, and Nordberg, 1972a), and never reaches the germinal epithelium in any detectable amounts, the changes observed in this epithelium most probably are secondary to the vascular damage.

6.6.1.5 Protective Measures

The necrotizing action of cadmium can be prevented by specific treatments. The effectiveness of the administration of zinc in this respect has already been mentioned. Totally 80 to 200 times the molar equivalent of zinc acetate given in three injections (5 hr before, simultaneously with, and 19 hr after the cadmium) afforded complete protection of the testicles of the rat against 0.04 mmol of Cd/kg (Parizek, 1956, 1957, and 1960). These observations have been confirmed by other investigators (Gunn, Gould, and Anderson, 1961) who used a total of 3 mmol/kg of zinc to prevent testicular damage from 0.03 mmol Cd/kg in Wistar rats. It was also shown that the duration of the prevention made possible by zinc was dependent upon the breeding of the animals. Animals bred immediately after injection of Cd + Zn showed a sharp decrease in fertility 4 weeks later, whereas animals bred 8 weeks after injection were still fertile 12 weeks after injection.

Selenium (0.04 mmol/kg) also protects the rat testicles against the necrotizing action of cadmium (0.02 mmol/kg) as shown by Kar, Das, and Mukerji, 1960, and Kar and Das, 1963. The same effect of selenium was studied in the mouse by Gunn, Gould, and Anderson, 1968b. The last mentioned authors also showed in an earlier communication (Gunn, Gould, and Anderson, 1966) that thiols such as cysteine and dimercaptopropanol (BAL) prevented cadmium-induced necrosis in the testicles. Cobaltous chloride given to rats in a dose of 30 mg $CoCl_2$/kg 17 hr prior to administration of 6.60 mg $CdCl_2$/kg also protected the testicles from damage (Gabbiani, Baic, and Deziel, 1967a).

Ito and Sawauchi, 1966, reported that injection of ½ to ¼ of a testis-destructive dose of cadmium chloride (0.1 ml of a 0.1% $CdCl_2$ solution per mouse) 2 days prior to the injection of the normally testis-destructive dose prevented testicle destruction in more than half of the animals. This protective effect of pretreatment with the harmful agent itself has later been confirmed in our own laboratory (Nordberg, 1971a), as well as by others (Gunn and Gould, 1970).

6.6.2 Chronic Effects

6.6.2.1 In Animals

From the point of view of environmental exposure, acute effects of large doses are of limited interest. Histological changes in the testicles of rats as a result of long-term administration of cadmium in the food were reported by Pindborg, 1950, and Ribelin, 1963. Pindborg gave doses of 0.025 to 0.1% $CdCl_2$ (=150 to 610 μg/g Cd) and Ribelin, 50 to 1,270 μg/g cadmium in the diet. The changes reported by Ribelin were different from those developing after a single injection but were not different from those produced by several other toxic substances tested by Ribelin. On the whole it is difficult to judge the frequency of the changes from the paper published by Ribelin, as no data are given concerning number of animals, frequency of testicular alterations, or the appearance of toxic reactions in other organs. In a brief report by Richardson and Fox, 1970, testicular development was studied in Japanese quail which had received 75 μg/g Cd in the diet up to 4 weeks of age. Seminiferous tubules lined with undifferentiated cells were said to have occurred in more than half of the tubules. Though it is impossible to judge the nature of the testicular damage from the short communication, it seems that no total testicular necrosis occurred.

Piscator and Axelsson, 1970, reported on histological and electron microscopic examinations of testicles from rabbits exposed by repeated daily subcutaneous injections of cadmium for as long as 24 weeks and followed for another 30 weeks before killed. The investigators did not observe any pathological changes in the testicles in spite of the presence of kidney damage. They stated that the absence of testicular changes could be explained either by rabbits' comparative insensitivity to the necrotizing action of cadmium or by the protection of the testicles by metallothionein formed in the liver. In order to shed some light on this question, Nordberg, 1971a, performed studies on mice of the CBA strain, which is very sensitive to cadmium-induced testicular necrosis. By repeated injections considerable amounts of cadmium were accumulated in the organs, including the testicles, but no histologically evident changes were found in this organ. These data evidently were in favor of the second alternative, suggesting a protective effect of metallothionein against the necrotizing action on the testicles. Such a metal-binding protein was also found in the livers of the CBA mouse (Nordberg, Piscator, and Lind, 1971). Injection of cadmium bound to metallothionein did not produce any testicular necrosis in doses effective in this respect when injected alone (Nordberg, 1971a). The earlier mentioned observations by Ito and Sawauchi, 1966, of a protective action of small doses of cadmium are consistent with the finding that no testicular necrosis has occurred in animals exposed repeatedly to cadmium.

Another observation which might be relevant when discussing long-term effects on the testicles is the decrease in proteinuria demonstrated in mice repeatedly injected with $CdCl_2$ for several months (Nordberg and Piscator, 1972). Male mice and rats normally excrete a high concentration of protein in the urine. The major urinary protein is synthesized in the liver under the influence of testosterone (Finlayson et al., 1965, Roy and Neuhaus, 1966, and Roy, Neuhaus, and Harmison, 1966). The observed decrease in proteinuria during cadmium exposure, therefore, might be a result of an action of cadmium on the production of testosterone in the interstitial cells of the testicles. The observation that cadmium accumulates in the interstitial tissue (around the capillaries) of the testicle (Nordberg, 1972a) favors such a possibility.

6.6.2.2 In Human Beings

The reports of acute testicular necrosis after systemic administration of cadmium salts in a great number of animal species strongly suggest that a similar effect is likely in human beings. However, such testicular necrosis has not been reported to be a result of cadmium exposure. Smith, Smith, and McCall, 1960, found high values of cadmium in the testicles of men industrially exposed to cadmium fume. These authors also reported some histological changes in the testicles at postmortem examination. These changes were of a rather unspecific nature, however, and the authors therefore ascribed them to the terminal illness. It is difficult, however, to exclude the possibility that these histological changes had some association with the cadmium exposure. Favino et al., 1968, investigated the fertility of ten cadmium workers and also analyzed androgens in urine. In this investigation one case of impotency with abnormally low testosterone levels in urine was found. Further studies similar to those made by Favino et al. are necessary before any conclusion can be drawn concerning the possible effect of cadmium on the endocrine function of the testicles of people exposed to cadmium.

6.6.3 Dose-response Relationships

The lowest dose effective for elicitation of acute testicular changes in calves is 0.2 mg $CdCl_2$/kg (0.12 mg Cd/kg) given by the intravenous route (Pate, Johnson, and Miller, 1970). A subcutaneous injection of cadmium salts in a dose of 0.01 mmol/kg (1.1 mg/kg) can produce a total testicular necrosis in some animals (e.g., certain strains of mice). For other animals (e.g., rabbits) higher doses are required. Some animal species or certain strains of a species are so resistant to the acute necrotizing action of cadmium that even doses in the lethal range do not cause alterations. The dose interval between no-effect and total testicular necrosis is very narrow for a given kind of animal. Testicular necrosis can be prevented by zinc, selenium, cobalt, thiol compounds, metallothionein, or pretreatment with a smaller dose of cadmium.

All data on testicular necrosis are from animal experiments. As many mammalian animal species, including monkeys, are susceptible to the action of cadmium on the testicles, it is highly probable that a similar effect would occur also in human beings. However, as cadmium is not used as a drug, people cannot be exposed by injections. If the absorbed dose necessary for elicitation of testicular changes is proportionally the same as the lowest subcutaneous dose in animals (1 mg Cd/kg), this means that 70 mg should be absorbed in a short time in a 70-kg man, an unrealistic exposure.

For chronic exposure no conclusive data are available concerning effects on the testicles. Animal experiments suggest that there might be a possibility for effects of a somewhat different nature from those seen after a single injection. As yet no evidence on human exposure has proven an effect on the testicles. Accordingly no dose-response relationships for chronic exposure can be discussed.

6.7 OTHER EFFECTS

6.7.1 In Human Beings

Over the years many different symptoms have been reported in men exposed to cadmium. These include loss of appetite, loss of weight, fatigue, increases in the E.S.R., etc. These unspecific symptoms can be related to the systemic effects. More specific effects have been a yellow coloring on the proximal part of the front teeth (Barthelemy and Moline, 1946, and Friberg, 1950) and anosmia (Friberg, 1950). The last mentioned effect was found in about one third of a group of workers with a mean exposure of 20 years to cadmium oxide dust. That anosmia is common in workers exposed for long periods of time to cadmium oxide dust was also noted by Baader, 1951. Suzuki, Suzuki, and Ashizawa, 1965, and Tsuchiya, 1967, did not find an increased prevalence of anosmia in workers exposed to cadmium stearate and cadmium oxide fumes, respectively.

Pancreatic function has attracted very little interest, which is regrettable, as the high concentrations of cadmium in the pancreas found at autopsy of cadmium workers (Section 4.3.2) have indicated the possibility of a disturbed function. According to Murata et al., 1970, a decrease in pancreatic function is a feature of the Itai-itai disease.

Gastrointestinal effects have been a common finding in earlier reports on acute cadmium poisoning due to ingestion of food highly contaminated with cadmium. The contamination was usually caused by storage of food in cadmium-plated containers, cans, etc. Nausea, vomiting, abdominal pains, and diarrhea have been reported symptoms, (Public Health Reports, U.S. Public Health Service 1942, Lufkin and Hodges, 1944, and Cole and Baer, 1944). Even if the outbreaks of food poisoning due to cadmium have decreased considerably since cadmium has been prohibited in utensils for cooking and storing food after the Second World War, some reports during the later years indicate that this problem still exists (Rème and Peres, 1959, Baker and Hafner, 1961, and Nordberg, Slorach, and Stenström, 1973). The prognosis seems good but there have been no reports on detailed follow-ups (cf. Section 3.2).

Symptoms from the nervous system have been reported by Vorobjeva, 1957b, who investigated 160 workers in an accumulator factory. Subjective symptoms consisted of headache, vertigo, sleep disturbances, etc. Physical examination revealed increases in knee-joint reflexes, tremor, dermographia, and sweating. Special investigations on sensory, dermal, optic, and motoric chronaxia showed that the cadmium-exposed workers with subjective disturbances also had changes on these tests.

Cvetkova, 1970, studied 106 women (ages 18 to 48) employed in cadmium industries; 61 worked in an alkaline accumulator factory, 21 in a zinc smeltery, and 24 in a chemical factory. A control group consisted of 20 women. Levels of cadmium in air varied between 0.1 to 25, 0.02 to 25, and 0.16 to 35 mg/m^3, respectively. Exposure times were not given in the report. In the alkaline accumulator factory and the zinc smeltery groups, the weights of newborn children, 27 in each group, both boys and girls, were significantly lower than the weights of children born to members of the control group. In 4 of the 27 children born to women in the zinc smeltery group, signs of rachitis and delayed development of teeth were recorded. The author concluded that pregnant women should not work in a cadmium-contaminated environment.

6.7.2 In Animals

Large doses of cadmium are toxic to nervous ganglia, as has been shown by Gabbiani, 1966, who gave rats subcutaneous injections of cadmium chloride (2.5 to 28 mg Cd/kg) and found severe hemorrhagic ganglionic lesions. Gabbiani, Baic, and Deziel, 1967b, found that by giving rats as pretreatment five subcutaneous injections of cadmium chloride in a relatively small dose (2 mg/kg), the ganglionic lesions produced by a single intravenous dose of 8 mg/kg were prevented. Damage to the cerebrum and cerebellum was found in newborn rats and rabbits receiving cadmium (Gabbiani, Baic, and Deziel, 1967b), whereas in mature animals such changes were not found. Gabbiani, Gregory, and Baic, 1967, found that pretreatment with zinc acetate prevented the development of ganglionic damage.

There have not been any long-term experiments in which small doses of cadmium were administered in order to investigate nervous functions.

Cvetkova, 1970, exposed fem[illegible] the same via the respiratory route to cadmi[illegible] drawn f[illegible]about 3 mg/m^3) during pregnancy. On the [illegible] day, half of the group was killed and the embryos were taken out. There was the same number of embryos in exposed rats as in a control group, but the mean weight was lower in the exposed group. Analysis of liver from embryos showed a higher content of cadmium in the exposed group. In the exposed group in which pregnancies were allowed to be full term, the weight of the newborns was lower than in controls. After 8 months, the weights in the exposed group were still lower. The rats born to the exposed group also had an increased mortality during the first 10 days after birth.

Schroeder and Mitchener 1971, looked for teratogenic and reproductive effects of cadmium. Mice were exposed during two generations to cadmium in drinking water (10 μg/g) from the end of the weaning period until the end of the experiment. Cadmium was found toxic for breeding mice to such an extent that the strain did not survive beyond the second generation. Congenital abnormalities appeared at a much higher frequency than in controls. Cadmium and zinc were not determined in organs. Since cadmium does not transverse the placenta, it is conceivable that a teratogenic effect would be due to a secondary zinc deficiency in the fetuses; the mothers retain more zinc during exposure to cadmium, leaving less zinc available for the fetuses. During exposure to cadmium, more zinc than normal is stored in liver and kidneys and less in testes (Petering, Johnson, and Stemmer, 1971).

With regard to influence on endocrine glands it has been shown that a single subcutaneous injection of cadmium chloride (10 mg Cd/kg) given to male rats causes a prompt decline in pituitary FSH (follicle stimulating hormone) level and an increase in LH (luteinizing hormone) (Kar, Dasgupta, and Das, 1960). These changes were thought to be caused by the testicular damage, and it is not known if cadmium exerts a direct effect upon the pituitary. In this connection it is of interest that Berlin and Ullberg, 1963, noted an uptake in the pituitary gland after a single dose of radioactive cadmium. Influence on the thyroid has been demonstrated by Anbar and Inbar, 1964, who gave mice (weight not given) 0.12 mg Cd by the intraperitoneal route. They found a decrease in the uptake of iodine, but as the dose was large and other metals, both essential and nonessential, had [illegible] effects, no certain conclusions can be [illegible] experiment. A decrease in the uptake of [illegible]ine was also found by Balkrishna, 1962, who gave rats single intramuscular injections of cadmium chloride (10 mg Cd/kg). Influence on the parathyroids was discussed in Section 6.4.

Baum and Worthen, 1967, found amyloid deposits in the glomeruli of rabbits given intramuscular injections of cadmium chloride, 1 mg Cd/kg, a total of 68 injections in 22 weeks, or the same dose for 9 weeks, a total of 47 injections. Similar findings have also been described by Vigliani, 1969, in rabbits exposed for 7 to 13 months to cadmium (0.25 mg Cd/kg subcutaneously, 5 days a week). He attributed the amyloid deposits to disturbances in immunoglobulin metabolism.

Biochemical effects of cadmium, especially with regard to enzymes, have recently been treated in an extensive review by Vallee and Ulmer,1972.

6.8 EFFECTS POSSIBLY PARTLY ERRONEOUSLY ATTRIBUTED TO LEAD

There are some reports that may indicate effects of cadmium even though the investigators involved have attributed them to lead. This problem was first pointed out by Friberg, 1948b, when he commented on a report by Seiffert, 1897. The latter investigator examined workers exposed to lead, zinc, and cadmium in a mine and concluded that he had found 65 cases of lead poisoning. Friberg concluded instead that the main cause of the disease had probably been exposure to cadmium since Seiffert's results showed that 82% of the persons said to have undergone lead poisoning had proteinuria and 83% had emphysema.

Graovac-Leposavic et al., 1972, reported on the health status of Meza Valley inhabitants exposed for several years to lead from a lead smeltery producing annually about 22,000 tons of lead and about 11,000 tons of zinc (Djuric et al., 1971). Lead emissions from the chimneys were estimated to be more than 200 tons per year. No data on cadmium emissions were given but it is obvious that an annual production of 11,000 tons of zinc would bring about the production or discharge of large quantities of cadmium. In an examination of 210 adults in different age groups, the main complaint was pain in the bones. Such bone symptoms, which 47% of the examined persons described, are not common in lead intoxication, but are regularly looked for when screening for

Itai-itai disease in Japan. No data on possible damage were given.

Cooper, Tabershaw, and Nelson, 1972, reported on a study of a total of 308 workers in lead smelting and refining. It is obvious from the report that a zinc plant was also involved. Urinary protein was not analyzed. The most significant finding was a decrease in serum phosphorus as a function of time of employment. There was no correlation between effects and lead levels of blood. The decreased serum phosphorus was not associated with an increase in serum calcium, indicating that the hypophosphaturia was due to tubular dysfunction. Since lead usually does not cause such tubular damage in adults, it is conceivable that other metals, particularly cadmium, might have been responsible.

In none of the reports referred to above is there any clear-cut evidence that cadmium actually was the cause of the disorders disclosed. These comments have been offered only to point out the need to look for effects of metals other than lead. In the Meza Valley, home of thousands of people, it is especially worthwhile to perform tests to find effects not only of lead but also of cadmium.

6.9 CONCLUSIONS

Acute manifestations in animals have been studied mainly by injection of cadmium. It has been possible to show effects from most organ systems, including hypertension and testicular, liver, and renal damage. Doses have varied between 1 to 20 mg Cd/kg. The results show that cadmium is an extremely toxic metal, but they have only limited bearings on the effects produced by long-term exposure to cadmium.

In human beings acute poisonings with local symptoms from the gastrointestinal tract are known to have occurred due to ingestion of cadmium-contaminated food or beverages. Systemic effects, however, are not known.

Long-term exposure gives rise to renal tubular damage in animals. Workers with prolonged exposure to cadmium oxide dust or cadmium oxide fumes will develop renal tubular dysfunction. A characteristic sign of the renal dysfunction is proteinuria of the so-called tubular type. This is considered to be due to a decreased tubular reabsorption of filtered proteins. Cadmium seems to be transported to the kidney with a low molecular weight protein, metallothionein. The reabsorption of this protein will also decrease and the result will be an increased excretion of cadmium. Other signs of the renal tubular dysfunction, usually appearing after the proteinuria, are glucosuria and aminoaciduria and changes in the metabolism of calcium and phosphorus.

Evidence from animal experiments and from analyses of cadmium in kidneys from workers exposed to cadmium indicates that renal dysfunction may become manifest at levels around 200 μg/g wet weight in renal cortex. In calcium-deficient animals exposed to cadmium, urine studies indicate that this level may be even lower under certain circumstances.

Osteomalacia has been observed in workers industrially exposed to cadmium. It is considered to be due mainly to a disturbance in calcium and phosphorus metabolism caused by the renal tubular dysfunction. Animal experiments have shown that cadmium exposure may cause demineralization of bone.

There is evidence that hypertension can be induced in animals either by single injections or by repeated exposure to cadmium in drinking water. In human beings who have died after a history of hypertensive disease, cadmium levels in kidneys have been higher than in people who have died from other diseases. The cause of the association found is not clear. Epidemiological investigations have indicated a relationship between cadmium levels in air and cardiovascular disease. There is reason to believe that this association is spurious, and in any case, no causal association has been proven.

Other long-term effects that have been noted are anemia and liver dysfunction, as shown in both animal experiments and investigations on exposed workers. Animal data have indicated that the activities of enzymes engaged in carbohydrate metabolism might be changed at concentrations of cadmium in liver within the range found in livers of normal human beings. The significance of this finding for man is not known.

Testicular changes have been noted during long-term exposure. These changes are milder than the pronounced testicular damage caused by acute doses of cadmium. It is not known if the testicles will be affected in man during long-term exposure.

Teratogenic effects have been reported in long-term studies on mice, and further studies, also on other species, are needed.

Chapter 7

CARCINOGENIC AND GENETIC EFFECTS

7.1 CARCINOGENIC EFFECTS

Information in regard to a possible carcinogenic effect of cadmium and cadmium compounds is given primarily from animal experiments. Only a very few reports dealing with such effects in human beings have appeared.

7.1.1 In Animals

In 1961 Haddow, Dukes, and Mitchley reported that repeated subcutaneous injections of rat ferritin produced sarcoma at the site of injection in rats. Testicular atrophy and benign Leydig cell tumors were also seen. The ferritin contained considerable amounts of cadmium because it had been prepared from rat liver protein by precipitation with a cadmium salt. The findings prompted further studies using cadmium alone and different investigators showed that cadmium and cadmium compounds injected subcutaneously or intramuscularly in rats could induce sarcoma. In 1964 Haddow et al. induced sarcoma at the site of subcutaneous injection with cadmium sulfate in 14 out of 20 rats but observed no tumors in mice after similar injections. In the same study, Roe et al., 1964, found testicular atrophy and Leydig cell hyperplasia and neoplasia. Gunn, Gould, and Anderson, 1963a and 1964, induced interstitial cell tumors in the testes of rats as well as subcutaneous sarcoma with a single subcutaneous injection of cadmium chloride. Both forms of cadmium-induced tumors were inhibited by subcutaneous administration of zinc acetate. Later, the same authors (Gunn, Gould, and Anderson, 1967) showed that a single subcutaneous or intramuscular injection of cadmium chloride in amounts as low as 0.17 to 0.34 mg of cadmium induced sarcoma. No tumors arose in skin, liver, salivary glands, prostate, or kidney. The authors concluded that cadmium may be one of the most potent metallic carcinogens yet known.

Kazantzis, 1963, and Kazantzis and Hanbury, 1966, showed that sarcomata could also occur after injection of a single dose of cadmium sulfide. Metastatic tumors were seen in regional lymph nodes and lungs. Heath et al., 1962, showed that cadmium given as a metal powder intramuscularly to rats induced sarcomata. Such tumors could be maintained by transplantation. At the time of publication, the cadmium tumors had been maintained for 6 years and 75 transplants (Heath and Webb, 1967). Most of the inducing metal was incorporated intracellularly by the primary tumors and bound by the nuclear fraction. It has subsequently been shown that at least 50% of the nuclear fraction is within the nucleoli. The remainder is bound by the nuclear sap and the chromatin (Webb, Heath, and Hopkins, 1972).

Schroeder and collaborators (Schroeder, Balassa, and Vinton, 1964 and 1965, and Kanisawa and Schroeder, 1969a) reported on life-term studies on the effect of trace elements on tumors in rats and mice. The rats given cadmium had more tumors than their controls while the opposite was true for mice. Schroeder concluded that the oral ingestion of cadmium cannot be considered carcinogenic in the doses given.

Malcolm, 1972, reported preliminary data from a 2-year study on rats and mice. The rats were exposed at weekly intervals subcutaneously (0.2 to 0.05 mg cadmium sulfate) or via stomach tube (0.8 to 0.2 mg). The mice were exposed perorally at weekly intervals to 0.02 to 0.005 mg cadmium sulfate/5 g body weight. No detailed data were given on any of the animals but Malcolm stated that sarcomata were found in a few cases at the site of the subcutaneous injection. No other cadmium-related tumors were reported.

7.1.2 In Human Beings

Potts in 1965 reported on the prevalence of cancer among workers exposed to cadmium oxide dust in the production of alkaline batteries. Among 74 men with at least 10 years' exposure to cadmium, 8 had died, 3 of them from cancer of the prostate and 2 from other forms of cancer. That the workers had been exposed to toxic amounts of cadmium was obvious from the high prevalence of proteinuria and anosmia. Potts did not examine any control group and did not draw any firm conclusions. He recommended as a matter of urgency, however, that the possibility of some association between cancer in man and cadmium as used in industry must be explored fully.

Kipling and Waterhouse, 1967, briefly mentioned the incidence of cancer in a group of 248 workers exposed for a minimum of 1 year to cadmium oxide. The type of work was not stated. From annual incidence data supplied by the Birmingham regional cancer registry, it was possible to compute expected values and to compare with observed values. For cancer of the prostate, 4 cases were observed vs. an expected number of 0.58.

Out of the 58 male workers in an alkaline battery plant examined by Friberg, 1950 (see Section 5.2.1.1.1 and 6.1.2.1.1), 17 had died as of 1972 (Friberg and Kjellstrom, unpublished data). The cause of death in three cases was cancer (urinary bladder, primary pulmonary, and large bowel, respectively). No control groups have been studied and no conclusions have been drawn as yet.

Humperdinck, 1968, reported on a follow-up of eight cases of chronic cadmium poisoning described by Baader in 1951. Four deaths had occurred, one due to primary cancer of the lung. Out of all workers (536) who during 1949 to 1966 had had any contact with cadmium, 5 had contracted carcinoma (including the above mentioned case of lung cancer). The author concluded that the data did not support any causal relationship between cadmium exposure and cancer. It should be added, though, that the majority of the workers were relatively young with relatively short periods of contact with cadmium (269 for less than 1 year, only 4 for 7 to 11 years). Thus, the study does not elucidate the question of a possible association between cadmium exposure and cancer. It should be of value to study the group on a longitudinal basis, particularly if more detailed information on exposure levels can be made available.

Holden, 1969, very briefly reported one case of carcinoma of the prostate and one of the bronchus among 42 men exposed to cadmium fumes from 2 to 40 years.

Winkelstein and Kantor, 1969, reporting data from the Erie County and Nashville Air Pollution Studies, found an association between prostatic cancer and concentrations of suspended particulate air pollutants in both studies. The authors were cautious in view of the small numbers but drew attention to earlier reports of an association between cadmium exposure and cancer of the prostate.

Lewis, Lyle, and Miller, 1969, found an increase in hepatic water-soluble protein-bound cadmium in emphysema and chronic bronchitis (Section 5.2.1.2). Morgan, 1969, studied the renal and hepatic cadmium levels for different disease groups at autopsies. In cancer patients she found such a wide variation in concentrations of cadmium that a further study was warranted. In two later investigations she pursued similar studies of bronchogenic carcinoma and emphysema. Cadmium was analyzed by atomic absorption spectroscopy with good recovery and repeatability. In the first of these two further investigations (Morgan, 1970) she compared the cadmium concentrations in a group of male patients who had died from bronchogenic carcinoma with a control group and with a group who had died from other forms of neoplasia. The results are shown in Table 7:1. The average cadmium concentrations in the group who had died of bronchogenic carcinoma were statistically higher than those in the other groups.

Morgan, 1971, reexamined her material and compared the controls with three groups: one

TABLE 7:1

Concentrations of Cadmium in Liver and Kidneys (μg/g ash) in Patients (Average Age: 60) Dying from Cancer of the Lung and Other Forms of Cancer and in Controls

	Number examined	Liver		Kidneys	
		Mean	S.D.	Mean	S.D.
Cancer of the lung	55	254	133	3,513	1,587
Other forms of cancer	50	179	140	2,937	2,065
Controls	47	182	99	2,406	1,299

From Morgan, 1970.

group of patients with bronchogenic carcinoma with emphysema, one group with bronchogenic carcinoma without emphysema, and one group with only emphysema. The purpose was to examine whether both emphysema and bronchogenic carcinoma were associated with high cadmium levels. She found that "renal cadmium content was elevated in both groups with cancer. Hepatic cadmium was increased in both emphysema alone and in combination with lung cancer as well as lung cancer alone." In 36 controls, the average value for the liver (μg/g ash) was 170 (S.D.: 95) against 290 (S.D.: 155) in the group with only emphysema. Corresponding values for the kidney were 2,512 (S.D.: 1,427) and 2,921 (S.D.: 1,252).

Morgan interpreted her findings with great caution. She pointed, for example, to earlier observations (Olson et al., 1954, and Teitz, Hirsch, and Neyman, 1957) in which tissue trace metal concentrations were found to be abnormal in a wide variety of neoplastic diseases. Furthermore, she added that there is no convincing evidence that bronchogenic carcinoma has been associated with industrial exposure to cadmium. Concerning emphysema she was also cautious. In addition, Morgan pointed to the possibility of an increased exposure due to smoking. This question is dealt with in detail in Sections 3.2 and 4.3.4.

There seem to have been no studies made concerning any possible association between cadmium exposure and cancer of the gastrointestinal tract. The incidence of stomach cancer in Japan is extremely high (Doll, Muir, and Waterhouse, 1970). This and the obvious exposure to cadmium via the gastrointestinal route in several areas should motivate epidemiological studies.

7.2 GENETIC EFFECTS

7.2.1 In Animals

Little information is available concerning possible genetic effects of cadmium and cadmium compounds. Preliminary genetic experiments with $CdCl_2$ on *Drosophila melanogaster* were performed by Ramel and K. Friberg (personal communication). Larvae were treated with 62 mg/l substrate, which was the maximum dose in the toxicity test giving a delay of larval development without causing an excessive lethality. As an indication of chromosome breakage the frequency of sex chromosome loss was used. One experiment was also performed on the combined effect of $CdCl_2$ and 3,000 R X-irradiation on sex-linked recessive lethals. The treatment with cadmium compound was given as mentioned above. The results of these two experiments (Tables 7:2 and 7:3) do not indicate any significant effect of $CdCl_2$ on chromosome breakage, induction of recessive lethals, or chromosome repair mechanism. It must be stressed, however, that the material is small and further investigations are needed.

7.2.2. In Human Beings

Shiraishi, Kurahashi, and Yosida, 1972, reported on chromosomal aberrations in cultured human leukocytes induced by cadmium sulfide (0.062 μg/ml culture fluid). The results are seen in

TABLE 7:2

Sex Chromosome Loss (XO-Males) after Larval Treatment with 62 mg $CdCl_2$ /l in the Substrate

Egg laying period in days after treatment	$CdCl_2$		Control	
	Total number counted	Percent sex chromosome loss (XO males)	Total number counted	Percent sex chromosome loss (XO males)
0–2	1,708	0.5	2,573	0.3
2–4	3,858	0.4	5,007	0.1
4–6	4,833	0.1	5,609	0.3
6–8	4,331	0.2	5,365	0.3
8–10	3,913	0.3	4,423	0.2
10–12	4,717	0.4	5,166	0.3
0–12	23,360	0.3	28,143	0.2

From Ramel and K. Friberg, unpublished data.

TABLE 7:3

Effect of $CdCl_2$ (Larval Treatment with 62 mg $CdCl_2$/l in the Substrate) on Recessive Sex-linked Lethals in Muller-5 Test after 3,000 R X-rays to Males

Egg laying period in days after treatment	$CdCl_2$ Number of X chromosomes tested	$CdCl_2$ Percent recessive lethals	Control Number of X chromosomes tested	Control Percent recessive lethals
0–2	235	10.6	149	11.3
2–4	445	7.7	190	5.8
4–6	80	17.5	98	6.1
6–8	285	4.2	305	3.6
0–8	1,045	8.1	742	6.1

From Ramel and K. Friberg, unpublished data.

TABLE 7:4

Chromosomal Aberrations and Their Frequencies in Human Leukocytes after Treatment with Cadmium Sulfide

Duration of treatment (hr)	Total number of cells observed	Number of cells with chromatid breaks (%)	Number of cells with isochromatid breaks (%)	Number of cells with translocation (%)	Number of cells with dicentric chromosomes (%)
0 (control)	50	0	0	0	0
4	50	17 (34)	2 (4)	0	7 (14)
8	50	7 (14)	14 (28)	6 (12)	3 (6)

From Shiraishi, Y., Kurahashi, H., and Yosida, T. H., *Proc. Jap. Acad.*, 48, 133, 1972. With permission.

Table 7:4. It was stated that it seems highly possible that cadmium sulfide has some mutagenic action. Shiraishi and Yosida, 1972, studied peripheral leukocytes from seven patients with Itai-itai disease. They found a higher frequency of chromosomal abnormalities compared with controls. Chromatid breaks were particularly common. The results are seen in Table 7:5.

7.3 CONCLUSIONS

The studies on animals show without doubt that cadmium and cadmium compounds can give rise to malignant tumors in rats at the site of injection. The *carcinogenic* evidence of cadmium in human beings is by no means conclusive but fully motivates further and intensified studies in groups exposed industrially as well as through food and ambient air. Very few *genetic* studies have been performed. The results are to some extent contradictory. One study on *Drosophila* did not show any genetic effects while studies on humans, both in vitro and in Itai-itai patients, have shown chromosomal abnormalities. There is a great need for further studies, including studies in people with excessive cadmium exposure but without complicating effects as seen in Itai-itai disease.

TABLE 7:5

Types and Frequencies of Chromosomal Aberrations in Itai-itai Patients and in Controls

Case no.	Age and sex	Total number of cells observed	Number of cells with single chromatid breaks	Number of cells with isochromatid breaks or gaps	Number of cells with chromatid translocations	Number of cells with dicentric or dicentric-like chromosomes	Number of cells with acentric fragment	Total number (%) of cells with chromosome abnormalities
	Controls							
C-1	58-F	50	1	0	0	0	0	1 (2)
C-2	62-F	50	0	0	0	0	0	0 (0)
C-3	67-F	50	0	0	0	0	0	0 (0)
C-4	70-F	50	1	0	0	0	0	1 (2)
C-5	70-F	50	0	0	0	0	0	0 (0)
C-6	78-F	50	0	0	0	0	0	0 (0)
	Itai-tai disease							
I-1	52-F	50	15	2	8	2	3	30 (60)
I-2	66-F	50	18	3	6	0	1	28 (56)
I-3	69-F	50	14	3	6	0	0	23 (46)
I-4	69-F	50	6	1	0	0	0	7 (14)
I-5	70-F	50	19	3	7	1	2	32 (64)
I-6	70-F	50	12	2	3	2	1	30 (60)
I-7	73-F	50	15	2	6	1	3	27 (54)

From Shiraishi, Y. and Yosida, T. H., *Proc. Jap. Acad.*, 48, 248, 1972. With permission.

Chapter 8

HEALTH EFFECTS OF CADMIUM IN THE GENERAL ENVIRONMENT IN JAPAN

8.1 INTRODUCTION

Mines with sulfide ores of copper, lead, and zinc are very common in Japan (Figure 8:1). The ores in most cases also contain certain amounts of cadmium. At least ten large zinc refineries are in operation in Japan, as shown in Figure 8:1. Zinc and cadmium production is even higher than the ore production would indicate, depending upon imports. Other industries as well are using cadmium, and the possibilities for people working at or living around such industries of being exposed to cadmium are manifold.

Arable land in Japan is extensively used for rice planting. Sometimes, river water polluted from mining or smelting operations is used for irrigation of the fields. In addition, crops may be planted on fields close to industries emitting cadmium via air. As a result, the cadmium content of rice in some areas is increased to more than ten times the estimated average in Japan. Since rice plays a large role in the Japanese diet, the daily cadmium intake will be elevated in these areas. Another factor which increases the likelihood of finding health effects of cadmium in polluted areas in Japan is that large populations (thousands of persons) generally are exposed.

In any population there is always an individual variation in sensitivity to a certain toxic substance. A certain daily cadmium intake might give renal tubular damage in one person but not in another because of variations in absorption rate, biological half-time, or sensitivity of tubular cells.

In areas with large populations, epidemiological studies should be able to establish accurate dose-response relationships. Studies from Japan are therefore of great importance for cadmium research. Their pertinence is accentuated by the significant and widely known finding of Itai-itai (ouch-ouch) disease in a cadmium-polluted area.

Itai-itai disease and cadmium pollution of the general environment of Japan have been issues of great social, economic, and political impact in that country. If it can be proved that cadmium pollution over a certain degree occurs in an area or that an individual is cadmium poisoned, the farmers and the victims are legally entitled to damage compensation from the cadmium-emitting industry. Within the scientific community, discussion of cadmium in the Japanese environment and of its possible health effects is still very active and shows a polarization into two camps. Hence, completely disagreeing conclusions about, for example, the etiology of the Itai-itai disease have been published.

The following account of the Japanese data will be presented for the most part in the order in which the information emerged. The first part will thus consider the situation in Toyama Prefecture and above all focus on the Itai-itai disease. In a separate subsection the prevalence of proteinuria and glucosuria in this area in persons without signs of Itai-itai disease will be reported. The second part will cover studies in other cadmium-polluted areas. A large amount of data on proteinuria and glucosuria in particular is available and the emphasis will be upon the prevalence of such findings. Insofar as data on Itai-itai disease are available, they are referred to. In the final part the findings, especially proteinuria and the Itai-itai syndrome, are discussed with special reference to cadmium as an etiological factor.

8.2 STUDIES WITHIN THE ITAI-ITAI DISEASE AREA (TOYAMA PREFECTURE)

8.2.1 History of Itai-itai Disease

In 1946, Dr. Hagino, a general practitioner returning from the war, reopened a private clinic in operation in Fuchu town, Toyama Prefecture, since his grandfather's time (Figure 8:2). Dr. Hagino was visited by patients with a painful disease, which he called the "Itai-itai byo," meaning "ouch-ouch disease." According to Hagino, 1957, the disease had occurred endemically in the area for several years. Nagasawa et al., 1947, reported on a "rheumatic disease" in 44 patients living in Fuchu. Most of the patients were women, between 40 and 70 years old, but there were also 13 male patients. Nagasawa et al. included only blood cell counts and subjective pain in their description of symptoms, and it is therefore hard to say to what extent this "rheumatic disease" is similar to the Itai-itai disease.

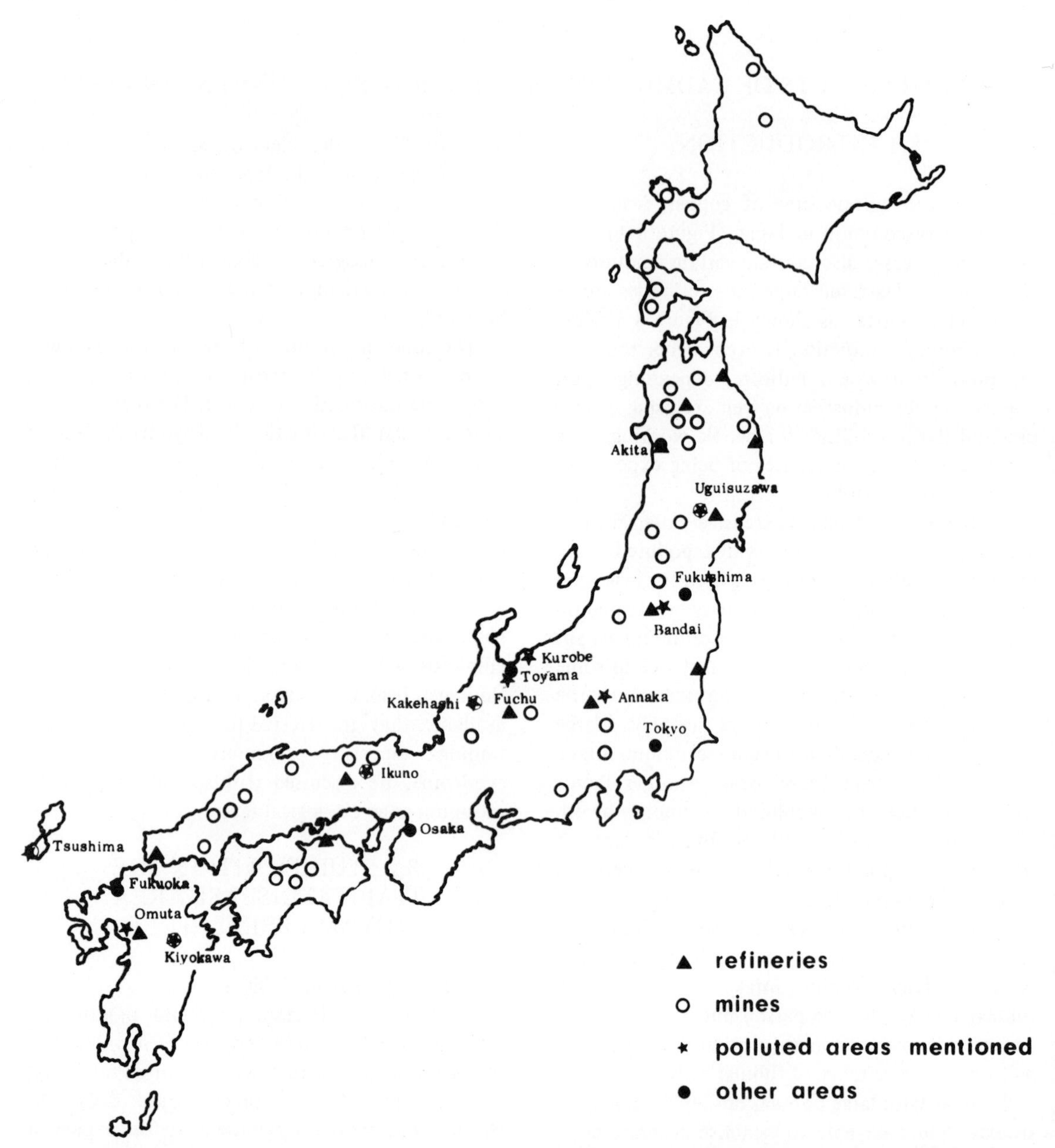

FIGURE 8:1. Location of sulfide ore mines and refineries in Japan as well as polluted areas mentioned in this report. (Redrawn from Yamamoto, 1972.)

In a review of the history of Itai-itai disease, Hagino, 1973, mentioned that in 1947 patients whom he referred to hospitals in Toyama City were diagnosed there as having "kidney trouble or diabetes." In 1948 osteomalacia was suspected and treatment with cod liver oil was tried without effect (Hagino, 1973). It was not until 1955 that Hagino and Kono brought about a more general awareness of the Itai-itai disease, making the first official use of the name, at the 17th Meeting of the Japanese Society of Clinical Surgeons. In the same year, treatment with vitamin D was started. After several months of this treatment, the bone symptoms improved (Section 8.2.6). During 1956 and 1957 a number of reports about the disease were written; at that time it was generally believed that Itai-itai disease was a vitamin D-deficient osteomalacia.

Osteomalacia among young adults and rickets among children had been rather common diseases in Toyama Prefecture during the early parts of this century, according to Takeuchi, 1973, citing pub-

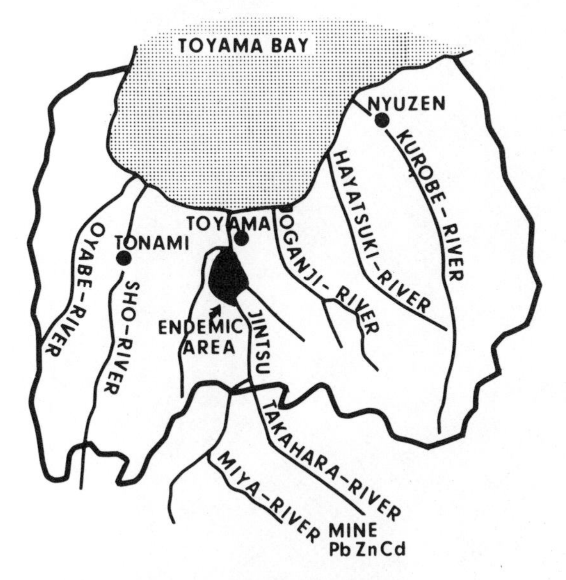

FIGURE 8:2. Location of population centers within Toyama Prefecture, with the location of the Kamioka mine. (From Ishizaki and Fukushima, 1968.)

lications from 1906 by Kinoshita, Hayashi, and Tashiro. However, the age distribution as well as the geographical distribution was completely different from those of the Itai-itai disease.

During World War II, damage to the rice crop in Fuchu had caused a severe dispute between the farmers and the Kamioka Mining Company (Kobayashi, 1971). The farmers argued that river water, used for irrigation of the rice fields, had been polluted by waste from the mine. They concluded that metals contained therein had damaged the crops. Kobayashi, 1943, made an investigation, but at that time no analysis of heavy metals in the soil or rice was performed. Moritsugi and Kobayashi, 1964, published data on cadmium analysis of rice performed in 1960. In 20 samples of rice from the endemic area, the average cadmium concentration (wet weight) was more than ten times higher (0.68 μg/g) than in about 200 samples from other areas of Japan (0.066 μg/g). In 1961 Hagino and Yoshioka reported about these data at the 34th Meeting of the Japanese Association of Orthopedics, and it was postulated that cadmium played an etiological role in the development of the Itai-itai disease.

Epidemiological and clinical studies were started on a larger scale in the 1960's by groups supported by the Japanese government. Based on the results of these studies, the Japanese Ministry of Health and Welfare declared in 1968, "The Itai-Itai disease is caused by chronic cadmium poisoning, on condition of the existence of such inducing factors as pregnancy, lactation, imbalance in internal secretion, aging, deficiency of calcium, etc." (Ministry of Health and Welfare, 1968a). The etiology of Itai-itai disease will be discussed in Section 8.4.2.

8.2.2 Clinical Features

8.2.2.1 Symptoms and Signs

The clinical course of the disease has been described by Nakagawa, 1960, Takase et al., 1967, Hagino, 1968a, 1968b, 1969, Ishizaki, 1969b, Murata et al., 1969, 1970, and Tsuchiya, 1969. The greatest number of the patients have been treated by Hagino and Murata. Figure 8:3 shows a patient with advanced skeletal deformities; Figure 8:4 shows multiple pseudofractures on X-ray of the ribs in one patient.

From Murata et al., 1970, the clinical manifestations can be summarized as follows. Most of the patients are postmenopausal women with several deliveries (average: six). The most characteristic features of the disease are lumbar pains and leg myalgia. Pressure on bones, especially the femurs, backbone, and ribs, produces further pain. Another characteristic of the disease is a ducklike gait. Such conditions continue for several years until one day the patient experiences a mild trauma and finds herself unable to walk. She is then confined to bed and the clinical conditions progress rapidly. Not only the bones of the extremities but also the ribs and other bones are susceptible to multiple fractures after very slight trauma such as coughing. Skeletal deformation takes place with a marked decrease in body height.

Tests have shown a tendency towards a more or less impaired pancreatic function in many cases (Murata et al., 1970). Examinations of the gastrointestinal tract have shown shortened ciliated epithelia in the small intestine, atrophy of the mucous membrane, and submucosal cell infiltration (Murata et al., 1970). In some cases ulceration of the mucous membrane of the small intestine has also been observed (Murata et al., 1970). Decreased fat absorption has been revealed using ^{131}I triolein and ^{131}I oleic acid methods, but this could have been partly dependent upon a pancreatic disturbance (Murata, personal communication). The changes in the gastrointestinal tract have been named "cadmium enteropathy" by Murata.

Nogawa and Kawano, 1969, reported that increased prevalence of hypertension was not

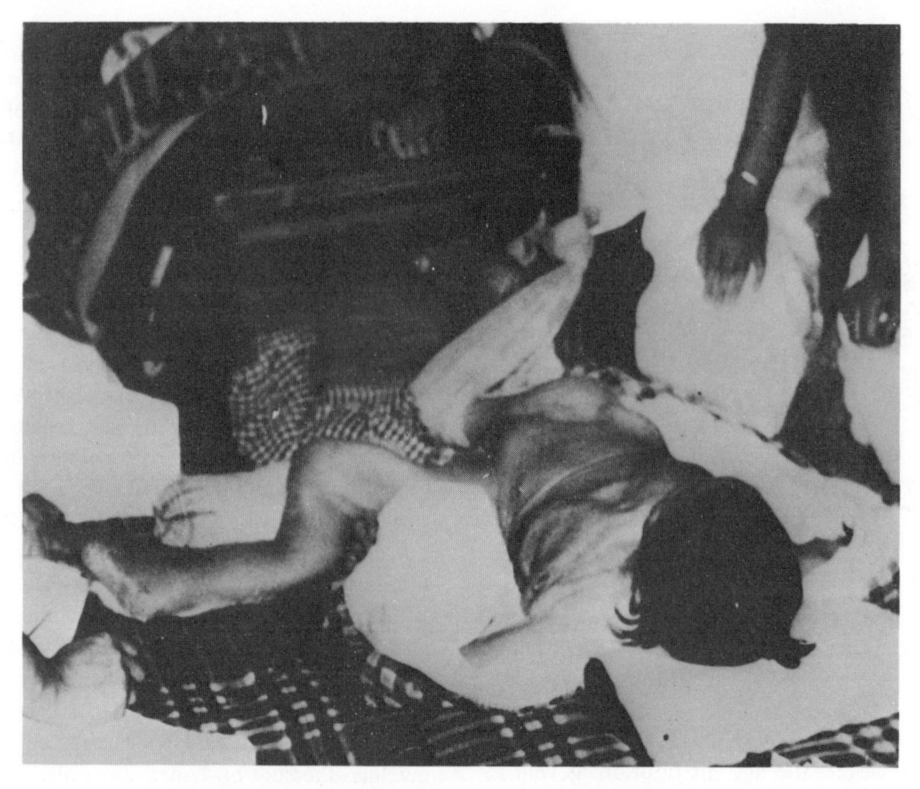

FIGURE 8:3. Advanced skeletal deformities in Itai-itai patient. (From Kobayashi, personal communication.)

found in patients with Itai-itai disease nor in persons living near the endemic district.

8.2.2.2 Laboratory Examinations

8.2.2.2.1 Blood Findings

According to Murata et al., 1970, blood examination has shown hypochromic anemia in most patients. Serum iron levels have been low in many, and the erythrocyte sedimentation rate has been elevated in some. Almost no deviation from the normal has been observed in serum protein and albumin/globulin ratios. The levels of calcium and inorganic phosphorus in serum are often low, while the alkaline phosphatase level is high. Data from studies by Murata et al., 1970, are shown in Figure 8:5.

8.2.2.2.2 Urinary Findings and Renal Function Tests

According to Hagino, 1973, proteinuria and glucosuria were common findings in the patients before any treatment had been given. Taga et al., 1956, reported bone symptoms among ten patients. Two of these patients had died in 1955 and were autopsied by Kajikawa. He reported that the patients had chronic pyelonephritis with arteriosclerosis (Kajikawa et al., 1957). It must be assumed that there had been severe kidney damage. Hagino, 1959, reported that he had found proteinuria in 82% and glucosuria in 32% of 71 surviving patients. The methods were not stated.

Nakagawa, 1960, described 30 patients from the Shinbo district (neighboring Fuchu) who were studied in 1955–1958. Vitamin D treatment had not been given to the patients prior to examination (Nakagawa, personal communication). All of the patients had proteinuria (sulfosalicylic acid method) and some had glucosuria as well (Nylander method). The tubular function was affected in a number of patients as judged by the PSP excretion test (phenolsulfonphthalein). The lower normal limit of excretion is considered to be 25% after 15 min. In seven out of ten patients the excretion was between 10 and 22.5% and in the other three, 25%. Kono et al., 1958, also reported low ratings (19% after 15 min) on the PSP excretion test in two patients. In four patients studied by Takeuchi et al., 1968, the PSP excretion was 6 to 12% after 15 min. A concentration test, dilution test, and phosphorus reabsorption test showed slight kidney damage in all four cases.

Ishizaki et al., 1968, reported that among 37

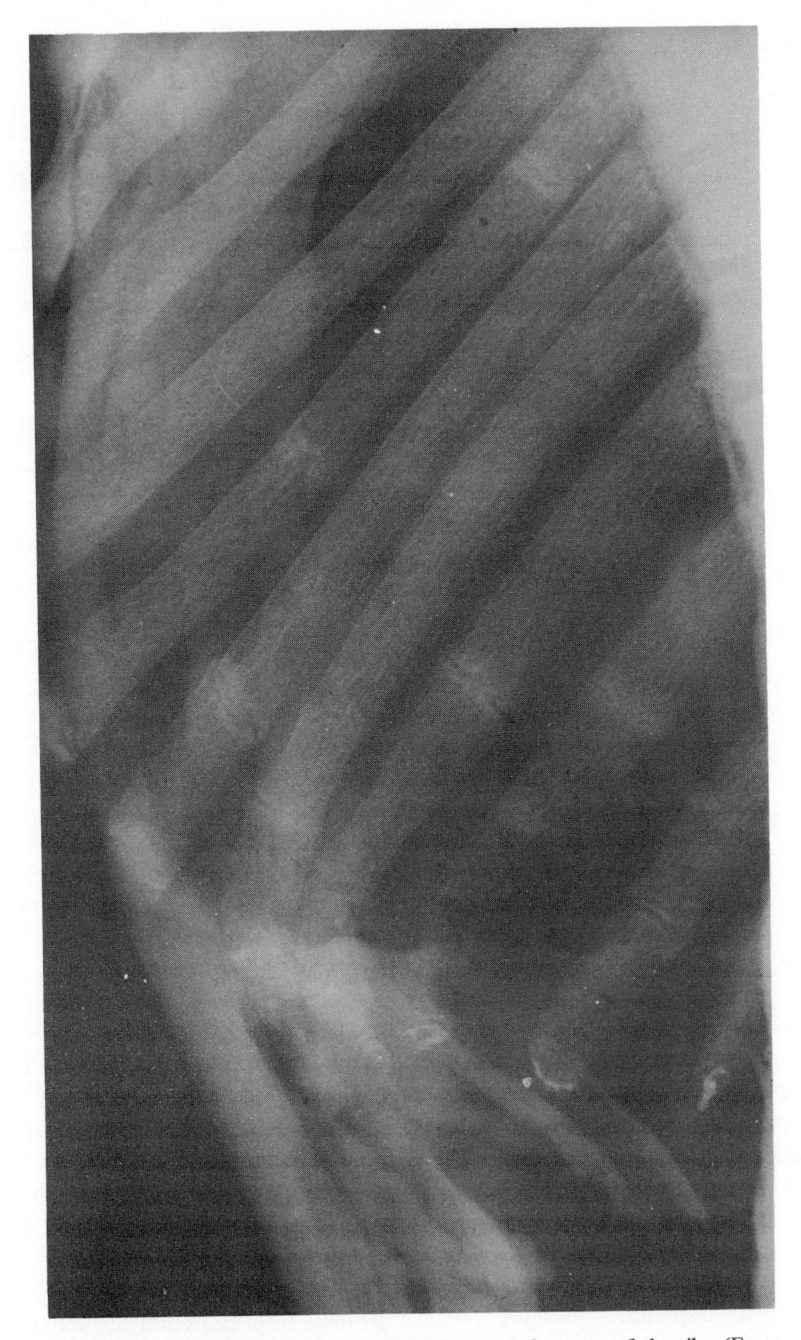

FIGURE 8:4. Itai-itai patient with multiple pseudofractures of the ribs. (From Murata, personal communication.)

Itai-itai patients (= I, Section 8.2.3) studied in the 1967 epidemiological investigation (Section 8.2.5.2) only 3 did not have glucosuria (= 1/32% or higher) and only 1 did not have proteinuria (= 10 mg/100 ml or higher).

While urinary calcium levels have been normal, urinary phosphate has been reduced. The urinary amino acid levels have been above normal (Takase et al., 1967, Takeuchi et al., 1968, Ishizaki, 1969b, Murata et al., 1969, Tsuchiya, 1969, and Fukushima, Kobayashi, and Sakamoto, 1973).

Quantitative determinations have shown an increase in cadmium concentration and a decrease in zinc concentration in blood and urine, in comparison with control subjects (see, for

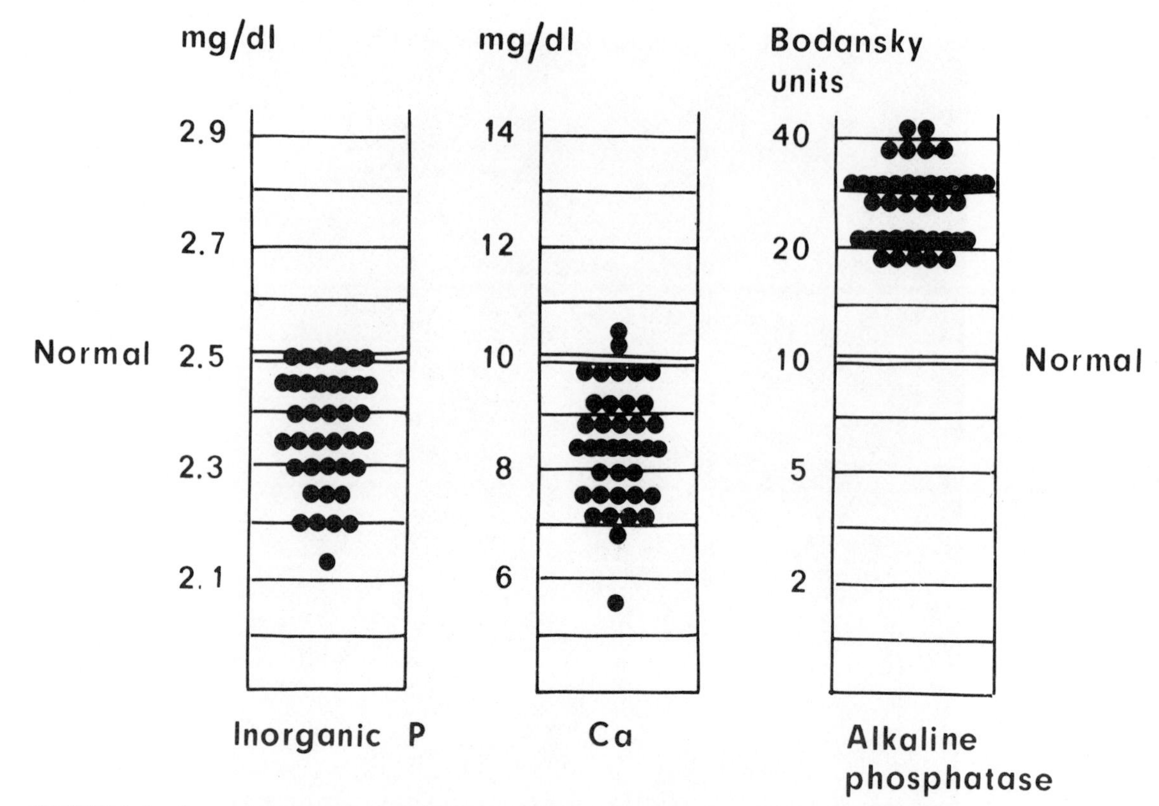

FIGURE 8:5. Inorganic phosphorus, calcium, and alkaline phosphatase in serum of patients with Itai-itai disease. (From Murata et al., 1970.)

example, Murata et al., 1970, and Figure 8:9 from Ishizaki, 1969b).

In Itai-itai disease, proteinuria, glucosuria, and aminoaciduria are considered signs of renal dysfunction. On separation by paper electrophoresis, a typical tubular protein pattern occurs in cases of Itai-itai disease and also in the preceding stages (proteinuria without bone changes), as shown by Piscator and Tsuchiya, 1971. They examined 10 female patients with Itai-itai disease and 12 persons (10 women and 2 men) from the endemic area with only proteinuria and glucosuria. The pattern found was similar to that observed in chronic occupational cadmium poisoning. In Figure 8:6 some of these patterns are shown (compare with Figure 6.2).

Further evidence that the proteinuria observed in women in the endemic district is similar to the one seen in workers exposed to cadmium and distinctly different from the one seen in glomerular disease is shown in Figure 8:7 (Piscator and Tsuchiya, 1971). The figure shows patterns of urinary proteins separated by electrophoresis in polyacrylamide gel (disc electrophoresis). Similar comparisons of disc electrophoretic patterns among cadmium workers, Itai-itai patients, and patients with other kidney diseases have been performed by Tsuchiya, 1972, and Harada, 1972. Cadmium workers and Itai-itai patients showed similar, even if not quite identical, patterns, as witnessed by Figure 8:8. It must be kept in mind that in these comparisons it has so far not been possible to find comparable sex and age groups of Itai-itai patients and cadmium workers.

Still further evidence that the proteinuria is of a tubular type has been obtained by using gel filtration on Sephadex G-100® (Piscator and Tsuchiya, 1971). The main part of the urine proteins from cases and suspected cases of Itai-itai disease had molecular weights less than that of albumin. Among these low molecular weight proteins, immune globulin chains and β_2-microglobulin were found, as has earlier been demonstrated in cadmium workers by Piscator, 1966a–c.

Electrophoretic examinations of urine proteins also have been performed by Fukushima and Sugita, 1970. In eight patients, they found a relatively small percentage of albumin and a dominance of α_2- β- and γ-proteins, also indicating a tubular type of proteinuria.

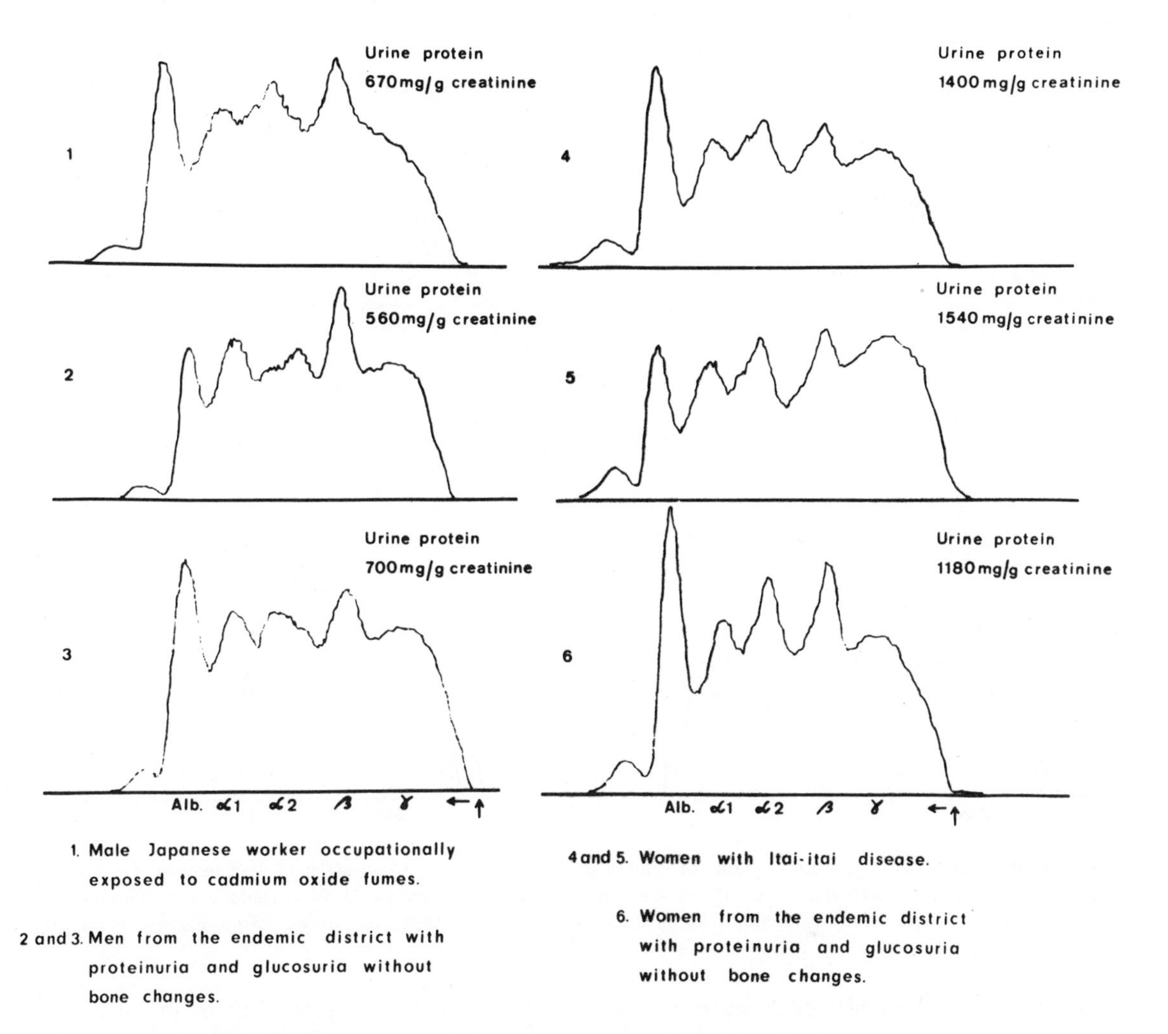
Urine protein
670mg/g creatinine
1
Urine protein
1400 mg/g creatinine
4
Urine protein
560mg/g creatinine
2
Urine protein
1540 mg/g creatinine
5
Urine protein
700mg/g creatinine
3
Urine protein
1180mg/g creatinine
6
Alb. α1 α2 β γ
Alb. α1 α2 β γ
1. Male Japanese worker occupationally exposed to cadmium oxide fumes.
2 and 3. Men from the endemic district with proteinuria and glucosuria without bone changes.
4 and 5. Women with Itai-itai disease.
6. Women from the endemic district with proteinuria and glucosuria without bone changes.

FIGURE 8:6. Paper electrophoretic patterns in an occupationally exposed male Japanese worker and male and female patients from the endemic district. (From Piscator and Tsuchiya, 1971.)

Fukuyama, 1972, and Sano, Iiguchi, and Kawanishi, 1973, compared the patterns of urinary proteins after gel filtration (G-100 and G-75, respectively) among Itai-itai disease patients, cadmium workers, and patients with various kidney diseases. Cadmium workers, persons from cadmium-exposed areas, and Itai-itai patients had similar patterns.

Ohsawa and Kimura, 1973, have isolated β_2-microglobulin from the urine of Itai-itai patients by separating on a Sephadex G-100 column and subsequently on a DEAE-cellulose column. The fractions were studied with disc electrophoresis from which the authors concluded that β_2-microglobulin is excreted to a remarkable degree in Itai-itai patients.

Isoelectric focusing (method described by Vesterberg, 1972, and Vesterberg and Nise, 1973 and radioimmunoassay (method described by Evrin et al., 1971) have been employed for analyzing urinary excretion of β_2-microglobulin in ten Itai-itai patients (I, Section 8.2.3) and ten suspected patients (i, Section 8.2.3) (Shiroishi et al., to be published). The I patients had levels between 24 and 55 μg/ml and the i patients between 6 and 41 μg/ml, a 500- to 1,000-fold increase over the normal value. The average β_2-microglobulin level in normal Swedish women in the age range 57 to 76 years has been found by radioimmunoassay to be 78 μg/24 hr (S.D. 80), roughly corresponding to 0.078 μg/ml (S.D. 0.080) (radioimmunoassay, Evrin and Wibell, 1972). In another study, using isoelectric focusing on 88 women in the age group 50 to 59 years (Kjellström, Vesterberg, and Holmquist, unpublished data), the average was about 0.23 μg/ml with a range of 0.1 to 0.77 μg/ml.

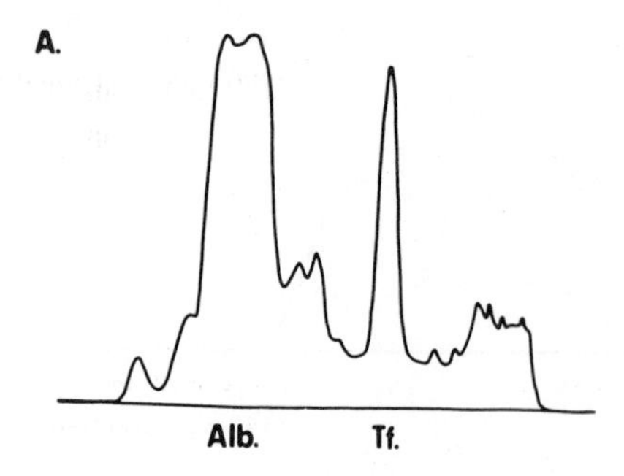

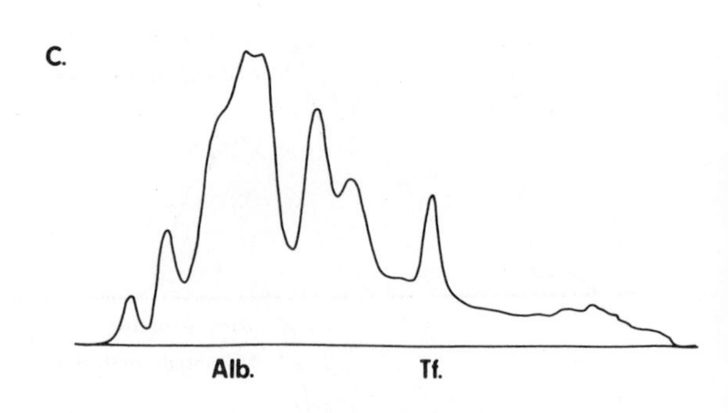

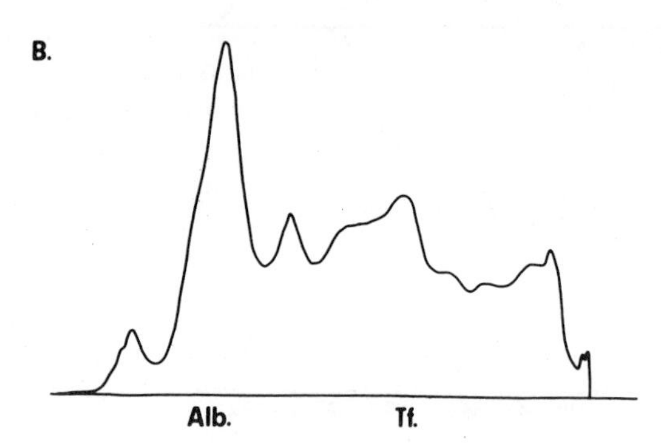

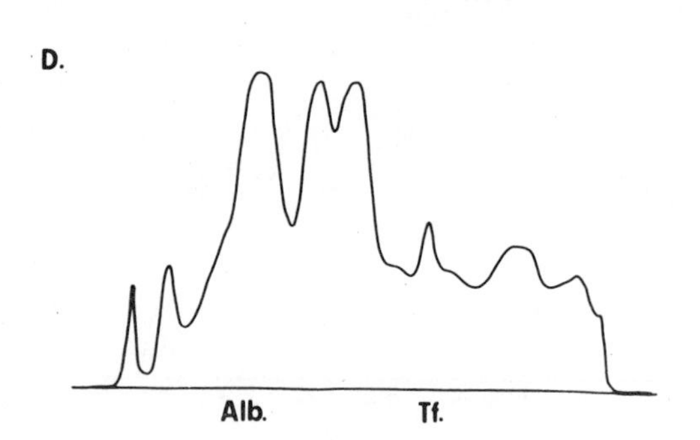

FIGURE 8:7. Disc electrophoretic patterns of urinary proteins from: A. 19-year-old man with nephrotic syndrome; B. 22-year-old cadmium-exposed worker; C. 67-year-old woman with Itai-itai disease; D. 74-year-old woman with proteinuria and glucosuria, but without bone disease, living in the endemic area for the Itai-itai disease. (From Piscator and Tsuchiya, 1971.)

Fukushima, Kobayashi, and Sakamoto, 1973, studied urinary amino acid excretion in two Itai-itai patients and in a control group. In each case, amino acid excretion was generally two to five times higher in the patients than in the controls. Arginine, citrulline, proline, and hydroxyproline were 20 to 50 times elevated in the patients. This is in accordance with what is generally found in patients with metabolic bone diseases. Increased amino acid excretion has also been found among cadmium workers without bone disease (Toyoshima, Seino, and Tsuchiya, 1973), in which cases the elevations of arginine and citrulline were especially high (Section 6.1.2.1.1) but not those of proline and hydroxyproline, which is in accord with the fact that these workers had no bone disease.

8.2.2.3 Histopathological Changes

The following, though very incomplete, are the only data on histological changes available to us (for cadmium concentrations in organs, see Section 8.2.4.2).

8.2.2.3.1 Kidneys

In two autopsies, Kajikawa et al., 1957, found senile arteriosclerotic contracted kidneys, pyelonephritis, and metastatic calcification. Takeuchi et al., 1968, found no marked changes in the glomerules in biopsy specimens from three patients. In the tubules, atropic changes with flattening of the epithelium were seen. Kajikawa, 1973, and Takeuchi, 1973, reported about six other patients. Five had "tubular nephropathy" (tubular atrophy and dilatation with eosinophilic casts and interstitial fibrosis) of varying degrees, and one had suppurative pyelonephritis. In all six patients, kidney soft tissue calcification of slight degree was reported. Since kidney stones have been a common finding among cadmium workers (Section 6.1.2.1.3), cadmium-induced disturbances in calcium metabolism may explain the soft tissue

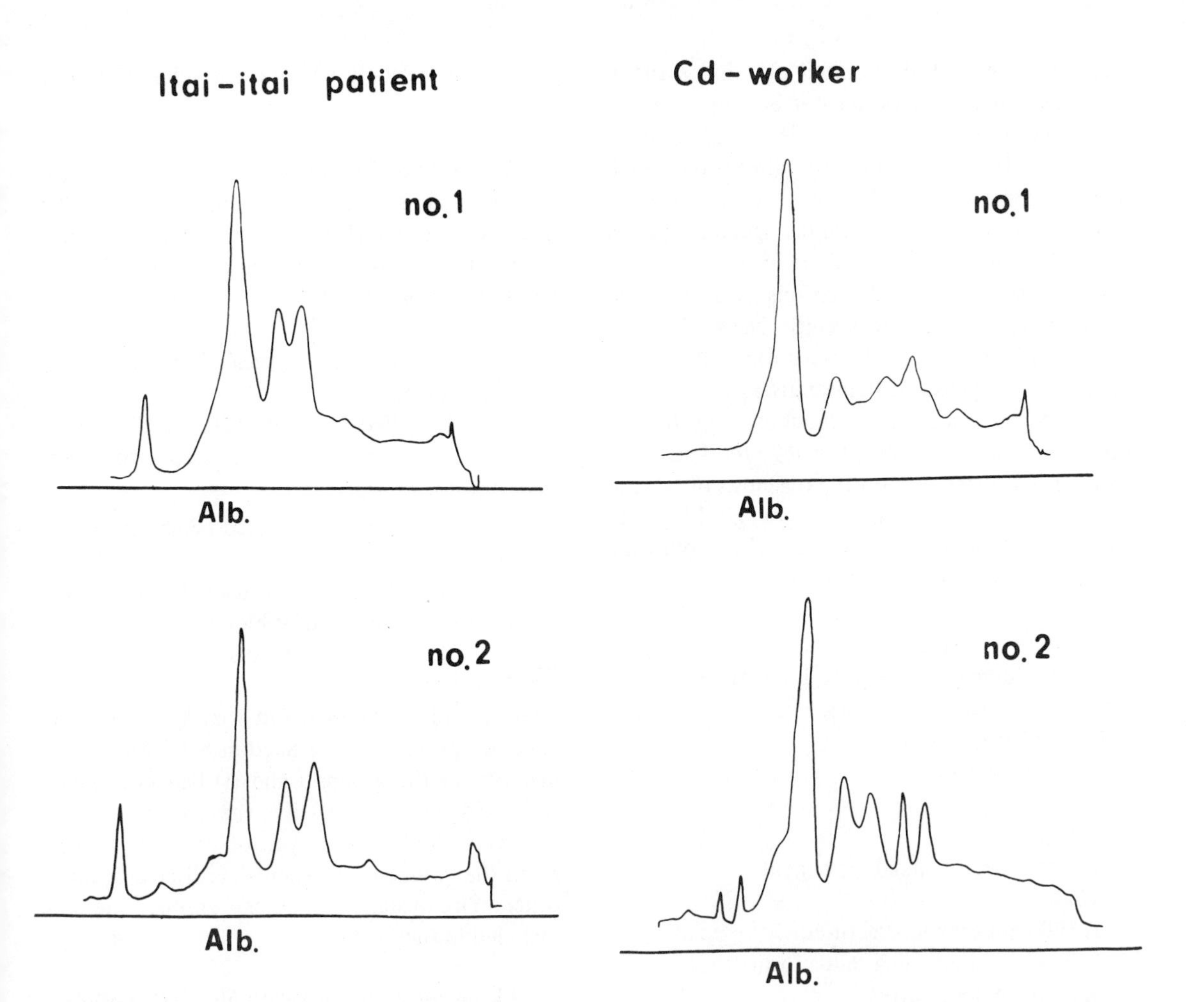

FIGURE 8:8. Comparison of disc electrophoretic patterns. (Modified from Tsuchiya, 1972, and Tsuchiya, personal communication.)

calcification. However, the Itai-itai patients were also exposed to large doses of vitamin D as a form of treatment for the illness (Section 8.2.6).

8.2.2.3.2 Bone

Ishizaki, 1969b, citing Kajikawa et al., 1957, and his own material, has described histological findings in bone from autopsies. The changes have been consistent with those found in osteomalacia. X-ray findings were also characteristic of osteomalacia. Nakagawa, 1960, on the basis of bone biopsies from seven patients, concluded that they had osteomalacia. Kono et al., 1958, found the Itai-itai disease very similar to common forms of osteomalacia. However, they found fewer osteoblasts in Itai-itai disease biopsies than are usually found in osteomalacia. Kajikawa, 1973, and Takeuchi, 1973, reported varying degrees of osteomalacia in combination with osteoporosis in eight patients.

8.2.3 Diagnosis of Itai-itai Disease

Ishizaki and Fukushima, 1968, discussed the clinical diagnosis of Itai-itai disease. They pointed out that the first and second stages could reasonably be considered to be renal dysfunction and bone changes, respectively. However, even if a specific diagnosis of renal impairment were possible, patients so diagnosed who can carry out their daily lives unhampered until bone symptoms develop cannot from a practical standpoint be included in the Itai-itai group. Because the subjective symptoms and the bone signs clinically observable through X-rays are very unspecific in the early stages, the diagnosis at that time could reasonably be verified by blood signs of osteomalacia, including increased alkaline phosphatase and decreased serum inorganic phosphate. These blood findings constitute the basis for the syndrome "latent osteomalacia" as introduced by Murata, Nakagawa, and Yoshimoto, 1958. Ishizaki and Fukushima, 1968, pointed out that a roentgenological diagnosis would not be helpful in the early stages because detectable X-ray changes would not be evident until the disease has developed to some extent. Moreover, as the disease is frequent in menopause and in old age, menopausal or senile osteoporosis could be mixed with the Itai-itai disease. To avoid such mistakes, it should be necessary to require specific signs of osteomalacia (Milkman's pseudofracture) for the roentgenological diagnosis of Itai-itai disease.

Based on elaborate discussions, the Itai-itai disease research group of 1962 to 1965 has worked out a standard for the diagnosis of Itai-itai disease, to be used in the epidemiological research (Kato and Kawano, 1968). The following observations should be noted:

A. Subjective symptoms
  1. Pain (lumbago, back pain, joint pain)
  2. Disturbance of gait (duck gait)

B. Physical examination
  1. Pain by pressure
  2. "Dwarfism"
  3. Kyphosis
  4. Restriction of spinal movement

C. X-ray
  1. Milkman's pseudofractures
  2. Fractures (including callus formation)
  3. Thinned bone cortex
  4. Decalcification
  5. Deformation
  6. Fish-bone vertebrae
  7. Coxa vara

D. Urine analysis
  1. Coinciding positive tests for protein and glucose
  2. Protein (+)
  3. Glucose (+)
  4. Decreased phosphorus/calcium ratio

E. Serum analysis
  1. Increased alkaline phosphatase
  2. Decreased serum inorganic phosphate

The subjects are classified into the following five groups:

I: Definitely Itai-itai patients
i: People deeply suspected for Itai-itai disease
(i): People suspected for Itai-itai disease
(O): People needing follow-up
O: People with no suspicion of Itai-itai disease

The methods for diagnosis differed a little with each year, as stated in Kato and Kawano, 1968, and a sixth group (I) is sometimes named. In some reports the groups are defined as follows (Ishizaki and Fukushima, 1968):

I: Patient (typical bone manifestations on X-ray)
(I): Person cured of Itai-itai disease
i : Person deeply suspected (bone signs but not typical)
(i): Person suspected (slight bone signs on X-ray
0_{ob}: Person with need for continuous observation (urinary and/or blood signs)
0: Person with no suspicion

It should be pointed out that because of the selection procedure (see Section 8.2.5.2), the vast majority in the groups i and (i) had both glucosuria and proteinuria, like the patients (= I). Sometimes only the I type of patients is meant when the "number of patients" is discussed, but in other circumstances, other groups are also included in the "patients."

8.2.4 Concentrations of Heavy Metals in Tissues

8.2.4.1 Concentrations of Heavy Metals in Urine

The 1962 to 1965 research group analyzed urine from inhabitants of different parts of the Toyama Prefecture. The method of sampling in the various patient categories cannot be known through literature available to us. Values reported, however, showed that cadmium concentrations in urine from persons in the endemic area were higher than from persons in other parts of the Toyama Prefecture (Ishizaki et al., 1965, Tsuchiya, 1969, and Ishizaki, 1969a). Lead and zinc values, on the other hand, showed no geographical differences (Ishizaki et al., 1965, and Tsuchiya, 1969). The 1967 research group, as referred to in Ishizaki's report, 1969b, also analyzed urine concentrations of cadmium (probably by atomic absorption after extraction). Increased concentrations were found in persons coming originally from the endemic area (see Figure 8:9).

8.2.4.2 Concentrations of Cadmium in Organs

Few values of cadmium in organs from Itai-itai

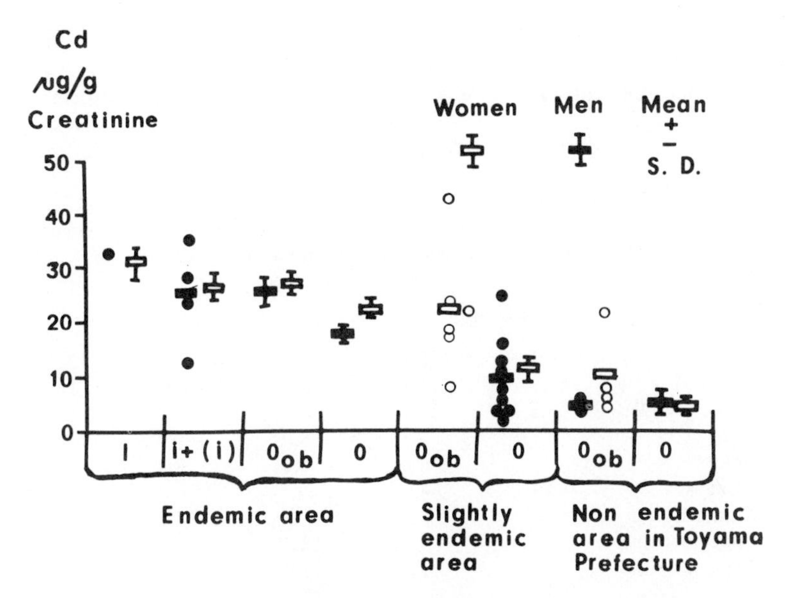

FIGURE 8:9. Cadmium concentration in urine of persons from different areas in Toyama Prefecture. (From Ishizaki, 1969b.)

patients have been reported. Some early data from one patient described high concentrations in bones. However, in this case the different organs had been stored together in formaldehyde for several years (Kobayashi, personal communication) so that the values must be considered unreliable.

Ishizaki, Fukushima, and Sakamoto, 1970b, and Ishizaki, 1972b, reported organ values (atomic absorption) from autopsies on five Itai-itai patients. The results are given in Table 8:1. Formaldehyde preservations were not used. At the same time cadmium analyses were performed on organs from 38 controls who died in hospitals in Kanazawa. All of them had been living in non-endemic areas in or near Kanazawa. Results are shown in Figure 8:10. As can be seen by comparing Table 8:1 with Figure 8:10, the cadmium content of the liver was five to ten times higher in Itai-itai patients than in the controls in the same age group. The patients' kidney values, however, were lower than those of the controls. The authors stated that the low kidney values in the Itai-itai patients could be explained by the advanced kidney damage. The values given should also be compared with "normal values" from Japan and other countries and with data from workers industrially exposed with signs of cadmium poisoning (see Sections 4.3.2.1 and 4.3.2.2). Such a comparison stresses the accumulation of cadmium, particularly in the liver of the Itai-itai patients.

8.2.5 Epidemiology of the Itai-itai Disease

Epidemiological studies regarding the Itai-itai disease will be treated in this section, while data on the prevalence of proteinuria and glucosuria in the general population, collected in connection with these studies, will be presented in Section 8.2.7. The limited research on the epidemiology of the disease prior to 1963 must be classified as preliminary or pilot studies. Since no diagnostic standard had been agreed upon, the different studies defined the disease differently.

There is no study known to us which would show that symptoms similar to those of the Itai-itai disease occurred in persons living outside the endemic area in the Jintsu River basin, at the time at which the disease was discovered there. Murata, Nakagawa, and Yoshimoto, 1958, reported that "latent osteomalacia" (Section 8.2.3) occurred outside the "Shinbo area" in 13 out of 38 persons studied. Whether this meant that "latent osteomalacia" occurred outside the Jintsu River basin is not clear since the exact location of the patients was not defined.

8.2.5.1 1962–1965 Epidemiological Research

An epidemiological survey made between 1962 and 1965 in Fuchu-machi, Toyama City, Osawano-machi, Yatsuo-machi, Nyuzen-machi, and Tonami City was evaluated according to a diagnostic standard from 1963 (Kato and Kawano, 1968). All of these places are in the Toyama

TABLE 8:1

Cadmium Content (μg/g Wet Weight) in Organs from Female Itai-itai Patients

	Patient 1	Patient 2	Patient 3	Patient 4	Patient 5
Age at death	79	71	60	73	67
Cause of death	Stomach cancer	Broncho-pneumonia	Endocar-ditis verrucosa	Stomach cancer	Uremia
Organ					
Liver	94.1	118.1	63.3	89.0	132
Renal cortex	41.1	31.8	19.8*		12
Renal medulla	39.5	26.1	*		10
Lung		2.5	8.0		2.1
Spleen		6.8	6.2		6.0
Pancreas	45.1	64.7			5.2
Stomach				4.8	
Small intestine		3.0	12.5	5.7	9.9
Large intestine		1.7	11.9		
Ribs			2.8		2.6
Bone cortex		1.6			
Bone marrow		1.1			
Skin		4.6	5.1	3.9	
Muscles		14.1			8.3
Brain	0.6				

*Cortex and medulla not separated.

Patients 1 to 4 from Ishizaki, Fukushima, and Sakamoto, 1970b; Patient 5 from Ishizaki, 1972b.

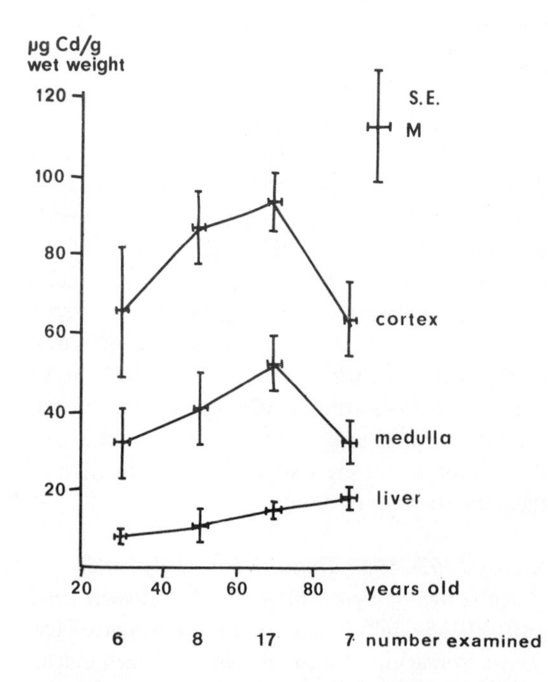

FIGURE 8:10. Cadmium content (μg/g wet weight) in controls who died in hospitals in Kanazawa. (From Ishizaki, Fukushima, and Sakamoto, 1970b.)

Prefecture (Figure 8:2). All females over 40 years of age were selected, except in Tonami City, where males over 40 years of age were also included in the 1,100 subjects examined. By examination of a total of 3,645 subjects, 28 patients, 14 deeply suspected subjects, and 19 suspected subjects had been found by 1965 in the endemic districts located along the Jintsu River (see Table 8:2). On the other hand, in Ota (Toyama City), located along another river in the endemic district, neither patients nor suspected patients were found among the 508 females over 40 years of age, 334 of whom were examined in 1964 in the same way as in the endemic districts. Also in Tonami City and Nyuzen-machi, located along other rivers in Toyama Prefecture, neither patients nor suspected patients were observed. These studies made clear that the Itai-itai disease was found only in specific districts in the Jintsu River basin. Further epidemiological research, therefore, was concentrated along this river.

8.2.5.2 1967 Epidemiological Research

In 1967, a larger, better-structured epidemio-

TABLE 8:2

Results of 1962–1965 Epidemiological Investigations on Women Over 40 Years of Age

District	Number Examined	Number of patients and suspected* I(I)	i	(i)
Kumano, Fuchu-machi†	259	9	8	13
Other villages in Fuchu-machi†	9	2	0	1
Shinbo, Toyama City†	352	15	5	3
Osawano-machi†	411	2	1	2
Ota, Toyama City	334	0	0	0
Nyuzen-machi	1,180	0	0	0
Tonami City‡	1,100	0	0	0

*Decided mainly by X-ray
†Jintsu River Basin
‡Also men over 40 years of age
I: Patient (typical bone signs on X-ray)
(I): Person cured of Itai-itai disease
i: Deeply suspected (bone signs but not typical)
(i): Suspected (bone signs, slight)

From Ishizaki and Fukushima, 1968.

logical study was performed by the Toyama Prefecture Health Authorities in cooperation with Kanazawa University in an attempt to find and treat all Itai-itai patients (the data in this section are from Kato and Kawano, 1968, and Ishizaki. personal communication, unless otherwise stated).

8.2.5.2.1 Groups Studied and Selection of Subjects for Final Diagnostic Procedure

A total of 6,717 subjects, all inhabitants over 30 years of age and of both sexes, of Fuchu-machi, certain parts of Toyama City, Osawano-machi, and Yatsuo-machi, were included in the first screening made by a questionnaire (shown in Table 8:3) and by semiquantitative determination of protein and glucose in urine. This screening was performed on 6,114 persons, 91% of the 6,717 subjects originally selected for the study.

Based on the results of the first screening, 1,911 persons were selected for the second examination in the following way (Kato and Kawano, 1968). A ranking of the responses to the questionnaire as well as of the results of urine analysis was performed. Those who had a ranking of 1 on either the questionnaire or urine analysis were selected for the second examination. Included were also those who had a rank sum of less than 5 and those who said that they had been told that they were Itai-itai patients. Those who answered "yes" more than ten times on the questionnaire were ranked as 1, those giving six to nine affirmative answers were ranked as 2, those giving one to five affirmative answers were ranked as 3, and those giving no "yes" responses were ranked as 4. The ranking of the urine analysis is depicted in Figure 8:11.

Out of the 1,911 people selected for the second examination, 1,400 underwent the examination, which included a medical interview, general clinical examination, and X-ray of the right upper arm and shoulder. As a result of the second examination, 451 persons were selected for a final examination. The exact basis for this selection has not been possible to elucidate, but among important criteria were bone changes.

8.2.5.2.2 Final Diagnostic Examination (Third Examination)

Out of the 451 persons selected for a final examination, 419 were examined. The exami-

TABLE 8:3

Questionnaire for First Screening Procedure in the 1967 Epidemiological Study

Please answer all of the following questions. Try to overcome any difficulties you might have in recalling the answers, "yes" or "no."

	Yes	No
1. Have you ever been told that you are an Itai-itai patient?		
2. Have you ever suffered from neuralgia or rheumatism?		
3. Have you ever suffered from bone or joint disease?		
4. Have you ever suffered from a low back pain?		
5. Have you ever suffered from a kidney disease?		
6. Have you ever suffered from diabetes?		
7. Do you have a low back pain now?		
8. Do you have an arm, leg, or foot pain now?		
9. Do you have a joint pain now?		
10. Do you have neuralgia now?		
11. Have you difficulty walking now?		
12. Have any of your family members (father, mother, brother, sister, husband, wife, or children) died of Itai-itai disease?		
13. Have any of your family members died of kidney disease?		
14. Have any of your family members died of diabetes?		
15. Have any of your family members suffered from rickets?		

From Kato and Kawano, 1968.

nation included a medical interview, X-ray of the pubic region, orthopedic examination, urine analysis for glucose, protein, calcium, phosphorus, creatinine, cadmium, lead, and zinc, as well as serum analysis of alkaline phosphatase and inorganic phosphorus. All of these analyses were not performed on all persons.

8.2.5.3 Total Number of Cases of Itai-itai Disease

It is not possible to state a precise figure on the incidence of Itai-itai disease over the years. The 1967 epidemiological survey is cross sectional, giving only prevalence data at the time of the study. It is estimated that nearly 100 deaths due to the Itai-itai disease occurred up until the end of 1965 (Yamagata and Shigematsu, 1970). The following numbers of persons with Itai-itai disease were found: I: 50, i: 17, (i): 31, and 0_{ob}: 136 (see Table 8:4 for the distribution of examined persons according to place of residence, sex, and age). The Toyama Prefectural Health Authorities now continuously register all living patients. Several new cases have been diagnosed since the 1967 study. Figures from September 1968 are given in Table 8:5.

The latest available figures on the total number

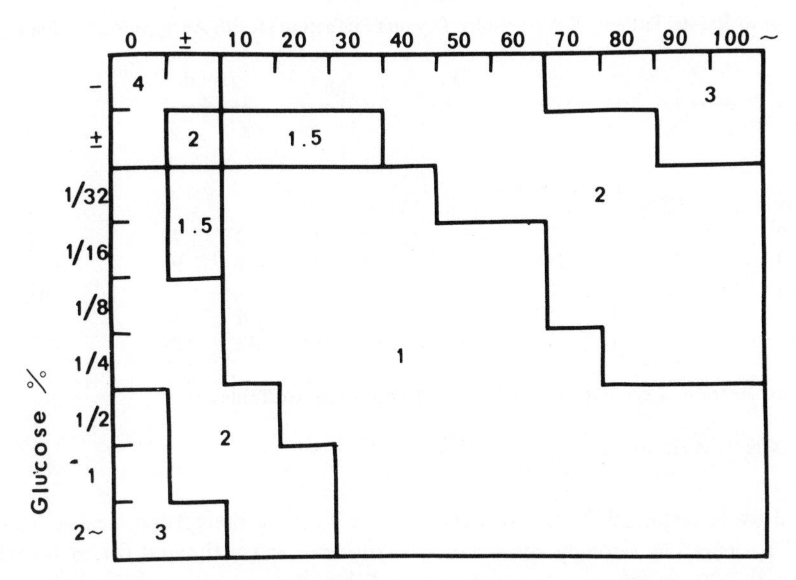

FIGURE 8:11. Ranking of subjects for selection to undergo the second examination from the results of urine analysis. (From Kato and Kawano, 1968.)

TABLE 8:4

Results from 1967 Final Examination

District	Total	I	i	(i)	O_{ob}	O
	419	50	17	31	136	185
Fuchu-machi	227	28	11	14	79	95
Toyama City	105	16	3	8	34	44
Osawano-machi	53	6	3	6	16	22
Yao-machi	34			3	7	24
Sex						
♂	122	1	1	3	37	80
♀	297	49	16	28	99	105
Age						
30–44	39	–	–	–	2	37
45–59	86	6	–	2	28	50
60–74	225	40	10	22	86	67
75+	69	4	7	7	20	31
Mean age:		66.6	72.2	68.3	66.1	58.0

I: Patient (typical bone signs on X-rays)
i: Deeply suspected (bone signs but not typical)
(i): Suspected (bone signs, slight)
O_{ob}: Need for continuous observation (urinary and/or blood signs)
O: No suspicion

From Ishizaki and Fukushima, 1968.

TABLE 8:5

Number of Itai-itai Patients Registered by Toyama Prefecture Health Authorities, September 1968

	I*		i		(i)		O_{ob}		Total		
Age	♂	♀	♂	♀	♂	♀	♂	♀	♂	♀	Total both sexes
30–39											
40–49		2				1		2		5	5
50–59		6		2			3	14	3	22	25
60–69		31		14		5	12	40	12	90	102
70–79		12		5		6	14	30	14	53	67
80–89		2	1	6		1	3	5	4	14	18
Total		53	1	27		13	32	91	33	184	217

*For explanation of symbols for the Itai-itai classification, see Table 8:4.

From Kubota et al., 1969.

of cases are stated by Ishizaki, 1971, as, until the end of 1968, 57 patients, 34 strongly suspected, 24 suspected, and 200 requiring observation. However, these figures are not directly comparable to those in Tables 8:4 and 8:5 because Ishizaki defined the groups somewhat differently than earlier. The "observation group" included subjects who had recovered.

8.2.5.4 Age and Sex

According to Table 8:5 from September 1968, only one male was placed in the group of patients. However, according to the 1967 epidemiological research, five male patients or suspected patients were found, as seen in Table 8:4. The reason for the registration of only one male patient in the table from September 1968 is not clear. Among the observation subjects there were an appreciable number of men.

Noting the age distribution in Table 8:4 from the 1967 report, one can conclude that the osteomalacia has been found only in women of 45 years of age or older. No patient was found in the age group below 44 years, but two observation cases were registered.

8.2.5.5 Incidence

Hagino (personal communication) had the impression that the frequency of disease onsets was higher in the 1950 to 1960 period than in the 1960 to 1970 period. At the same time he pointed out that a number of cases not known before are still diagnosed yearly.

By means of the 1967 epidemiological research and the subsequent registration of all patients by the Toyama Prefectural Health Authorities, a year of onset was estimated for each patient, as seen in Table 8:6. Great caution must be exercised in interpreting the data. As seen in the table, a majority of the *presently* registered patients first complained of symptoms in the period between 1945 to 1964 but 10 to 15% did not complain of symptoms until 1965 or later. Many Itai-itai patients who contracted the disease some years earlier must have died by 1968 (Section 8.2.5.3).

8.2.5.6 Geographical Distribution of Itai-itai Patients

The locations of the different population centers within the Toyama Prefecture are seen on the maps in Figures 8:2 and 8:12.

It has become evident from the research that the disease has been found only in a limited area around the parts of the Jintsu River near Toyama City, where its water is used for irrigation of rice fields. The prevalence of the Itai-itai disease in different parts of the endemic area has been best illustrated by the detailed survey of 1967, the results of which can be seen in Figure 8:13. The frequency of the disease is illustrated there by the percentage of women over 50 years of age having or suspected of having the Itai-itai disease. The geographical distribution of patients thus confirms the results from the 1962–1965 study (Table 8:2).

The 1967 research group also analyzed the cadmium concentration in the upper stratum of the soil in the paddy fields (see Section 8.2.5.10) and found a striking correlation with the prevalence of the Itai-itai disease, even within the endemic area (see Figures 8:12 and 8:13).

TABLE 8:6

Number of Patients with Itai-itai Disease According to Year of Onset of Disease Estimated by Patients' Complaints (Patients Registered by Toyama Prefecture, June 15, 1968)

		Starting year					
		Before 1934	1935–1944	1945–1954	1955–1964	1965–later	Not clear
Total	92	6	10	25	33	12	6
I*	51	4	7	16	16	5	3
i+(i)	41	2	3	9	17	7	3

*For explanation of symbols for the Itai-itai classification, see Table 8:4.

From Kato and Kawano, 1968.

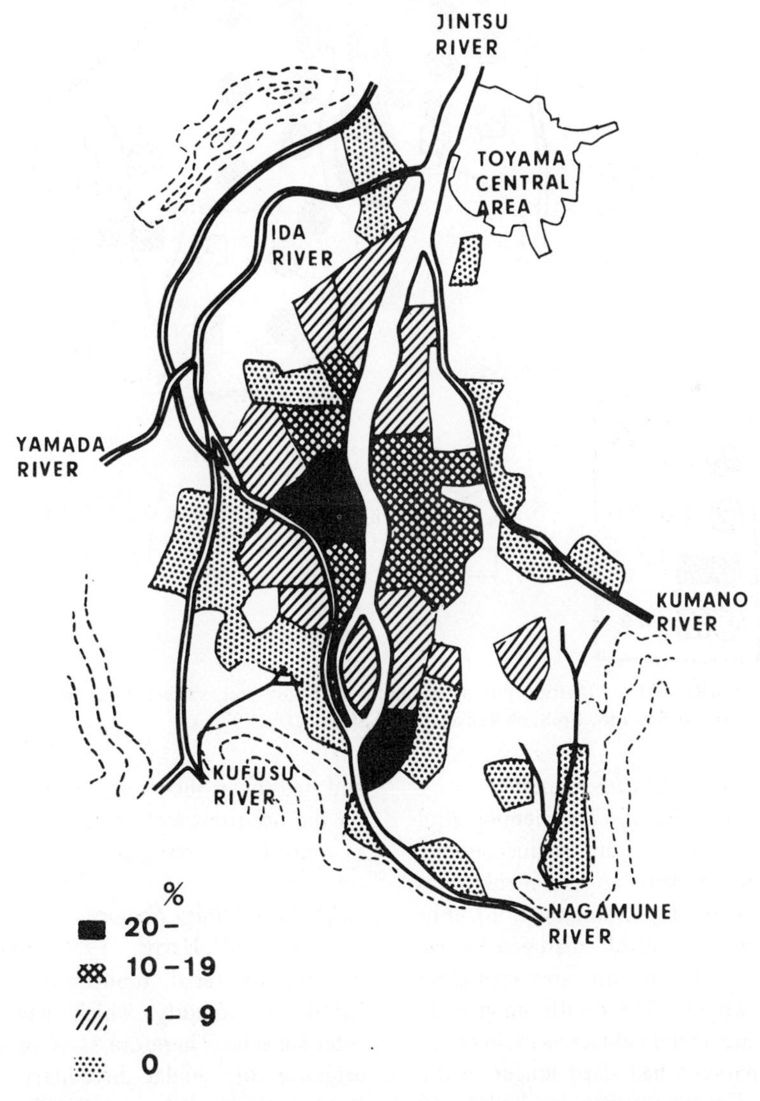

FIGURE 8:12. Percentage of women over 50 years of age with Itai-itai disease (I, i, or (i)) at the examination of 1967. (Redrawn from Ishizaki and Fukushima, 1968, Kato and Kawano, 1968, and Yamagata and Shigematsu, 1970.)

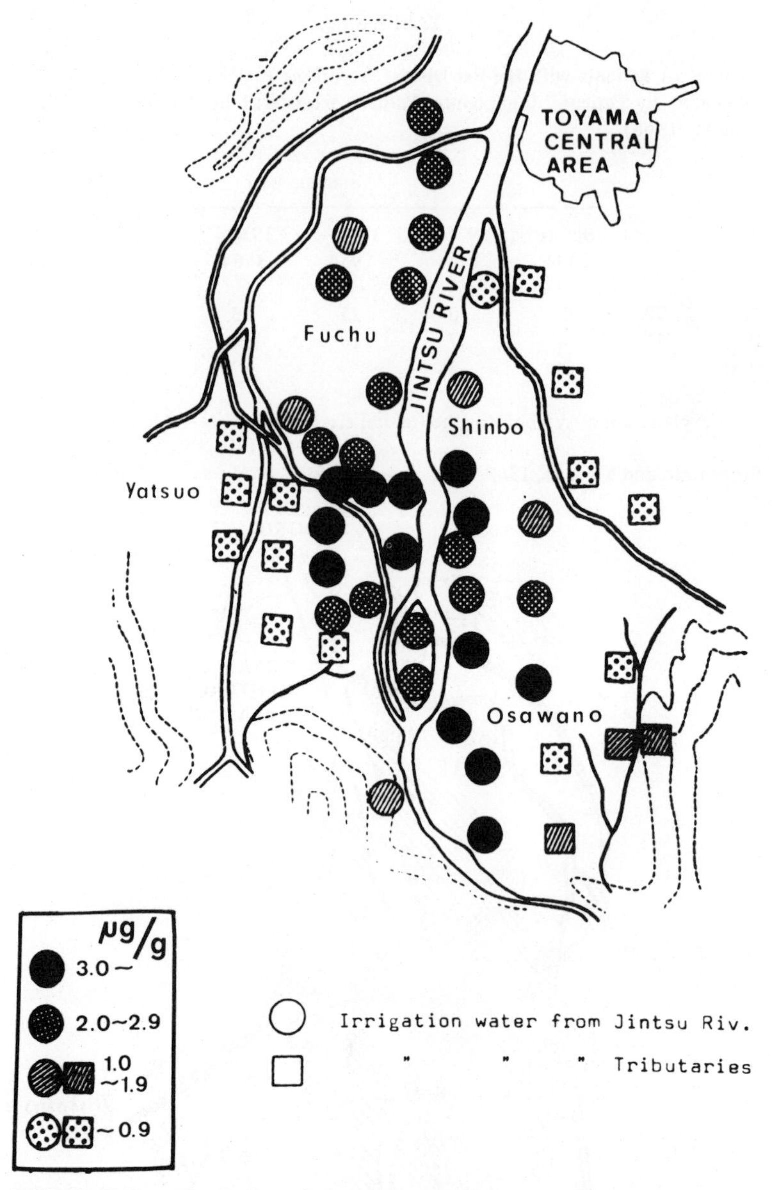

FIGURE 8:13. Distribution of cadmium in paddy soil, surface layer. (From Kato and Kawano, 1968, or Yamagata and Shigematsu, 1970.)

8.2.5.7 Profession and Social Characteristics

In connection with the 1967 epidemiological study, information on certain socioeconomic factors about Itai-itai patients and "controls" was obtained (Kato, personal communication). For each Itai-itai patient, a "control" matched for age and sex and living in the endemic area very close to the patient was chosen. The results are given in Table 8:7. There are certain differences between the groups. The patients had lived longer in the same area, had had more pregnancies, had drunk water from the Jintsu River to a greater extent, and had lower incomes. No information was given on the occurrence of proteinuria and/or glucosuria in controls, for example.

8.2.5.8 Hereditary Factors

Dent and Harris, 1956, have described a number of renal tubular reabsorption defects based on heredity which have given rise to osteomalacia. Therefore it is of interest to seek evidence of similar hereditary factors in the Itai-itai disease. Ishizaki, 1969b, reported that in many families in the endemic district, both the

TABLE 8:7

Socioeconomic Data Concerning Itai-itai Patients and Controls

Factors studied	Itai-itai patients* (percent)	Controls† (percent)
1. Lived in same area 40 years or more	96	77‡
2. Worked in agriculture	83	79
3. Six or more pregnancies	56	41‡
4. Drank water from Jinstu River	83	57‡
5. Total family income 1,000,000 yen or more	15	28‡
6. "Better" condition of living room	61	63
7. Refrigerator in house	56	64
8. Area of rice field (mean)	1 ha	1 ha

*124 women, 3 men; mean age: 67.6.
†124 women, 3 men, mean age: 66.3.
‡Difference statistically significant.

From Kato, personal communication.

farmer's mother and wife had contracted the disease. No cases had occurred in women from the endemic district who had married and moved to a family in the nonendemic district. These facts indicate that the etiology rests on environmental rather than hereditary factors.

8.2.5.9 Dietary Factors

The classical form of osteomalacia is based on a deficiency of vitamin D. In the beginning of the research on the Itai-itai disease, dietary and climatological factors were assumed to play an important role in its etiology.

The Toyama Prefectural Health Authorities performed a survey on the dietary conditions in 200 households in the endemic area during 1955 to 1956. The results are given in Table 8:8. For comparison, averages for Sweden in 1960 are also given. It is evident that the dietary standard in the Itai-itai patients' homes and in the endemic area as a whole is not much different from the mean in the Toyama Prefecture or from the Japanese average. It is not known to what extent the figures for the different areas in Japan are comparable to the figures for Sweden, but for certain dietary components the large differences noted are not likely to disappear, even if the figures are corrected for the different methods of calculation. As vitamin D is a known essential dietary factor when discussing osteomalacia, it is unfortunate that no Japanese figures for this vitamin are available. The vitamin A and fat intakes in the Itai-itai patients are lower than the Japanese average and considerably lower than the Swedish average. Therefore, it is probable that the vitamin D intake is also lower. If such is the case, a great exposure to sunlight could compensate for the low intake of vitamin D. However, since the endemic area has a rainy climate, and since the women dress in such a way that they screen off a great part of the sunshine, this compensation is not likely. A higher calcium intake could also partly compensate for a low level of vitamin D, but the calcium intake in Japan is very low. The figures from the investigation of 1955 to 1956 in the homes of the Itai-itai patients represented a mean intake for all family members. Several investigations in various countries have shown that the calcium intake is usually much lower in women; therefore, it is likely that the women in the endemic area consume significantly less calcium than indicated by the figures.

It must also be remembered that the nutritional situation in Japan during and immediately after the Second World War was poor. Insull, Oiso, and Tsuchiya, 1968, documented acute generalized malnutrition problems arising in 1936 and 1945. They cited evidence showing that the heights of children during and directly after the war were significantly retarded.

8.2.5.10 Concentrations of Heavy Metals in the Environment. Daily Intake of Cadmium

As has been discussed, most of the Itai-itai patients displayed their symptoms many years ago. If heavy metals caused the disease, even exposure several decades earlier must have been of importance. However, no quantitative information on such exposure is available.

After the theory about cadmium as a cause of the Itai-itai disease was advanced, the Jintsu River was suspected as the carrier of the cadmium from the Kamioka mine to the endemic area. Kobayashi, 1971, stated that the increased production at the mine together with faulty treatment of waste water polluted the Jintsu River heavily during World War II. He explained that particles carried by the river became deposited in the rice fields. Damage to the rice crop prompted a severe dispute between the farmers and the owners of the mine. Studies on cadmium content in year rings

TABLE 8:8

Nutritional Data From Japan and Sweden

Foodstuff		Itai-itai patients' homes*	Kumano district*	Mean in in rural District (Toyama Pref.)	Mean in Japan, 1955	Mean in Sweden, 1960
Calorie	(cal)	2,139	2,209	2,237	2,074	2,868
Protein	(g)	73	74	62	69	74
Fat	(g)	14	17	17	21	124
Ca	(mg)	408	374	331	364	938
P	(mg)	1,556	1,432	1,433	1,822	1,378
Fe	(mg)	34	15	23	6	14
Vitamin A	(I.U.)	1,325	1,754	1,376	2,814	3,267
Vitamin B_1	(mg)	0.8	1.1	1.0	1.1	1.8
Vitamin B_2	(mg)	0.6	0.6	0.7	0.7	1.8
Vitamin C	(mg)	69	93	64	66	75
Vitamin D	(mg)					5.1

*Total: 200 households.

Japanese data for 1955 and 1956 from Takase et al., 1967, or Tsuchiya, 1969; Swedish data for 1960 from Blix et al., 1965.

from cedar trees may also speak in favor of a higher concentration of cadmium in the Jintsu River several years ago (Ishizaki et al., 1970). On the other hand, Nitta, 1972, reported that his geological studies showed a "remarkably high" cadmium content in the soil of the Jintsu River basin. He stated that the distribution of cadmium indicated that it had been sedimentous in this specific area during hundreds of years.

In 1959, Kobayashi and Yoshioka, as cited in a report by Yamagata and Shigematsu, 1970, analyzed samples of rice from the endemic area. Values (μg/g in ash) in polished rice were given as 120 to 350 μg/g in exposed areas, compared with 21 μg/g in a control area. In the root of rice plants, corresponding figures were 690 to 1,300 and 35 μg/g, respectively. No information about the representativeness of the samples was reported. Moritsugi and Kobayashi, 1964, reported cadmium concentrations in over 200 samples of polished rice from throughout Japan. Analysis was performed with the dithizone colorimetric method as described by Saltzman, 1953. The average of 20 samples from the Itai-itai disease endemic area was 0.68 μg/g in rice "kept in an air-dried condition at room temperature" while the average of samples from other parts of Japan was 0.066 μg/g (wet or dry weight not stated).

In 1967, a study headed by Shigematsu on the distribution of cadmium in the endemic area was made under a grant from the Ministry of Health and Welfare. This study is included in the above mentioned report by Yamagata and Shigematsu, 1970. In water upstream from the Kamioka mine and tributaries of the Jintsu River, cadmium was detected in trace amounts or not at all in the samples taken. In water downstream from the mine a maximum value of 9 ng/g was found. Out of four samples from the drainage from the mine, three contained 5 to 60 ng/g (pH about 7 to 8) and one contained 4,000 ng/g (pH 2.8) of cadmium. Near the drainage area of the mine, cadmium concentrations of 363 and 382 μg/g were found in suspended materials.

Samples of paddy soils irrigated with water from the Jintsu River and from its tributaries have been analyzed by Kato and Kawano, 1968, and Yamagata and Shigematsu, 1970. The results, revealing much higher concentrations in river-irrigated soil, are shown in Figure 8:13. Figure 8:12 shows the prevalence of sure and suspected cases of Itai-itai disease in different areas around the Jintsu River and its tributaries. Unfortunately the maps cover only a limited area; comparable data from more distant districts would have been of great value. However, when Figures 8:12 and 8:13 are considered together, the association between high soil concentrations of cadmium and the Itai-itai disease is obvious. The variation in cadmium concentration among individual rice samples (Figure 8:14) taken close to each other is larger than that among soil samples. This fact and the lack of a subdivision of the concentrations in rice under 0.5 μg/g make it hard to evaluate the rice data. It is possible that cadmium concentration in soil is a better indicator of long-term exposure than cadmium concentration in rice.*

Ishizaki, Fukushima, and Sakamoto, 1968, recorded values of over 2 μg/g for cadmium concentrations in rice from polluted districts, very seldom finding values under 0.5 μg/g in such districts. The results are shown in Figure 8:14.

Based on the National Nutrition Survey of 1966 and available data on cadmium content in food, a representative value for the daily intake of cadmium in the Japanese diet has been estimated as 60 μg (Ministry of Health and Welfare, 1969b). A breakdown of the 60 μg indicated that 23 μg of cadmium should come from an intake of 335 g of polished rice with a concentration of 0.07 μg/g (estimated from data by Moritsugi and Kobayashi, 1964; see also Section 8.3.2.1). The daily intake of cadmium via food in the endemic area was calculated at 600 μg by assuming an average cadmium concentration in rice of 1 μg/g and a concentration in other foodstuffs produced and consumed locally of about ten times the value for Japan as a whole. An additional intake from river water used as drinking water has also been assumed. The cadmium content of river water as reported by Yamagata and Shigematsu, 1970, is too low to contribute significantly to the daily intake. In a calculation of daily intake in the endemic area the Ministry of Health and Welfare, 1968b, used an assumed water concentration during World War II of about 500 ng/g to reach a cadmium intake from water of about 1 mg. No data have been reported in support of such a high estimate, so the actual average long-term daily intake in the area is not known.

The concentrations of lead in the soil are about 100 times higher than the concentrations of cadmium.** In rice samples from the Itai-itai disease endemic area, on the other hand, Tsuchiya, 1969, reported average values for lead of about 0.2 μg/g and for cadmium of about 0.8 μg/g. In the same report, he noted that the average urinary lead and zinc excretion was the same for Itai-itai patients, other persons in the endemic area, and persons in the control area. On the other hand, cadmium excretion was lower in persons in the control area than in those in the endemic area. No studies on lead exposure nor on the possible interaction effects of lead and cadmium have been performed.

8.2.6 Treatment of Itai-itai Disease

No objection-free study on the effects of treatment of Itai-itai disease has been published. There are, however, some data which elucidate to what extent vitamin D has been successful in improving overt osteomalacia symptoms. As such information has some bearing when discussing the etiology of Itai-itai disease, available data are briefly included here. Further, Takeuchi, 1973, in a recent discussion of the etiology of Itai-itai disease, states that the effects of vitamin D were satisfactorily prompt and sufficient, and the patients were never resistant to therapy. The data available do not support such a conclusion.

When Hagino discovered the first cases, he believed the disease to be vitamin D-deficient osteomalacia, and during 1949 to 1955 tried treatment with cod liver oil (5 to 10 g/day corresponding to about 1,000 I.U. vitamin D) on 30 patients, but this was without effect (Hagino, 1973).

Taga et al., 1956, reported on the treatment of ten patients with vitamin D. After 3 months of treatment with vitamin D, serum inorganic phosphate had increased and the patients were able to move. There was an improvement of subjective symptoms; X-ray pictures of bones showed improvement. It was not stated how large the doses were or which route of entry was used. Since Hagino was one of the authors of the paper, one may assume that treatment was similar to his method. He reported (Hagino, 1973) that during 1955, 1956, and 1957, he administered 50,000 to 100,000 I.U. vitamin D/day to about 50 patients. The route of entry was not mentioned.

Hagino, 1957, and Nakagawa and Furumoto,

*Compare with Section 8.2.7, Figure 8:19.

**See Figure 8:19.

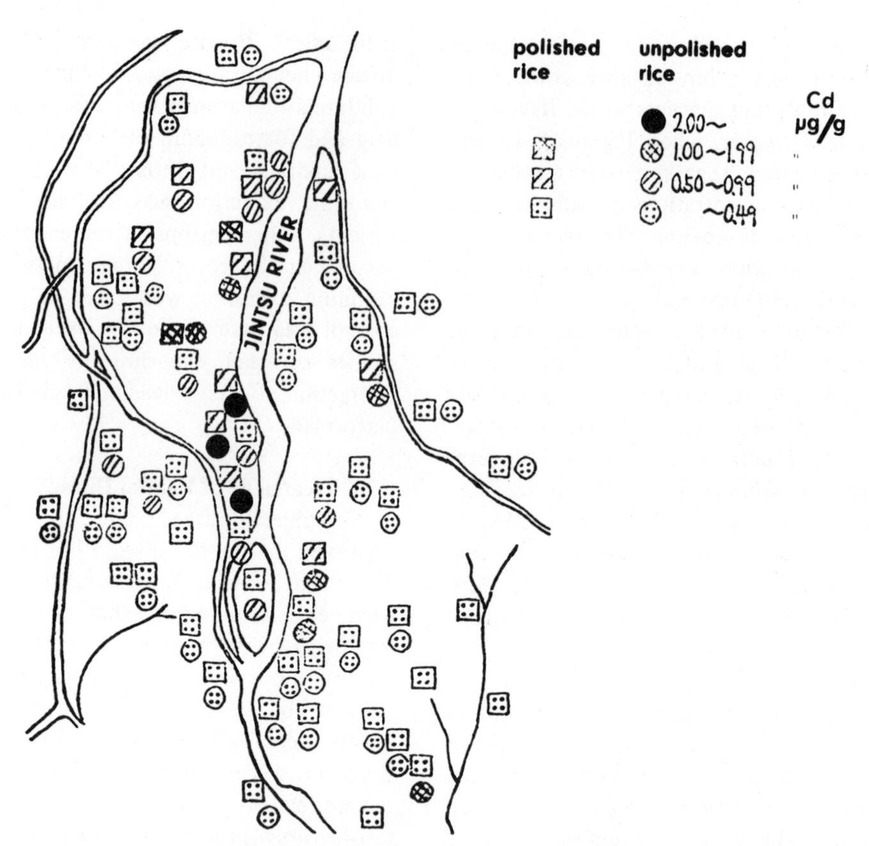

FIGURE 8:14. Concentration of cadmium in rice in the Fuchu area in 1967. (From Ishizaki, A., Fukushima, M., and Sakamoto, M., *Cadmium Content of Rice Eaten in the Itai-itai Disease Area,* Annual Meeting of the Japanese Association of Public Health, Kyoto, 1968. With permission.)

1957, reported that large doses of vitamin D were needed to reverse the bone symptoms. Toyoda et al., 1957, studied three patients who had been treated first for 2 months with a diet containing high amounts of calcium and phosphorus; in two cases, the pains were relieved after 2 months. At that time vitamin D treatment (100,000 I.U./day intramuscularly) was started and after 3 months of this treatment the pain had completely disappeared. Two patients displayed the "duck gait" from the beginning but after 3 and 5 months, respectively of vitamin D treatment, these symptoms disappeared. Toyoda et al. stated that the findings on X-ray showed a very slow healing process.

Murata, Nakagawa, and Yoshimoto, 1958, reported that vitamin D was effective for both Itai-itai patients and persons with "latent osteomalacia." Kono et al., 1958, made balance studies on two Itai-itai patients. Vitamin D treatment brought a change from negative to positive phosphate balance, and the calcium balance became more positive. In one case the vitamin D treatment was ended after a certain time (how long was not reported) and 3 months later the pains resumed. The patient, who had spontaneous pains upon movement, could hardly walk 8 months after the end of treatment. Kawano et al., 1958, stated that the disease could not have been caused only by malnutrition, as the symptoms of this patient worsened when the administration of vitamin D and other medications was terminated but otherwise complete nutrients were given.

The most detailed account of treatment and its effect has been given by Nakagawa, 1960. He studied a total of 30 patients and divided them into three groups according to their capability of moving: slight symptoms (duck gait, some limitations in movement, otherwise few bone symptoms), medium symptoms (can crawl but hardly walk), and heavy symptoms (cannot move at all).

Many of the patients with slight symptoms could be treated as outpatients, while most of the other patients were treated as inpatients. Outpatients were given vitamin D tablets (20,000 I.U./day = calciferol 0.5 mg) and arrangements were made for better nutrition (more milk, egg, meat, vegetables) and more sunlight. Inpatients received intramuscular injections of 100,000 I.U. vitamin D 6 days/week (calciferol 2.5 mg) during 5 to 6 weeks. After that period treatment was withdrawn for a short time. It was taken up again in the form of injections of 100,000 I.U. every second day plus 10,000 I.U./day as tablets. The patients ate regular hospital food and were taken out in the sunlight as much as possible. After leaving the hospital the patients received tablets corresponding to 20,000 I.U./day and were given an injection of 200,000 I.U. every time they came for a check-up (one to two times each month). Check-ups were made of serum calcium level and subjective symptoms. Treatment was continued for 1 to 3 years.

Pain decreased after about 3 weeks of treatment for the inpatients and after 2 months of treatment for the outpatients (Nakagawa, 1960), but the impaired mobility remained for a long time. Persons with slight symptoms could return to everyday life after about 4 months, by which time the duck gait had disappeared. Patients with medium symptoms were relieved from pains after about 2 months, could stand erect after 5 months, could walk inside the hospital after 8 months, and returned to everyday life after 13 to 14 months. The effect of treatment was still slower in persons with heavy symptoms: after 5 months they could sit up, after 7 months they could stand up, after 15 months they could walk with canes, and after 20 months they could walk unaided a couple of meters. The erythrocyte sedimentation rate decreased and the leukocyte count increased after about 1 month of treatment. Serum inorganic phosphate increased to normal and alkaline phosphatase decreased, but these changes were slower. When the progress of treatment was studied by X-ray of bones, it was found that pseudofractures persisted after 1 to 3 years of treatment in some cases, but in others they disappeared after 3 to 6 months.

Murata, Nakagawa and Hirono, 1972, report on treatment with vitamin D (about 100,000 I.U./day) of five patients for 15 years. The amounts of vitamin D given varied from year to year, but the supplementation was never discontinued during a whole year. Murata, 1971, showed that in two patients the pathological fractures had healed after 6 and 24 months, respectively, when judging from X-ray pictures. According to Murata, Nakagawa, and Hirono, 1972, X-ray pictures of one patient showed pathological fractures in 1956 and 1968 but not in 1962. They argued that vitamin D treatment had first cured the patient and then by overdoses caused bone porosity. However, it was not explained why 10,000 to 30,000 I.U. vitamin D/day was given from 1962 to 1969.

Hagino, 1973, stated that the patients' symptoms returned when treatment was discontinued, a finding reported by Kono et al., 1958, as well. The very long treatment periods reported by Murata, Nakagawa, and Hirono, 1972, would have been unnecessary if such recidivism did not occur. Thus, there are clear-cut indications from studies referred to above that long-term administration with large doses of vitamin D was required in order to bring about improvement of the osteomalacia in Itai-itai disease. This would not be the case if the patients had had a classical vitamin D-deficient osteomalacia. Thus, all the evidence favors Itai-itai disease as being a form of vitamin D-resistant osteomalacia.

Vitamin D treatment with such large doses as were used in many cases (over 100,000 I.U./day) may give rise to side effects, which mainly are caused by hypercalcemia. It may well be assumed that the treatment generally was performed in such a way as to avoid acute side effects (nausea, headache, anorexia, etc.). Only incomplete data are available regarding serum calcium concentrations, but Murata, Nakagawa, and Hirono, 1972, reported that the patient with the recidivous fracture referred to above had concentrations over 11 mg/100 ml in 2 analyses out of 15 during a 15-year period.

Some persons diagnosed as in need of continuous observation regarding Itai-itai disease (0_{ob}, Section 8.2.3) were given "prophylactic treatment" with vitamin D during 1968 (Hagino, 1973, and Murata, Nakagawa, and Hirono, 1972). Average daily dose varied between less than 25,000 I.U. and more than 200,000 I.U. In 22 out of 48 cases serum calcium concentration exceeded 11 mg/100 ml (Murata, Nakagawa, and Hirono, 1972), but the data presented do not indicate any dose-response relationship between serum calcium and vitamin D dose or between proteinuria or glucosuria and vitamin D dose.

Murata, Nakagawa, and Hirono, 1972, presented data on the PSP excretion test, urea clearance, and urea-nitrogen from different time periods for five patients treated with large doses of vitamin D. A continuous aggravation of kidney function is clear. Takeuchi, 1973, referred to clinical data for three patients studied in 1955 and four patients studied in 1965 which show that the kidney function was damaged in the latter patients, whereas the others had normal or only slightly impaired kidney function. Takeuchi did not report to what extent the patients had been treated with vitamin D. As no control patients were used in these studies it is not clear whether differences in kidney function tests at different times were caused by continued cadmium exposure, vitamin D treatment, or other cause.

8.2.7 Prevalence of Proteinuria and Glucosuria in the Endemic Area

In addition to sure or suspected cases of Itai-itai disease, a great number of persons with proteinuria and/or glucosuria have been found in connection with the 1967 epidemiological stuides. The total prevalence of proteinuria (method: sulfosalicylic acid, trichloracetic acid, or test tape) and glucosuria (method: Benedict's reaction) in males and females and in different age groups is shown in Figure 8:15 from Ishizaki, 1969b. The number of persons examined is given in Table 8:9.

As can be seen, about 50% or more of inhabitants older than 60 in the endemic area had proteinuria. Among those over 70, 70 to 80% had proteinuria. The prevalence was somewhat higher among females, and for both sexes was considerably higher than in a nonendemic area of the Toyama Prefecture. However, even in such a nonendemic district the prevalence was high, especially in women. Thus, about 20% or more of women older than 60 and 30 to 40% older than 70 in nonendemic districts had proteinuria. From the methodological point of view, exactly comparable figures from other parts of Japan are not available (Section 8.3.2.2).

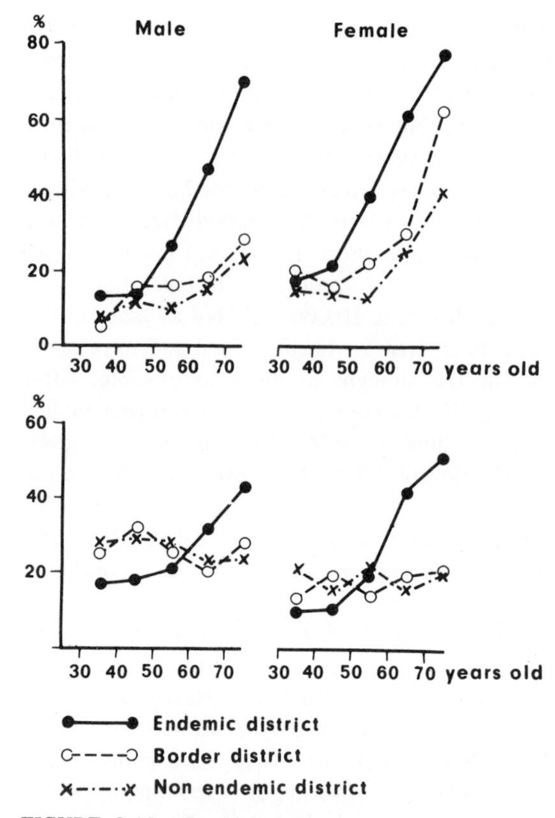

FIGURE 8:15. Prevalence of proteinuria (above) and glucosuria (below) in different districts in men and women in different age groups at the 1967 epidemiological study. (From Ishizaki, 1969b.)

Since either proteinuria or glucosuria can indicate a renal tubular impairment, the occurrence of

TABLE 8:9

Number of Persons Examined for Proteinuria and Glucosuria – Men and Women in Different Districts and Age Groups at the 1967 Epidemiological Study

	Men						Women					
Age group	30–39	40–49	50–59	60–69	70+	Total	30–39	40–49	50–59	60–69	70+	Total
Endemic district	394	382	291	255	160	1,482	503	392	352	289	134	1,670
Borderline district	97	95	73	66	32	363	109	99	95	78	47	428
Nonendemic district	306	220	202	178	85	991	349	245	265	184	116	1,159
Total	797	697	566	499	277	2,836	961	736	712	551	297	3,257

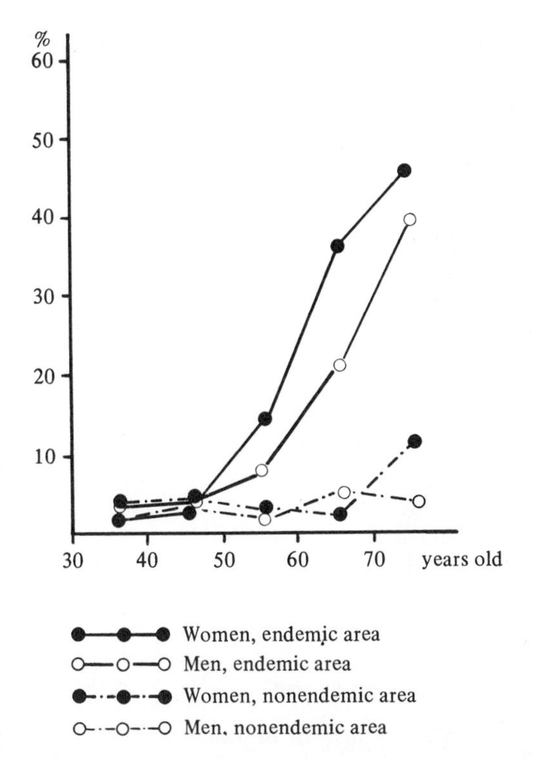

FIGURE 8:16. Concurrent prevalence of proteinuria and glucosuria in endemic and nonendemic districts in men and women in different age groups at the 1967 epidemiological study. (Modified from Ishizaki, 1969b.)

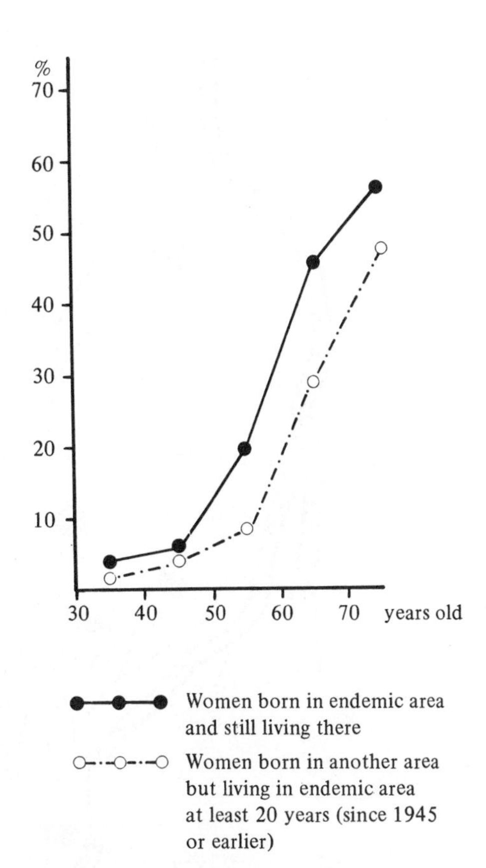

FIGURE 8:17. Concurrent prevalence of proteinuria and glucosuria in women in different age groups in the endemic district according to place born. (From M. Fukushima, personal communication.)

both in a person is a stronger indication of such a renal defect. The concurrent appearance of these two conditions in different age groups in different areas of the Toyama Prefecture is shown in Figure 8:16. Whereas the frequency in age groups up to 45 years has not been different from that found in a nonendemic area, a very clear difference can be seen in age groups of 55 years and older, in both women and men. No information on the number of persons examined is available. Concurrent glucosuria and proteinuria in women of the endemic area can also be seen in Figure 8:17. The women born and spending their lives in the endemic area have had a higher prevalence of proteinuria and glucosuria than women born elsewhere but living in the endemic area at least 20 years. It is not known whether or not the differences are statistically significant.

Further epidemiological studies have been performed in Fuchu (Fukuyama and Kubota, 1972a–1972c) on the relation between the level of heavy metal pollution and proteinuria or glucosuria. Data (Itai-itai Research Committee, 1968) on paddy soil were used for estimating exposure levels of cadmium, lead, and zinc in 12 villages along four different irrigation canals (Figure 8:18). The villages here mentioned were in some cases a combination of two to three hamlets, where populations were small. Fukuyama and Kubota, 1972c, examined women over 40 years of age and achieved a participation of 79.7%. As shown in Figure 8:19, proteinuria (sulfosalicylic acid method) and glucosuria (test tape) were reported to be well correlated to metal concentrations in soil. These analyses were performed on a blind basis (Kubota, K., personal communication).

8.3 EPIDEMIOLOGICAL STUDIES IN OTHER CADMIUM-POLLUTED AREAS OF JAPAN (1969–1973)

In order to map the extent of possible effects of cadmium, the Japanese Ministry of Health and Welfare and the authorities of the prefectures concerned have made during the last years a

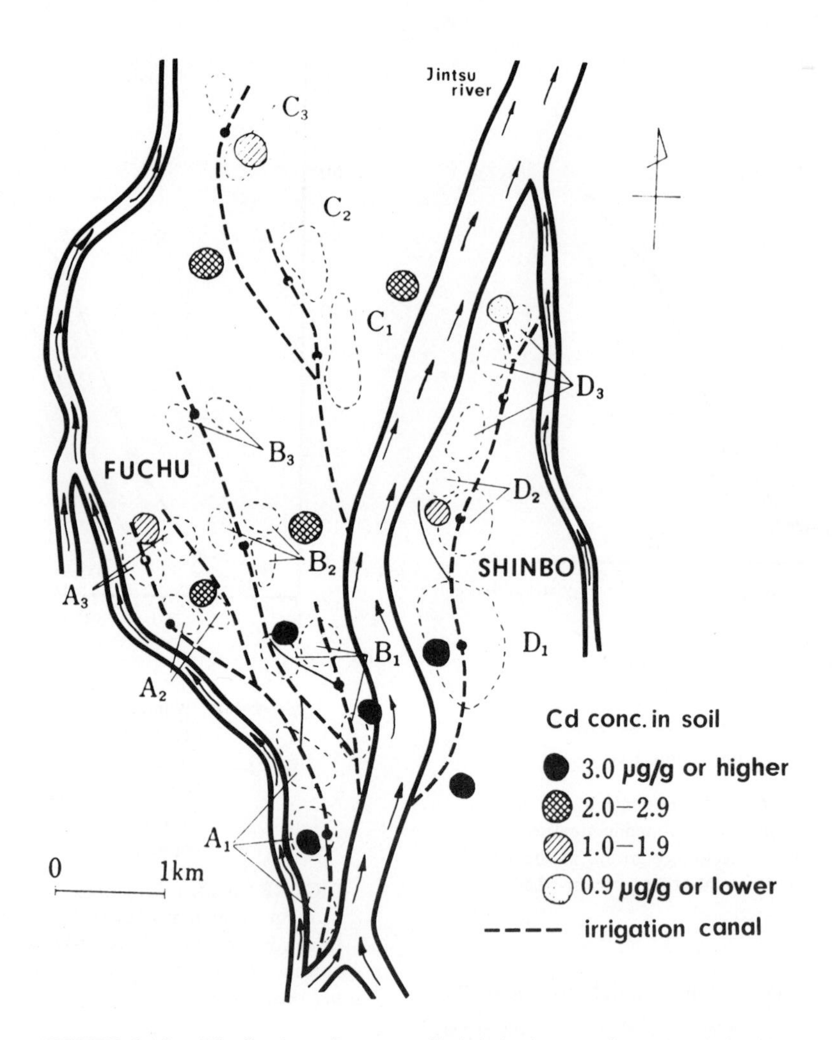

FIGURE 8:18. Distribution of areas studied, irrigation canals, and cadmium in paddy soil. Thin lines show borders of hamlets included in the study. Cadmium concentration in paddy soil is an average of samples from top soil at water inlet, center, and water outlet of paddy. (From Fukuyama, Y. and Kubota, K., *Med. Biol.*, 84, 305, 1972. With permission.)

number of epidemiological studies in at least 13 areas suspected of cadmium pollution.

Data on daily cadmium intake are scarce. However, cadmium concentrations in rice are usually available, since the Ministry of Health and Welfare has ordained that in the areas suspected of cadmium pollution yearly analysis of cadmium in rice should be undertaken. There are generally no data on past cadmium concentrations in rice. Possible changes often have to be inferred from production changes at the cadmium source.

Most of the data on medical effects presented below concern proteinuria, the only one of all possible indicators of an effect that has been measured and reported on consistently. Available data which have been considered of interest for discussing dose-response relationships have been included. The areas mentioned in this section can be spotted in Figure 8:1.

Health screenings in the polluted areas have generally been performed according to standard methods recommended by the Ministry of Health and Welfare. These methods as well as methods for estimating daily intake are presented and discussed in this section.

8.3.1 Methodology

8.3.1.1 Method for Selection of Areas Studied for Cadmium Pollution

The Ministry of Health and Welfare, 1969a, published a standard method for selection of "areas requiring observation of cadmium pollu-

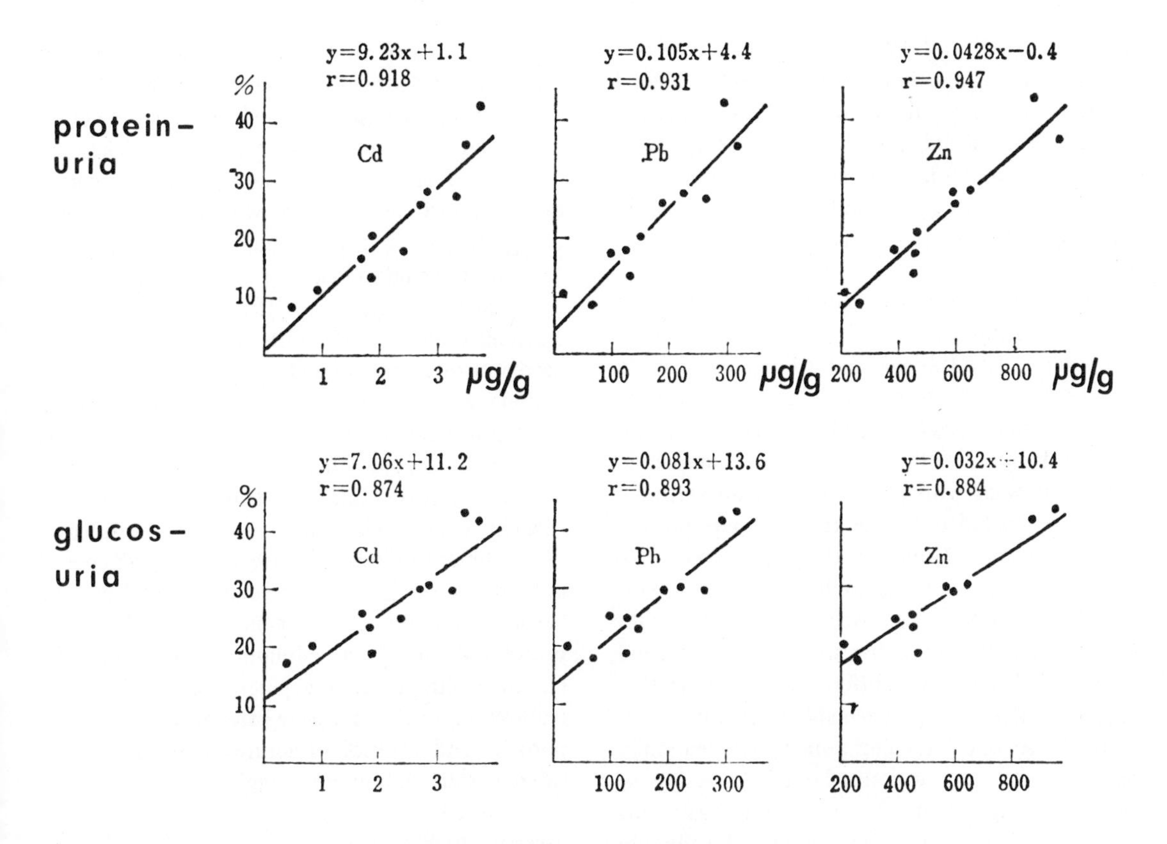

FIGURE 8:19. Village averages of proteinuria and glucosuria in relation to concentration of heavy metals in paddy soil. Regression equations and correlation coefficients. (From Fukuyama, Y. and Kubota, K., *Med. Biol.,* 85, 103, 1972. With permission.)

tion." If within a village values of cadmium concentration in drinking or irrigation water higher than 10 ng/g or values in unpolished rice higher than 0.4 μg/g are found, a closer investigation of environmental cadmium should be conducted. If the average daily cadmium intake in the area could be calculated to be more than 300 μg (Section 8.3.1.5), the village should be designated as an "area requiring observation." In most polluted districts a large number of villages are included in the "area requiring observation" (observation area). Decisions regarding to what extent studies on health effects of cadmium should be carried out are made at the prefectural level. In one study (Ikuno, Section 8.3.3.1) the selection of target areas for health screening was performed by preliminary urine analysis on pooled samples, using 9 μg/l as a limit.

It should be pointed out that in some of the areas discussed in Section 8.3.3 (e.g., Annaka) cadmium concentrations in rice over 0.4 μg/g have been recorded outside the observation area. On the other hand, in large parts of the observation area in Bandai cadmium concentrations in rice do not exceed 0.4 μg/g, according to recorded values.

Control areas "regarded as free from any man-made cadmium contamination" (Ministry of Health and Welfare, 1969a) are chosen on the basis of their similarity to the observation area with regard to population structure, living conditions, climate, etc. As "man-made" cadmium does not include natural cadmium some areas have been used as "controls" in spite of reported average daily cadmium intakes far above the Japanese average.

8.3.1.2 Standard Methods for Screening of Cadmium-related Disease

Two types of health screening have been used between 1969 and 1972. The old type of health screening, used until 1971, was divided into two principal stages, the first of which was a general screening of the population over 30 years of age in defined areas. Included in the first screening were interview, proteinuria analysis (trichloracetic acid), glucosuria analysis (test tape), and blood pressure measurement (Ministry of Health and Welfare, 1969a).

All persons confirmed as having proteinuria in the first screening should have been continued to the second. This rule was not strictly observed in all areas, which is obvious when going through the listings of the individual results (Japanese Association of Public Health, 1970a). For example, in Bandai, persons with glucosuria but not proteinuria were included in the target group for the second screening. Moreover, the participation rate in the second screening could not reach 100%, meaning a difference between the number with proteinuria and the number studied in the second screening. However, the difference is less than 10% in all cases that we have been able to check.

In the second screening a more detailed study on blood and urine was stipulated. The results of this screening would be used for evaluating whether a person had any effects of cadmium exposure.

In May 1971, the Ministry of Health and Welfare published a new standard method for screening cadmium-related disease (Ministry of Health and Welfare, 1971b). The target population was persons over 30 years of age. For qualitative proteinuria measurements both the trichloracetic acid method and the sulfosalicylic acid method were recommended. The new standard method called for measurement of cadmium concentration in urine at the first screening, as well as qualitative proteinuria analysis and blood pressure measurement. Glucosuria measurement was moved to the second screening, where quantitative measurement of proteinuria was to be made also, by any quantitative method as long as it was clearly described. Disc electrophoresis of urinary proteins was to be part of the second screening.

If the disc electrophoresis showed a tubular pattern, a third and final screening was performed. It included a number of blood and urine tests as well as renal function tests and X-ray of certain bones.

The study group for differential diagnosis of cadmium poisoning and Itai-itai disease, set up by the Japanese Association of Public Health, discusses the results of all persons undergoing the third screening and draws a final conclusion. Such a study group is also organized in each prefecture having an "area requiring observation." The conclusions of the national study group are mainly made on an individual basis with possible verdicts being "Itai-itai disease," "cadmium poisoning," or "neither cadmium poisoning nor Itai-itai disease." The diagnostic criteria for Itai-itai disease used in Toyama Prefecture (Section 8.2.3) have not been used in other areas (Tsuchiya, personal communication). No definite criteria for "cadmium poisoning" have been set but the study group consults various clinical specialists and discusses the clinical and laboratory findings from case to case (Tsuchiya, personal communication). As no cases of "cadmium poisoning" have ever been diagnosed as a result of the third screening it is clear that neither cadmium exposure in combination with proteinuria nor even cadmium exposure in combination with proteinuria, glucosuria, renal tubular reabsorption defects, severe osteomalacia or osteoporosis, and multiple fractures have been considered as "cadmium poisoning."

As noted before, the screening procedures were designed to find suspected cases of Itai-itai disease. As the screenings do not primarily aim at detecting the prevalence of proteinuria, the methods of analyzing and classifying proteinuria are not uniform from district to district and in some cases not even within districts. Comparisons among areas must be made with caution.

8.3.1.3 Standard Method for Qualitative Measurement of Proteinuria (Ministry of Health and Welfare, 1971b)

Either one of methods (a) or (b) may be used. It is preferable to combine them.

(a) Trichloracetic acid (TCA) method:

To 5 ml urine 5 ml of TCA solution is added (concentration = 35 g/100 ml). Foaming should be avoided when mixing, and the mixture is heated to 37°C for 10 min.

(b) Sulfosalicylic acid (SA) method:

Three milliliters urine is put into each of two test tubes of equal size. In one of the tubes a drop of a solution of sulfosalicylic acid (20 g/dl) is added. The other tube is used as a control for evaluation.

TABLE 8:10

Result of Comparison of Proteinuria Methods in Two Areas of Toyama

		Polluted area		Control area	
	Total number examined:	165*	(100.0)	225*	(100.0)
Sulfosalicylic acid method	+	21	(12.7)	3	(1.3)
	±	19	(11.5)	4	(1.8)
	–	125	(75.8)	218	(96.9)
Trichloracetic acid method	+	13	(7.9)	5	(2.2)
	±	24	(14.5)	20	(8.9)
	–	128	(77.6)	200	(88.9)
G-25 gel filtration method	+	63	(38.2)	3	(1.3)
	±	47	(28.5)	3	(1.3)
	–	55	(33.3)	219	(97.3)

*Numbers indicate how many people were studied in each group; numbers within parentheses are vertical relations (%).

From Fukuyama, Y., Shiroishi, K., and Kubota, K., *Med. Biol.*, 83, 85, 1971. With permission.

Evaluation: same in both (a) and (b):

Transparent = –
Slight opalescence = ±
Opalescence = +
Intensive opalescence = ++
Milky opalescence = +++
Precipitate = ++++

In the final reports proteinuria has usually only been classified as positive and negative, in some cases using + or higher as the criterion and in other cases ± or higher. Another drawback of most of the epidemiological studies on proteinuria is that the analyses are not made on a "blind" basis. The persons who made the analyses knew whether the urine sample to be analyzed was from an "exposed" area or a "control" area. When studying proteinuria on a nonblind basis with methods relying on subjective evaluations, obvious possibilities for systematic errors exist.

8.3.1.4 Comparisons of Proteinuria Methods

Various methods for detecting cadmium-induced proteinuria have been developed and studied during recent years (see Section 6.1.2.1.1). For epidemiological studies in Japan the trichloracetic acid method, sulfosalicylic acid method, gel filtration, and disc electrophoresis methods have mainly been employed.

It should be mentioned here that a large number of studies on the characteristics and methods for analysis of proteinuria in chronic cadmium poisoning have been performed in Japan by Nomiyama, Tsuchiya, Fukuyama, and M. Fukushima, among others. Such studies were discussed in Chapter 6; only proteinuria methods that have been used in epidemiological studies will be commented upon here.

Fukuyama, Shiroishi, and Kubota, 1971, compared the trichloracetic acid, sulfosalicylic acid, and G-25 gel filtration methods with disc electrophoresis of urine. The studies were made on a blind basis (Kubota, K., personal communication). In the G-25 method (Fukuyama and Kubota, 1970) urine samples are eluated (20 ml/hr in phosphate buffer) on a Sephadex G-25 column (300 × 9 mm), and 2-ml fractions are collected. Folin Lowry reagent is added to each fraction and the absorbance at 750 nm is measured. An elution pattern of Folin Lowry positive substance is achieved and usually three peaks (P, Q, and R) are found, the first of which has been found to consist mainly of proteins (Fukuyama and Kubota, 1970). The G-25 protein quotient is calculated as $P/(P + Q + R)$. The G-25 gel filtration method is rated as + if the G-25 protein quotient is higher than 15%, ± if the quotient is 10 to 14.9% and – if the quotient is less than 10% (Fukuyama, Shiroishi, and Kubota, 1971).

In Table 8:10 results of three methods used in two areas in Toyama are compared. In the polluted district the prevalence of proteinuria rated as + according to the sulfosalicylic acid

TABLE 8:11

Comparison Between Disc Electrophoresis and Other Proteinuria Methods

	Disc (+)		Disc (±)		Disc (–)		Total	
	N	%	N	%	N	%	N	%
	50	30	42	26	73	44	165	100
Sulfosalicylic acid method ⩾ ±	33	83	5	13	2	5.0	40	100
Trichloracetic acid method ⩾ ±	25	68	8	22	4	11	37	100
SA or TCA ⩾ ±	36	68	11	21	6	11	53	100
G-25 gel filtration method +	47	75	11	18	5	7.9	63	100

N = number of persons examined.
% = horizontal relations.

Calculated from a table by Fukuyama, Shiroishi, and Kubota, 1971.

method was slightly higher than that rated + according to the trichloracetic acid method: 12.7 and 7.9%, respectively. The prevalence according to the G-25 method was considerably higher (38.2%). Taking persons with the ratings + and ± together, there would be no difference between the sulfosalicylic acid and the trichloracetic acid methods. On the other hand, in the control district the prevalences of proteinuria (+) according to the sulfosalicylic acid method, trichloracetic acid method, and G-25 method were 1.3, 2.2, and 1.3%, respectively; when the prevalences of ± or higher are considered the figures would be 3.1, 11.1, and 2.7%, respectively. In this area there is a tendency for the trichloracetic acid method to give the highest results. Data on specific gravity are not given, which would have been of value for interpretation of the results. The G-25 method is not dependent on specific gravity, whereas the result of acid tests is. It is conceivable that there are persons with decreased concentrating ability in urine in the polluted area.

Fukuyama, Shiroishi, and Kubota, 1971, also studied urines of persons from the polluted district with the disc electrophoretic method, in which a disc rated + means that the pattern of the electrophoresis is of the tubular type. As can be seen in Table 8:11 about one third (30%) of those studied had a tubular pattern. Among those with proteinuria, as related above, about 70 to 80% had tubular patterns. The G-25 method could "detect" the largest number of those with discs rated + and the trichloracetic acid method the least.

Another study comparing proteinuria methods has been furnished by Kakinuma et al., 1971, in which 2,436 persons from Annaka were studied with both the trichloracetic acid and sulfosalicylic acid methods. A good correlation between the two methods was found (Table 8:12). When using only one of the methods for calculating the prevalence of proteinuria (+ or higher) in an area, the result may be expected to arrive at about 10% lower prevalence figures than if both methods are used (+ or higher in either of them). When setting the limit for proteinuria at ± or higher a combined use of both methods will give 10% higher figures than the trichloracetic acid method alone and less than 1% higher figures than the sulfosalicylic acid method alone (Table 8:12). If there is a bias in the examiner such that he includes all those in the ± group in the + group, the studies by Fukuyama, Shiroishi, and Kubota, 1971, and Kakinuma et al., 1971, indicate that the number of persons registered in the + group would be doubled.

In relation with studies in the separate areas treated in the sections below (especially Ikuno, Section 8.3.3.1), methodology of proteinuria analysis will also be discussed.

8.3.1.5 Methods for Estimating Daily Intake

Cadmium concentration in rice has had to be used as an indicator of daily intake when no other data are available. In some cases more elaborate methods have been employed.

The *National Nutrition Census Method* (Japanese Association of Public Health, 1970b) is

TABLE 8:12

Results of Comparison of Proteinuria Methods in Annaka

	Rating	Trichloracetic acid method (TCA) +	±	–	Total (SA)
Sulfosalicylic acid method (SA)	+	257	39	17	313
	±	31	233	48	312
	–		2	1,809	1,811
	Total (TCA)	288	274	1,874	2,436

Figures indicate the number of persons with different combinations of ratings according to the trichloracetic acid method (35% solution, standard method) and the sulfosalicylic acid method (3% solution, modified method).

Modified from Kakinuma, K. et al., *About the Prevalence of Proteinuria as well as a Result of a Health Examination of the Inhabitants Along the Usuigawa River,* Japanese Association of Public Health, Tokyo, November, 1971.

based on data on the average daily intake of food divided into 22 different groups and the average cadmium concentration in each food group. A questionnaire is sent to a sample of families in the polluted and control areas. They are instructed to record the amount of each of the 22 food groups consumed in the family for 3 days. It is not defined exactly how the families are to be selected. Using figures on the relative calorie requirements in different age groups the family food intake data are adjusted to an "average adult" in the area. From some families included in the questionnaire survey, samples of the different food groups are collected, a procedure as yet systematized only for stored rice and "miso" soybean paste. Cadmium concentrations in the food samples are analyzed with atomic absorption spectrophotometry after extraction in an organic solvent. However, especially in the control areas, only samples from a few food groups were collected in the studies on intake presented in Japanese Association of Public Health, 1970b. Instead, the calculations of daily cadmium intake in control areas were mainly based on values from a control area used in a study in Oita Prefecture (Ministry of Health and Welfare, 1969c).

The *Total Diet Collection Method* (Japanese Association of Public Health, 1970b) requires that the persons participating put a "twin" sample of each foodstuff they eat during 1 day in a container. The material in the container is homogenized and the cadmium concentration is analyzed with atomic absorption spectrophotometry after extraction in an organic solvent.

The *Standard Diet Method* (Yamagata et al., 1971) bases the daily cadmium intake calculation on a specific standard menu recommended by the National Institute of Nutrition. Samples of seven major food groups included in the menu are collected from the area to be studied and analyzed for cadmium concentration. The daily intake is then calculated by multiplying the standard amount of every food group with the measured cadmium concentration.

The *Rice Method*, proposed by us and sometimes used in this review, is an estimation of daily intake based on the assumption that on the average half the daily cadmium intake comes via rice, which corresponds to the figure found by M. Fukushima, 1972, regarding farmers. From the data published in Japanese Association of Public Health, 1970b, it can be calculated that in eight different areas 14 to 71% (average 44%) of the calculated daily intake came via rice. The daily intake of rice is assumed to be the same as the estimated Japanese national average, 300 g (Association of Health Statistics, 1972). The method for selecting rice samples to be collected has varied among areas. A standard method was published by the Ministry of Agriculture, 1971, stating that one sample should be taken in the central part of every 2.5 ha.

The variation in estimated daily intake within an area, as indicated by the range, is a crucial factor when assessing dose-response relationships. The individual intakes in polluted and control

TABLE 8:13

Average Daily Cadmium Intake for Adults

Area		Average rice conc. of Cd (μg/g)	National nutrition census		Total diet collection		
			N	Average μg Cd	N	Average μg Cd	Range μg Cd
Uguisuzawa	Polluted	0.30	513	244.7	6	180	119–227
(Miyagi Pref.)	control[a]	0.11	158	84.5	6	108	78–141
Annaka	Polluted	0.38	676	211.0	6	281	148–485
(Gumma Pref.)	control	0.30	138	113.4	6	99	71–132
Tsushima	Polluted	0.37	567	215.2	8	213	49–533
(Nagasaki Pref.)	control	0.08	284	59.1	8	104	58–192
Kiyokawa	Polluted	0.18	666	221.6	6	391	117–120
(Oita Pref.)	control	0.03	224	79.2	6	–	–

[a]Not the control area used in the epidemiological studies.
N = number of persons examined.

From Japanese Association of Public Health, 1970b. With permission.

areas overlap in Uguisuzawa and Tsushima (Table 8:13). For example, Yamagata et al., 1971, studied the average adult daily cadmium intake with the Total Diet Collection Method in nine families living along the same river, and it varied between 39 and 257 μg.

A further study of the data in Japanese Association of Public Health, 1970b, showed that the daily intake of the foodstuffs included in the National Nutrition Census as well as their cadmium concentration varied to a great extent between the different areas.

8.3.2 Average Daily Intake and Proteinuria Prevalence in Nonpolluted Areas of Japan

8.3.2.1 Cadmium Intake

Using the calculated daily intakes of various foodstuffs according to the 1966 Nutrition Census as well as cadmium concentrations in these foodstuffs (National Nutrition Census Method, Section 8.3.1.5), the Ministry of Health and Welfare, 1969b, estimated that the average daily intake of cadmium would be 60 μg.

Moritsugi and Kobayashi, 1964, analyzed over 200 samples of polished rice from throughout Japan (Section 8.2.5.10) in which they found an average cadmium concentration of 0.066 μg/g wet weight. Using the Rice Method (Section 8.3.1.5) it can be calculated that the corresponding average daily cadmium intake would be 40 μg. The highest prefectural average cadmium concentration in rice was 0.26 μg/g (Tokyo, five samples) and the lowest was 0.010 μg/g (Wakayama, five samples). As seen in Table 8:13, in the control areas used in the epidemiological studies the daily intake was estimated to be 60 to 120 μg using the National Nutrition Census Method and 99 to 108 μg using the Total Diet Collection Method. The rice concentrations of cadmium were 0.03 to 0.30 μg/g, corresponding to 18 to 180 μg daily intake (Rice Method).

M. Fukushima, 1972, analyzed cadmium concentrations in a large number of food samples from stores in Kanazawa City, Ishikawa Prefecture. Using these results, he estimated the average daily intake (National Nutrition Census Method) to be 47 μg. The intake for female farmers was estimated to be 68.5 μg and for male farmers 84.2 μg.

Yamagata et al., 1971, used the Total Diet Collection Method (Section 8.3.1.5) and found an average cadmium intake of 43 μg in Ishikawa Prefecture (eight samples) and 25 μg in seven samples from other prefectures. Yamagata and Iwashima, 1973, reported on a study covering each prefecture of Japan. The total diet collection method was modified so that during 1 day all food from five adults, one from each of five families, was mixed and analyzed. The families were from nonpolluted areas. Out of 45 samples analyzed, with the exception of two which had 149 and 147 μg/g, the range was 3 to 57 μg/g. The calculated average cadmium intake in Japan came to 31 μg. Apart from cadmium, zinc concentration in the ashed samples was analyzed, showing an average zinc intake of 8.1 mg. The correlation coefficient between cadmium and zinc intake was 0.27, which is not significant. The average cadmium intake in 16 farming villages was 25 μg. A large variation in intake among sampled families may occur when using the Total Diet Collection Method (Section 8.3.1.5).

As discussed in Section 3.2, feces analysis may well be a useful method for estimating daily intake. Tsuchiya, 1969, referred to Japanese data from a nonpolluted area where an average of 57 μg/day was found in four persons.

Figure 4:25 shows that reported kidney cortex concentrations of cadmium in three separate studies from nonpolluted areas of Japan are considerably higher than in the U.S.A. and Europe. The reason for the difference is not clear since the average daily intake in nonpolluted areas in Japan does not seem to be considerably higher than in nonpolluted areas elsewhere. Further, due to a lower average body weight in Japan, the daily dose in micrograms per kilogram body weight will be higher. A probable factor of importance is difference in absorption.

8.3.2.2 Proteinuria

In a treatise covering each polluted area (Section 8.3.3), relevant studies on proteinuria in control areas will be discussed. Data from studies not related to polluted areas will be presented in this section.

The Ministry of Health and Welfare, 1964, reported on a study of cardiovascular disease performed in 1961 and 1962. During the latter year, 4,541 persons over 40 years of age were studied for proteinuria. The sulfosalicylic acid method was used for detecting proteinuria, but this was not the standard method as described in Section 8.3.1.3. In a test tube 2.5 ml of urine was mixed with 7.5 ml of sulfosalicylic acid (3 g/100 ml). After 10 min the mixture was checked against a black background; if it displayed "opalescence" the person was rated as having proteinuria.

The prevalence of proteinuria in relation to age, sex, and systolic blood pressure is illustrated in Figures 8:20 and 8:21. Between 161 and 703 persons are included in each group. Prevalence of proteinuria increases with age and is higher among women and persons with high blood pressure. Similar prevalences of proteinuria were found by Freedman et al., 1967, in an epidemiological study on about 5,000 persons from Hiroshima and Nagasaki, which comprised a portion of the clinical sample under study at the Atomic Bomb Casualty Commission. Positive reactions to the methods applied, sulfosalicylic acid or test tape, were confirmed with nitric acid. Results of the study are shown in Table 8:14.

Shiroishi, Tanii, and Kubota, 1972, studied proteinuria in 161 persons (aged 6 to 81 years) from "O-area," a nonpolluted area of Toyama Prefecture, and 105 persons (aged 59 to 92 years) from a home for the aged in Toyama City. The latter persons were also considered to have lived in nonpolluted areas prior to admission to the home. Proteinuria was measured with the Kingsbury-Clark Method, and disc electrophoresis was also performed. The persons from O area were divided into two groups according to age, over and under 60 years. The distribution of different protein concentration levels in urine is shown in Figure 8:22. A protein concentration over 10 mg/100 ml was found in about 5% of both the older and younger persons from O area but in 20% of the persons from the home for the aged. Table 8:15 shows the prevalence of different patterns in disc electrophoresis. Among the young group from O area, 95% had normal patterns as compared with 87 to 89% in the other groups. About 7% had glomerular patterns, both among the old in O area and those from the home for the aged in Toyama. In this latter group 2% Itai-itai patterns were found, but none were found in the other two groups. These data indicate that disc electrophoresis patterns of the tubular type (Itai-itai type) are uncommon among "normal" elderly people. It was not mentioned whether or not the study was performed on a blind basis.

8.3.3 Results of Epidemiological Studies in Cadmium-polluted Areas (1969–1973) (Excluding Toyama Prefecture)

8.3.3.1 Ikuno Area of Hyogo Prefecture

The source of cadmium in this area is a copper and zinc mine which had been in operation for several hundred years and which had had a fairly constant production during the last decades. The

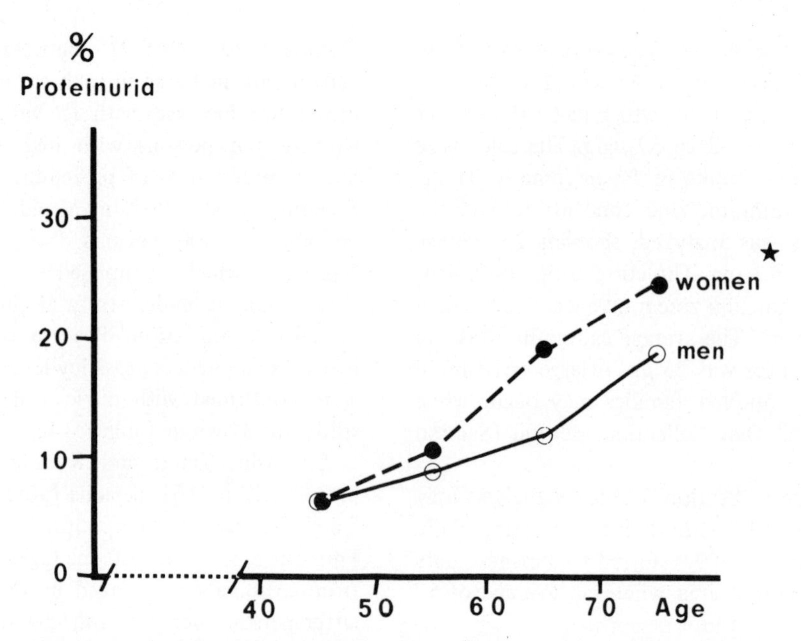

FIGURE 8:20. Age-related prevalences of proteinuria in Japan as recorded in a study on cardiovascular disease. ★: About two thirds of the women were selected for urinary examination. (From M. Fukushima, personal communication, based on data from Ministry of Health and Welfare, 1964.)

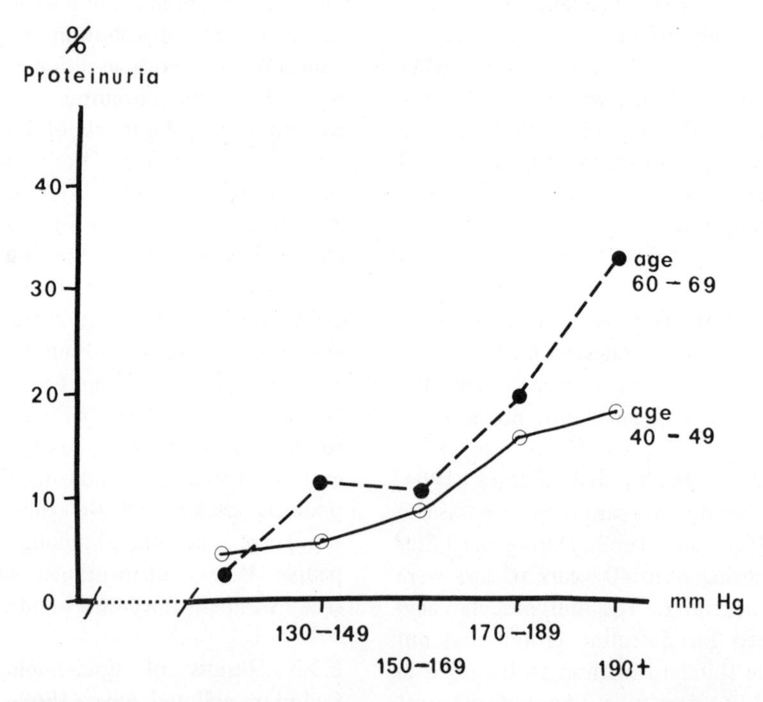

FIGURE 8:21. Systolic blood pressure. Prevalence of proteinuria in relation to blood pressure in Japan as recorded in a study on cardiovascular disease. (From M. Fukushima, personal communication, based on data from Ministry of Health and Welfare, 1964).

TABLE 8:14

Prevalence of Proteinuria Among Men and Women of Different Ages in Hiroshima and Nagasaki (1960–1962)

Age at examination (years)	Male		Female	
	Number tested	Positive tests (percent)	Number tested	Positive tests (percent)
Hiroshima				
10–19	118	6.8	163	3.1
20–29	329	2.4	524	4.4
30–39	708	2.8	1,576	2.5
40–49	501	3.6	1,085	3.5
50–59	678	7.1	1,245	7.2
60–69	616	11.4	837	8.8
70+	219	19.2	275	24.0
Total	3,169	6.8	5,705	5.9
Nagasaki				
10–19	60	3.3	63	1.6
20–29	188	1.6	273	1.1
30–39	321	1.6	769	1.4
40–49	211	3.3	294	3.1
50–59	233	3.4	234	4.3
60–69	174	3.4	134	2.2
70+	33	12.1	47	10.6
Total	1,220	2.9	1,813	2.3

From Freedman et al., 1967.

TABLE 8:15

Prevalence of Different Patterns in Disc Electrophoresis among Persons from Nonpolluted Areas in Toyama Prefecture

Group	Total studied (100%)	N	K	I	X
"O" area Age: 6–59 years	109	104 (95.4%)	2 (1.8%)	–	3 (2.7%)
"O" area Age: 60–81 years	52	46 (88.5%)	4 (7.6%)	–	2 (3.8%)
Toyama City home for aged 59–92 years	105	91 (86.6%)	7 (6.6%)	2 (1.9%)	5 (4.7%)

N = Normal pattern
K = Glomerular pattern
I = Itai-itai pattern
X = Other pattern

From Shiroishi, K., Tanii, M., and Kubota, K., *Med. Biol.*, 85, 197, 1972. With permission.

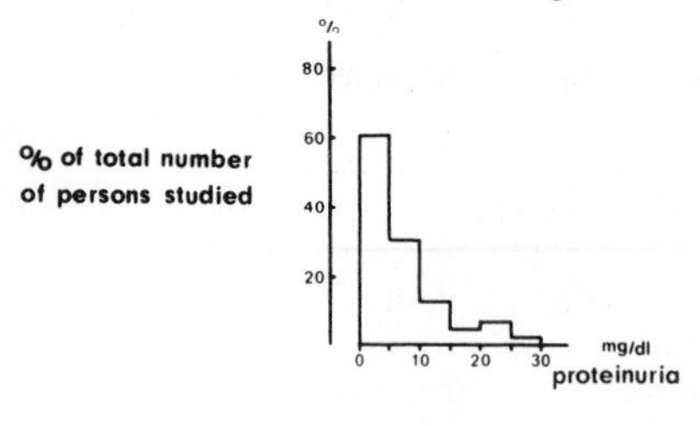

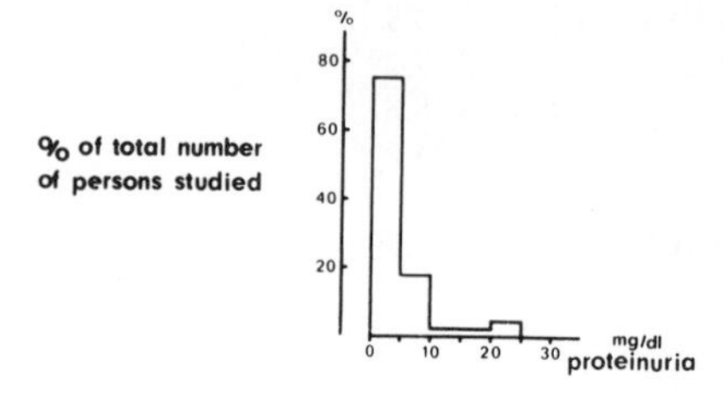

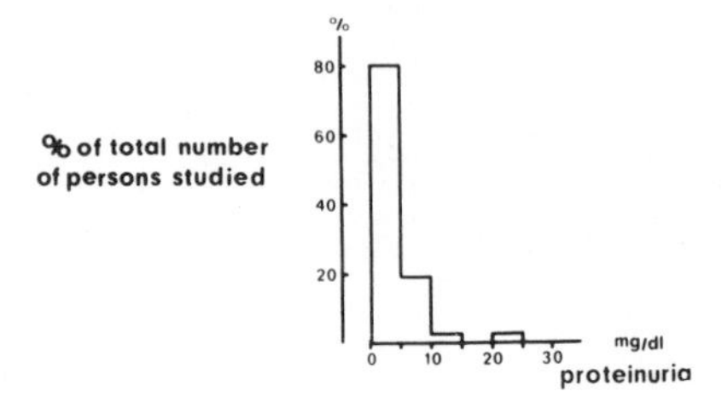

FIGURE 8:22. Distribution of urinary protein concentrations among people who have lived in nonpolluted areas in Toyama Prefecture. Present residence: Toyama City (home for the aged) or "O area." "O area": a city in Toyama Prefecture. (From Shiroishi, K., Tanii, M., and Kubota, K., *Med. Biol.,* 85, 197, 1972. With permission.)

mine ceased operation in 1973. All studies referred to here as authored by Hyogo Prefecture were carried out under the responsibility of Dr. Watanabe, Director of the Hyogo Prefecture Institute of Hygiene.

The first detailed epidemiological study was started in 1971 on persons over 30 years of age living in polluted villages along the Ichikawa River (Hyogo Prefecture, 1972a). The study was performed in three stages. First, a preliminary study was made in April 1971 with the principal aim of establishing average cadmium concentration in urine within villages; 15 villages were designated as "areas requiring health examinations" as the cadmium concentration in pooled urine samples from 30 adults was more than 9 μg/l. The second stage and the final stage consisted of two screenings, one in August–September 1971 and another in October–November 1971. No control areas were included.

In June 1972 the same area was studied once more as well as three other areas with lower cadmium exposure (Hyogo Prefecture, 1972b; Watanabe, 1973, and Watanabe, unpublished data).

In the early spring of 1971 cadmium pollution was detected in the Ikuno area, spurring the Hyogo Prefecture Agricultural Bureau to check rice produced in 1970 and 1971. They reported the average cadmium concentration in 68 samples of stored rice harvested during 1970 in Ikuno town to be 0.85 μg/g. The average concentrations of cadmium in fresh rice from 1971 grown in six towns along the Ichikawa River are given in Table 8:16 and in Figure 8:23 (Hyogo Prefecture, 1972c). The samples were chosen according to a standard method by the Ministry of Agriculture (Section 8.3.1.5). Watanabe, 1973, when analyzing the results of the different studies, divided the area into a highly polluted part having an average of the maximum cadmium concentration in unpolished rice within each one of the included villages of 0.6 to 1.0 μg/g and a less polluted part having an average maximum cadmium concentration in unpolished rice of about 0.3 μg/g.

For the second epidemiological study, June 1972, three areas were tentatively chosen as control areas (Watanabe, personal communication) based on their similarity to the polluted area (Section 8.3.1.1). The cadmium, copper, and zinc concentrations in rice from the three areas are given in Table 8:17. The reported average cadmium concentrations within these areas are below 0.066 μg/g, the Japanese average (Moritsugi and Kobayashi, 1964).

In Ikuno no studies on daily cadmium intake have been performed. Based on the Rice Method (see Section 8.3.1.5) applied to data in Table 8:16, it can be calculated that in Iluno (I) town the average daily intake would be around 410 μg (range = 90 to 860 μg) and in Kamisaki (K) town 290 μg (range = 90 to 600 μg). As mentioned in Section 8.3.1.5, these estimates are rather rough.

The observation area in the 1971 epidemiological study (Hyogo Prefecture, 1972a) was selected according to a modified standard method (Section 8.3.1.1) as indicated earlier. In these villages 1,344

TABLE 8:16

Cadmium Concentration in Rice from the Ikuno Area, 1971

Area	Number of rice samples	Cadmium concentration (μg/g wet wt) Maximum	Minimum	Average
Ikuno town (I)	23	1.43	0.15	0.69
Ookochi town (O)	57	1.15	0.01	0.35
Kamisaki town (K)	30	0.98	0.15	0.48
Ichikawa town (IC)	20	0.49	0.01	0.19
Fukusaki town (F)	5	0.37	0.03	0.15
Kooji town (KO)	10	0.72	0.06	0.27

From Hyogo Prefecture, The General Situation as Regards Countermeasures Against Pollution of Cadmium Etc. in the Vicinity of the Mine in Ikuno, June 1972. With permission.

TABLE 8:17

Heavy Metal Concentrations (Mean and Range) in Rice from Three Areas Tentatively Chosen as Control Areas in Hyogo Prefecture

Area	Number of samples	Cadmium (μg/g)	Copper (μg/g)	Zinc (μg/g)
"A"	14 unp.[a]	0.057	2.67	17.3
Kasuga town		(0.020–0.138)	(1.95–3.11)	(7.8–23.3)
	5 p.[b]	0.038	1.95	11.7
		(0.024–0.071)	(1.23–2.39)	(9.1–14.6)
"B"	21 unp.	0.055	2.31	16.6
Nishiki town		(0.022–0.142)	(0.92–4.47)	(10.8–24.8)
"C"	1 unp.	0.029	1.48	16.1
Nikata town	6 p.	0.017	1.77	15.7
		(0.005–0.029)	(1.04–2.20)	(12.2–17.7)

[a]unp. = unpolished rice
[b]p. = polished rice

From Watanabe, unpublished data.

of the 1,700 persons over 30 years of age were examined in August–September 1971 with a modified first screening for Itai-itai disease (cf. Section 8.3.1.2) consisting of interviews, measurement of blood pressure, and analysis of glucosuria and proteinuria (trichloracetic acid method). On the trichloracetic acid test 22.7% had a rating of + or higher in the first screening.

The 1972 epidemiological study was performed on the same target population (Hyogo Prefecture, 1972b). This time three other areas were also examined. Area "A," Kasuga town, coincides with the control area used by Watanabe, 1973, in his evaluation of the studies in 1971 and 1972. The areas "B" and "C" were not included as control areas for administrative reasons (Watanabe, personal communication). Proteinuria in the two areas was analyzed by one laboratory technician and proteinuria in area "A" and Ikuno by another (Hyogo Prefecture, 1972b, and Watanabe, personal communication). The method for proteinuria analysis in 1972 was the sulfosalicylic acid method, which gave the results shown in Table 8:18 (Hyogo Prefecture, 1972b). The polluted area (Ikuno) has a strikingly higher prevalence of proteinuria than areas "B" and "C" and the prevalence is also considerably higher than in area "A." The studies were not carried out on a blind basis. Within area "A" a number of very small open pit manganese mines have been in operation (Watanabe, personal communication). Watanabe (unpublished data) studied 313 persons living in the "mine village" to see if they had a higher prevalence of proteinuria than others. He found

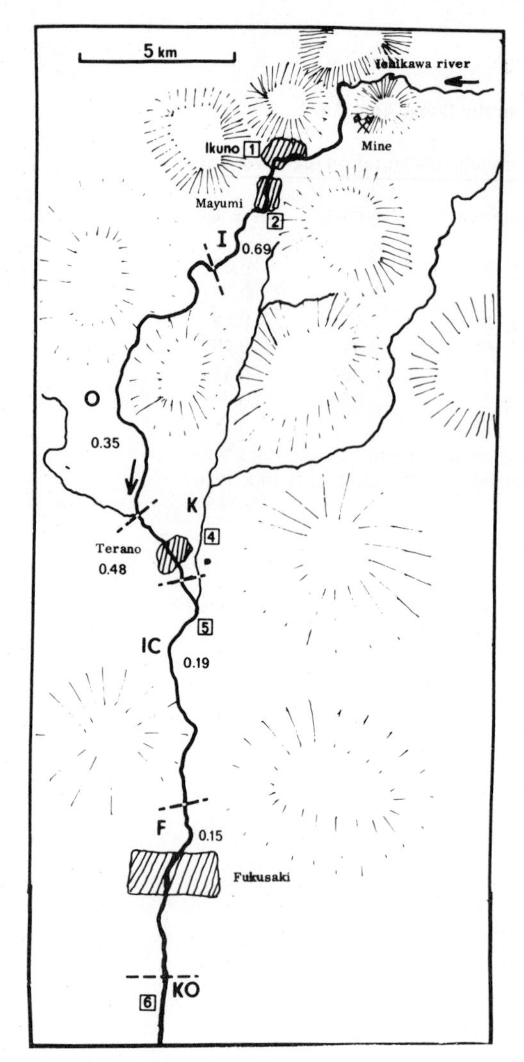

FIGURE 8:23. Ichikawa River and polluted places in Ikuno area with indications (□) of residence of suspected Itai-itai patients and concentrations of cadmium (μg/g wet weight) found in rice. ------: town boundaries (I, O, K, IC, F, KO). Patient 4 is Patient I in Table 8:15. See Table 8:16 for explanation of area name abbreviations. (Data from Hyogo Prefecture, 1972c and Ishizaki, unpublished data).

that 35% had + or higher by the sulfosalicylic acid method and 9% ±, as compared to 33 and 8%, respectively, in area "A" as a whole.

In order to further evaluate differences in prevalence of proteinuria between areas, Watanabe, 1973, studied an age- and sex-matched sample of 376 persons from the Ikuno area and 250 persons from the Kasuga area. It is not clear how the sample was selected. Proteinuria analysis was performed on fresh urine and thawed frozen urine (Table 8:19). The higher prevalence of proteinuria in the polluted area compared to the nonpolluted area when analyzing fresh urine is obvious. This study was not carried out on a blind basis. No clear-cut difference is seen between polluted and control areas when frozen samples are compared. This latter comparison was carried out on a blind basis.

Watanabe, 1973, also studied proteinuria with the Biuret method, using trichloracetic acid for precipitation, on the same age- and sex-matched sample. The results can only be judged as semiquantitative since values under about 10 mg/100 ml are unreliable with this method of precipitation. However, the distribution of different protein concentrations was depicted as in Figure 8:24. The geometrical average protein concentration in urine was calculated as 2.2 mg/100 ml in the polluted Ikuno area and 2.4 mg/100 ml in the Kasuga control area. Because of the semiquantitative character of the method these general averages cannot be used for epidemiological analysis. If Watanabe's data as seen in Figure 8:24 are further scrutinized, differences between the two groups are revealed. In the range under 15 mg/100 ml no obvious differences are seen. However, above 15 mg/100 ml a larger proportion of persons from the polluted area is found, i.e., about 11% as estimated from the figure, compared to about 5% from the nonpolluted area. This difference is statistically significant.

Watanabe et al., 1972, have shown that the globulin/albumin ratio as measured with disc electrophoresis of urine can be found to be elevated among persons living in an area with high cadmium exposure. The ratio was calculated by dividing the total area of the globulin peaks in the disc electrophoresis pattern with the area of the albumin peak. An elevated globulin/albumin ratio reflects tubular proteinuria, and Watanabe et al., 1972, suggest it as an index for the screening of cadmium-induced proteinuria.

Urinary cadmium and globulin/albumin ratios in persons, probably with proteinuria, living in six villages with different levels of cadmium exposure have been compiled in Table 8:20 from Watanabe. He did not state the age distribution of the persons involved. There was a significant difference in globulin/albumin ratios between persons living in villages with 0.33 to 0.35 μg/g average cadmium concentrations in rice and in those with 0.61 to 1.10 μg/g. Urinary cadmium concentrations increased distinctly with increasing cadmium in rice.

The relation of the globulin/albumin ratio to

TABLE 8:18

Prevalence (%) of Different Degrees of Proteinuria in Areas of Hyogo Prefecture. Persons of Both Sexes, All Over 30 Years. Sulfosalicylic Acid Method

Area	Target population	Persons examined	Rating		
			+	±	–
Ikuno	1,699	1,560	58	5	37
Area "A" Kasuga	1,935	1,574	33	8	59
Area "B"*	2,514	2,002	4	5	91
Area "C"*	771	638	9	2	89

*Laboratory technician performing these analyses was not the same as in Area "A" and Ikuno.

From Hyogo Prefecture, Results of a Study on 13 Inhabitants of the Surroundings of the Ikuno Mine. Report to the Committee for Differential Diagnosis of Itai-itai Disease and Cadmium Poisoning, September 6, 1972. With permission.

TABLE 8:19

Prevalence (%) of Different Degrees of Proteinuria in an Age- and Sex-matched Sample from Ikuno and Kasuga*

Method	Area	Rating			
		–	±	+	++ or higher
Sulfosalicylic acid method on fresh urine	Ikuno	39	4	34	25
	Kasuga	62	8	24	6
Sulfosalicylic acid method on frozen urine after thawing†	Ikuno	60	7	18	16
	Kasuga	66	10	18	7
Trichloracetic acid method on frozen urine after thawing†	Ikuno	77	6	10	8
	Kasuga	86	2	9	3

*Number of persons studied: Ikuno – 376; Kasuga – 250.
†Blind tests.

Modified from Watanabe, 1973.

age and level of pollution has been studied with regard to people with proteinuria using the sulfosalicylic acid method (Table 8:21). The ratio increased significantly with age and was in each age group about twice as high in the highly polluted area (towns 3 to 6 in Table 8:20) as in the slightly polluted area (towns 1 and 2 in Table 8:20). The differences in the globulin/albumin ratio between the areas with different pollution levels were also significant in all age groups. According to Watanabe (personal communication) some people in the control area, Kasuga, had a globulin/albumin ratio up to about 1.5, but the majority had a G/A ratio under 0.5.

In June 1972, Watanabe (unpublished data) studied the retinol binding protein (RBP) reaction in urine in the highly polluted area and in the control area (Kasuga). Among 829 persons with proteinuria studied in the former area, 51 showed increased excretion of RBP; among 507 persons with proteinuria from the control area only 3 showed such an increase.

Watanabe, 1973, reported the average urinary cadmium excretion in different villages within the polluted area (Figure 8:25). Among school children, no increase in urinary cadmium concentration related to an increase in average cadmium concentration in village rice was found. On the other hand, average urinary cadmium concentration among adults over 30 years of age was seen to

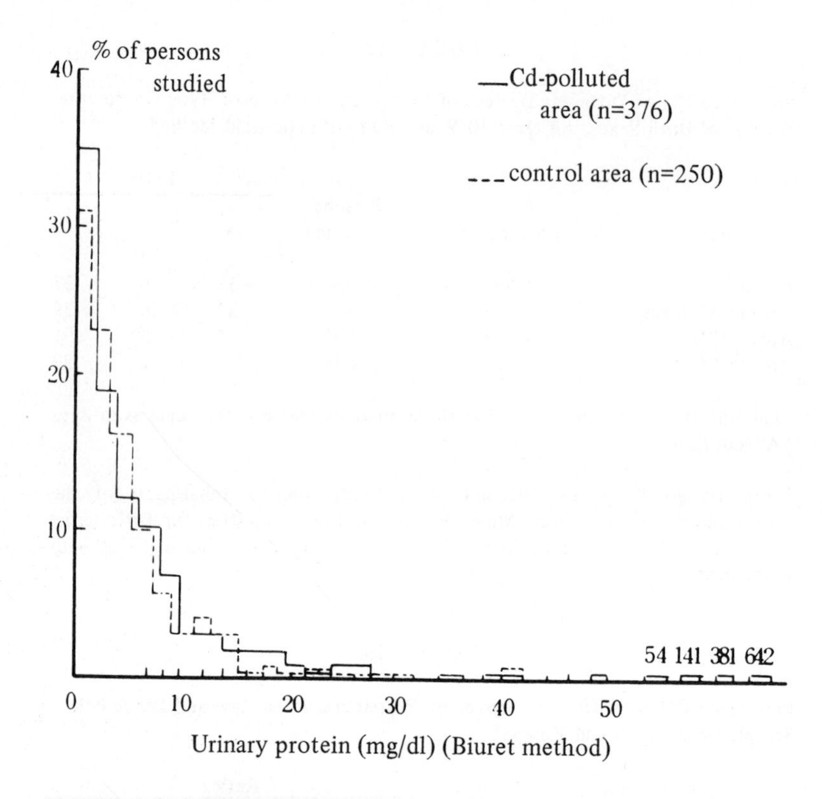

FIGURE 8:24. Distribution (%) of persons according to protein concentration in urine in polluted and control districts in Ikuno area. (From Watanabe, H., A Study of Health Effect Indices in Populations in Cadmium-polluted Areas, presented at Meeting on Research on Cadmium Poisoning, Tokyo, March 25, 1973. With permission.)

TABLE 8:20

Relationship Between Cadmium Concentration in Rice, Urinary Cadmium Concentration, and Globulin/Albumin Ratio

Towns	Cadmium in rice (μg/g)	Number of urine samples	Urinary Cd (μg/l); mean ± S.D.	Globulin/albumin ratio; mean ± S.D.
1 (S.P.*)	0.33	35	7.5 ± 5.2	0.75 ± 0.61
2 (S.P.)	0.35	34	8.0 ± 5.1	0.81 ± 0.58
3 (H.P.)†	0.61	49	9.4 ± 5.3	1.80 ± 0.91
4 (H.P.)	0.88	50	12.9 ± 6.1	1.79 ± 0.84
5 (H.P.)‡	0.91	27	13.3 ± 9.0	1.41 ± 0.95
6 (H.P.)	1.10	55	13.8 ± 10.7	1.97 ± 0.82

*S.P. = slightly polluted.
†H.P. = highly polluted.

‡Town 5 is along another river polluted by the same mine but with a shorter exposure time.

From Watanabe, unpublished data.

increase strongly with an increase in average cadmium concentration in village rice. No such correlation for prevalence of proteinuria (trichloracetic acid method) was reported.

Cadmium concentration in urine among persons with proteinuria was in each age group higher in the highly polluted area (age group averages between 11.50 and 14.04 μg/l) than in the slightly polluted area (age group averages between 6.00 and 10.29 μg/l) (Watanabe, 1973, and Watanabe, unpublished data). No increase with age was found.

Watanabe (unpublished data) also compared the prevalence of glucosuria between the highly polluted area and the control area (Table 8:22). There was a tendency for slightly higher prevalence of glucosuria in highly polluted areas, a difference statistically significant only in women of ages 60 to 79.

In April 1971 certain actions were taken to lessen the amount of cadmium ingested in the polluted area (Watanabe, unpublished data). The globulin/albumin ratio was studied in a large number of subjects both in October 1971 and June 1972 (Tables 8:23 and 8:24). The number of persons with a globulin/albumin ratio higher than 1.55 decreased during that interval. Watanabe, 1973, stated that this might be an effect of decreased cadmium exposure.

The epidemiological studies by Hyogo Prefecture, 1972a and 1972b, resulted in the selection of

TABLE 8:21

Comparison of Globulin/Albumin Ratio by Age Between Highly and Slightly Polluted Areas in Protein-positive Persons Examined in October 1971

Age	Slightly polluted area		Highly polluted area	
	N	G/A ratio (mean ± S.D.)	N	G/A ratio (mean ± S.D.)
30–39	8	0.47 ± 0.17	16	0.93 ± 0.17
40–49	11	0.62 ± 0.46	19	1.24 ± 0.88
50–59	8	0.55 ± 0.26	23	1.56 ± 1.07
60–69	18	0.70 ± 0.48	47	1.75 ± 0.72
70–79	17	0.90 ± 0.60	62	2.21 ± 0.67
80–89	6	1.43 ± 0.47	13	2.02 ± 0.74

From Watanabe, unpublished data.

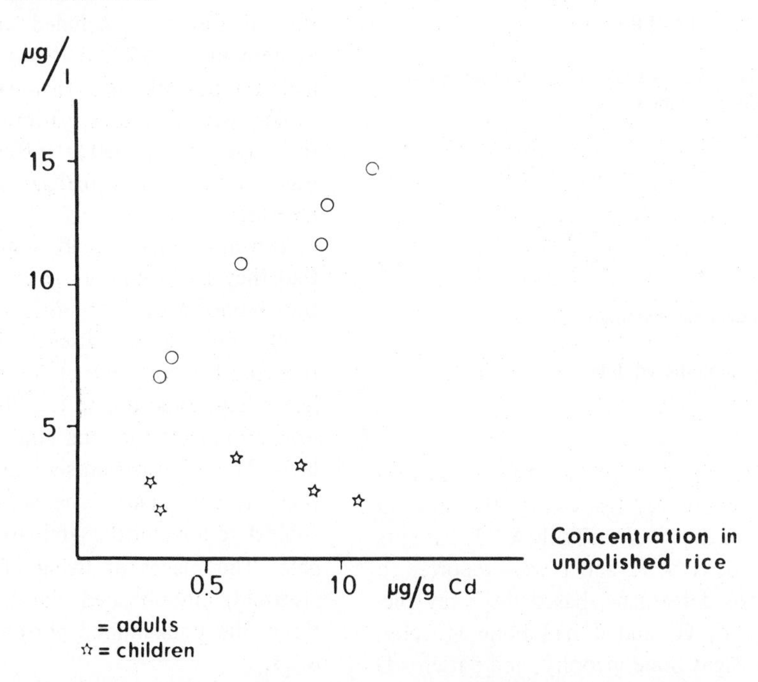

FIGURE 8:25. Village average urinary cadmium concentrations among children and adults in relation to village average rice concentrations. (Redrawn from Watanabe, H., A Study of Health Effect Indices in Populations in Cadmium-polluted Areas, presented at Meeting on Research on Cadmium Poisoning, Tokyo, March 25, 1973. With permission.)

TABLE 8:22

Comparison of Positive Glucosuria Rates (%) by Age and Sex Between Highly and Nonpolluted Areas. Persons Examined June 1972

		Age			
		40–59		60–79	
Sex	Area	N	Positive rate	N	Positive rate
Male	Highly polluted	348	5	202	9
	Nonpolluted	340	4	245	7
Female	Highly polluted	405	2	252	11
	Nonpolluted	445	2	244	6

From Watanabe, unpublished data.

TABLE 8:23

Follow-up Study of Prevalence of Different Globulin/Albumin Ratios at Two Different Times

		Globulin/albumin ratio		
Time of examination	N*	1.55 and over	1.54 – 1.0	Below 1.0
October 1971	164	111 (68%)	15 (9%)	38 (23%)
June 1972	164	44 (27%)	13 (8%)	107 (65%)

*N = Number of persons examined.

From Watanabe, unpublished data.

TABLE 8:24

Comparison of Prevalence of Globulin/Albumin Ratios ⩽ 1.55 at Two Different Times

Time of examination	N*	%
October 1971	1,295	6.8
June 1972	1,560	3.9

N* = Number of persons examined.

From Watanabe, unpublished data.

13 persons for the third screening (Section 8.3.1.2). For reference, laboratory findings on these persons are depicted in Table 8:25. Patients A, B, C, D, F, G, H, J, K, and L were reported to have spondylosis deformans, based on X-ray pictures. Patients C, G, and J had bone atrophy, patient K had slight bone atrophy, and patients D and I had considerable bone atrophy. Case I also had a history of multiple fractures. The number of pregnancies and other socioeconomic data on the persons were not reported. The study group for differential diagnosis of cadmium poisoning and Itai-itai disease concluded at their meeting on September 6, 1972, that none of the 13 persons had Itai-itai disease. However, concerning one subject with fractures, patient I, some members of the study group had a different opinion, which was noted in the official statement from the meeting.

Ishizaki, Hagino, and Shibata, 1973, reported that they had found an additional five patients in the Ikuno area with a clinical syndrome very similar to Itai-itai disease. Detailed data were reported for only one of the patients. The patient had a high exposure to cadmium, tubular proteinuria, aminoaciduria, and clinical signs of osteomalacia. The other patients are difficult to evaluate at present, particularly one who has not yet been subject to a medical examination, which is remarkable. The places of living of suspected patients (Ishizaki, unpublished data), excluding number three, the unexamined person, are seen in Figure 8:23.

In summary, several patients who have displayed a clinical picture similar to that of the Itai-itai disease have been found in the Ikuno area. Whether or not these symptoms are cadmium induced is not possible for us to evaluate but must

TABLE 8:25

Laboratory Findings on 13 Persons from Ikuno Area Studied in the Third Screening

Person identification	A	B	C	D	E	F	G	H	I	J	K	L	M
Sex	M	F	F	F	M	F	F	M	F	F	F	M	F
Age	81	74	72	69	82	57	80	73	75	76	74	56	61
Blood pressure	–	150/96	120/80	136/82	134/68	124/76	102/50	192/94	120/60	142/100	142/80	196/80	116/70
[a]Proteinuria	3/3	4/4	4/4	2/3	2/3	2/4	1/3	2/4	4/4	1/3	0/4	0/5	4/4
[b]Proteinuria (mg/day)	936	225	543	506	636	283	525	427	744	619	174	97	649
[b]Aminoaciduria (mg/day)	230	217	217	293	236	184	294	183	416	268	156	138	245
[a]Glucosuria	3/3	0/4	2/5	0/3	3/3	2/5	3/3	1/5	4/4	3/3	0/4	0/5	5/5
[b]Cd-uria (μg/day)	12.6	14.3	22.4	17.6	35.0	25.0	19.4	29.3	10.5	33.8	11.9	11.9	24.4
[c]Phosphaturia (g/day)	0.54	0.48	0.48	0.57	0.41	0.44	0.50	0.38	0.68	0.78	0.47	0.27	0.32
[c]Urine calcium (g/day)	0.17	0.13	0.20	0.24	0.14	0.33	0.27	0.19	0.06	0.40	0.16	0.18	0.09
Urine osmotic pressure mosm/l	440	672	588	552	435	538	566	513	380	515	415	360	280
PSP test 15 min / 30 min		15 / 40	15 / 30			15 / 35		15 / 30	5 / 10		15 / 25	10 / 15	10 / 25
60 min / 120 min		65 / 80	45 / 60			50 / 60		45 / 65	20 / 40		40 / 55	20 / 25	35 / 45
Phosphorus reabsorption test (%)		92.0	76.6			82.5		85.2	66.1		90.4	84.2	79.0
Serum calcium (mg/dl)		11.5	9.8	8.5		9.3	8.3	8.6	9.2	9.2	9.8	7.8	8.7
Alkaline phosphatase (Jan 72)	17.4	14.6	16.4	11.6	7.5	15.3	15.3	12.1	70.7	22.6	11.6	37.3	26.5
Alkaline phosphatase (King-Armstrong) (May 72)		9.6	10.0	10.5		12.0	12.0	8.3	42.0	12.0	9.6	18.5	18.5
Blood glucose (mg/dl)		69	63	86		64	73	84	98	74	69	76	76

[a]Frequency of tests + or higher. Proteinuria: trichloracetic acid or sulfosalicylic acid methods. Glucosuria: test tape.
[b]In some cases average of two measurements (Tsuchiya-Biuret method for proteinuria).
[c]Average of two or three measurements.

Modified from Hyogo Prefecture, 1972b.

be considered an open question. Independent of the etiology, the number of patients on a percentage basis must be very small.

8.3.3.2 Tsushima in Nagasaki Prefecture

After the theory that cadmium poisoning was the etiology of Itai-itai disease had been put forward, Kobayashi and Hagino, among others, visited other cadmium-polluted areas of Japan in search of similar patients. In 1964 they found two patients with similar symptoms in a village downstream from a mine on Tsushima Island (Kobayashi, 1971).

Preliminary medical studies were carried out by the local health authorities in 1965. The results showed that Kashine, downstream from a mine in the River Sasu basin, was a most suspect area. New studies were performed in 1968 and, on a more comprehensive basis, in 1969 (Nagasaki Prefecture, 1970). The medical examinations were carried out by the staff of the Section of Health, Izuhara. Urine analyses were made at the Laboratory of Hygienic Chemistry, Faculty of Pharmaceutical Sciences, Nagasaki University.

Mining in the area dates as far back as 674 A.D. but large-scale production did not start before World War II. The location of mines and villages studied is shown in Figure 8:26. The volume of zinc production increased from 1,000 tons in 1948 to 21,000 tons in 1958 and has been relatively stable (22,000 to 29,000 tons) since then (Terada, personal communication).

The area under investigation consists of a control area, the village Are in the basin of the Are River and the village Hikage-Kamiyama, and an observation area, villages in the basins of the rivers Sasu and Shiine, Shimobaru (including Tokoya and Nagon), Komoda (including Komoda-hama), Shiine (including Shiine-hama), and Kashine (including Uragochi). This area can be seen in Figure 8:26. In the area under observation, 653 people over 40 years of age had been living there for at least 5 years. The corresponding number of the control area was 176. Out of a total of 829 subjects, 541 were medically examined in 1969.

Similar investigations were performed in 1970, 1971, and 1972. Unfortunately, neither the methods nor the presentation of results were the same. Only parts of the control area used in 1969 were included in the later studies. In 1972 no control

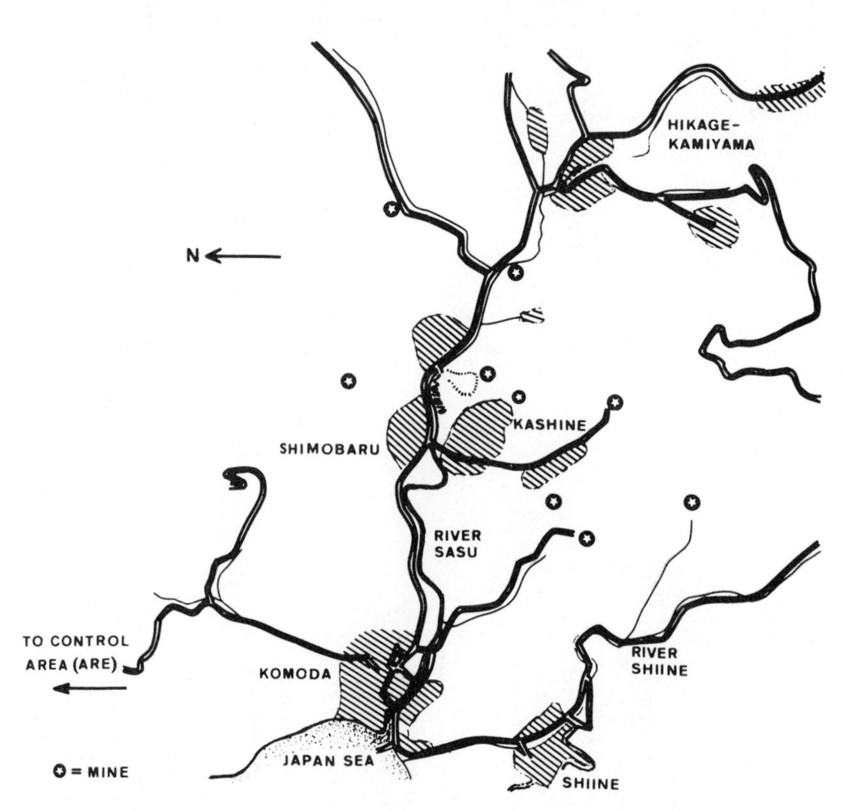

FIGURE 8:26. Area under investigation, Tsushima Island. (From Nagasaki Prefecture, 1970.)

areas came under consideration, according to the report available to us (Nagasaki Prefecture, 1972a). The studies in the Tsushima area were not performed on a blind basis (Terada, personal communication).

According to five different studies the rice levels of cadmium were at least six times higher in the polluted areas (Kashine, Shimobaru, Komoda, Shiine) than in the control area (Are) (Table 8:26). The villages Kashine and Shiine tended to have the highest levels, and the same tendency was found as regards urinary levels; they are therefore treated together in some tables on effects below. The average daily cadmium intake for adults was calculated (Japanese Association of Public Health, 1970b) with the National Nutritional Census Method to be 215 μg in the polluted villages and 59 μg in the control village Are (Table 8:13). The Total Diet Method on a sample of eight persons from each of the polluted areas and the control area gave, respectively, 213 and 104 μg. Thus this control area should have had a higher daily cadmium intake than the Japanese average (Section 8.3.2.1).

In the 1969 study only persons over 40 years of age were examined. Participation in the observation area was about 70%, and in the control area about 50%. It was not stated in the report (Nagasaki Prefecture, 1970) whether or not the examined and nonexamined persons differed from each other in any way.

The investigation included an interview, urinary protein (test tape) and glucose (test tape) determinations, X-ray examination of the chest, and analyses of urinary content of cadmium. The protein and glucose examinations were performed on 522 subjects; results are given in Table 8:27. There is a considerably higher prevalence of both proteinuria and glucosuria in the Kashine, Shiine area than in the other areas.

Data for calculating the age-related prevalences of proteinuria on a village basis (Figure 8:27) are available only as regards the study in 1969. It must be stressed, however, that the number of persons examined (Table 8:28) in some groups was extremely small, which invalidates any precise interpretation of the data. Figure 8:27 indicates a difference between the two villages with the highest pollution level (Kashine, Shiine) and the other villages inasmuch as the prevalence of proteinuria generally is higher and increases with age in Kashine and Shiine.

In Nagasaki Prefecture, 1970, individual data on urinary cadmium concentration (atomic absorption after extraction) were also reported (Figure 8:28) regarding persons with proteinuria. Cadmium levels were between 10 and 30 μg/l in the majority of cases in Kashine and elevated

TABLE 8:26

Concentration of Cadmium in Unpolished Rice from Different Villages of Tsushima

Village	Range of reported average cadmium concentrations from 1969–1970 (μg/g)[a]	Averages, October 1970[b]
Polluted		
Kashine	0.38–0.89	0.60 (58)[c]
Shimobaru	0.30–0.50	0.42 (39)
Komoda	0.36–1.32	0.50 (34)
Shiine	0.44–0.60	0.58 (52)
Control		
Are	0.05[d]	

[a]Results from five different studies.
[b]Average of results from a study by the Ministry of Agriculture and Prefectural Agriculture Center, October 1970, and a study by the Ministry of Health and Prefectural Pollution Bureau, October 1970.
[c]Numbers in parentheses indicate number of samples.
[d]Only studied in 1969.

Compiled from Nagasaki Prefecture, 1971.

TABLE 8:27

Prevalence of Proteinuria and Glucosuria

Area	Males				Females			
	Number examined	Percent with protein-uria	Percent with glucos-uria	Percent with protein-uria and/or glucos-uria	Number examined	Percent with protein-uria	Percent with glucos-uria	Percent with protein-uria and/or glucos-uria
Kashine, Shiine	73	14	12	23	96	14	10	17
Shimobaru, Komoda	99	3	3	6	101	5	4	7
Hikage-Kamiyama, Are	46	0	4	4	107	6	0	6

From Nagasaki Prefecture, 1970.

TABLE 8:28

Age Distribution of Examined Subjects (Actual Numbers) in Different Areas)

Area	Males				Females			
	Age: 40–49	50–59	60–69	70+	40–49	50–59	60–69	70+
Kashine, Shiine	30	22	15	7	36	30	14	17
Shimobaru, Komoda	46	29	16	11	60	24	17	10
Hikage-Kamiyama, Are	19	10	14	3	50	27	26	8

From Nagasaki Prefecture, 1970.

above normal (Section 4.3.3.1) in most cases in the other villages, too.

In 1970 547 persons over 30 years of age from the polluted area (participation 43% of the target population) and 120 persons from a control village (Hikage-Kamiyama participation 27%) were studied (Nagasaki Prefecture, 1971). The reported prevalence of proteinuria by the trichloracetic acid method showing + or higher, according to Terada (personal communication), was 4.9 and 0.8%, respectively.

Takabatake, Keshino, and Matsuo, 1972, reported on urinary cadmium concentrations in Kashine and Hikage-Kamiyama (Table 8:29) studied in connection with the 1970 epidemiological investigation. For both male and female farmers the average cadmium concentration was about five times higher in the polluted village than in the nonpolluted areas. Concerning nonfarmers, the average was only about two times higher in the polluted village. These differences were statistically significant for three of the four groups. No difference can be seen between men and women nor between farmers and nonfarmers within the control area.

In 1971 882 persons in the same age groups were examined in the polluted area (participation 81%), and 5.3% had proteinuria (Nagasaki Prefecture, 1972b). In the control area (Hikage-Kamiyama) 177 were examined (participation 43%) and 4.5% had proteinuria (trichloracetic acid method) (Nagasaki Prefecture, 1972b). In Table 8:30 a comparison is made between prevalence of proteinuria in the villages with the highest cadmium levels in rice (Kashine, Shiine), in villages with elevated levels (Shimobaru, Komoda), and in the control village (Hikage-Kamiyama). The first villages have consistently about twice as high a prevalence as Shimobaru and Komoda or the control village.

Age-related proteinuria prevalences from the studies in 1969 to 1972 as reported by Terada (personal communication) generally reveal an increase with age. Since the control groups studied

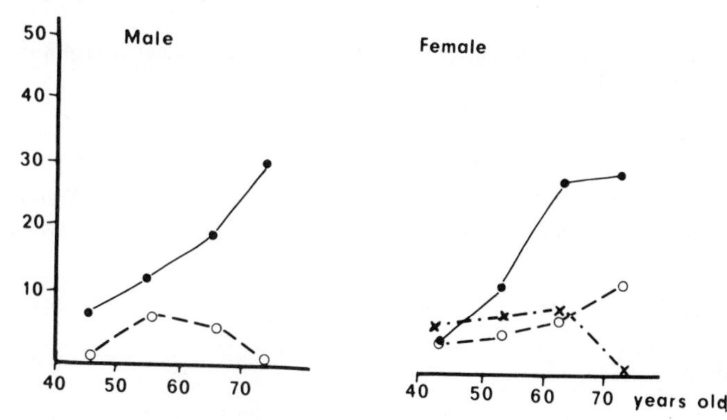

•——• most polluted villages, Kashine, Shiine

o– – –o less polluted villages, Shimobaru, Komoda

x·—··x control villages, Hikage-Kamiyama, Are (No cases of proteinuria found among males)

FIGURE 8:27. Prevalence of proteinuria in different areas in men and women in different age groups. Number of persons studied is small in some age groups, as shown in Table 8:28. (Based on data from Nagasaki Prefecture, 1970.)

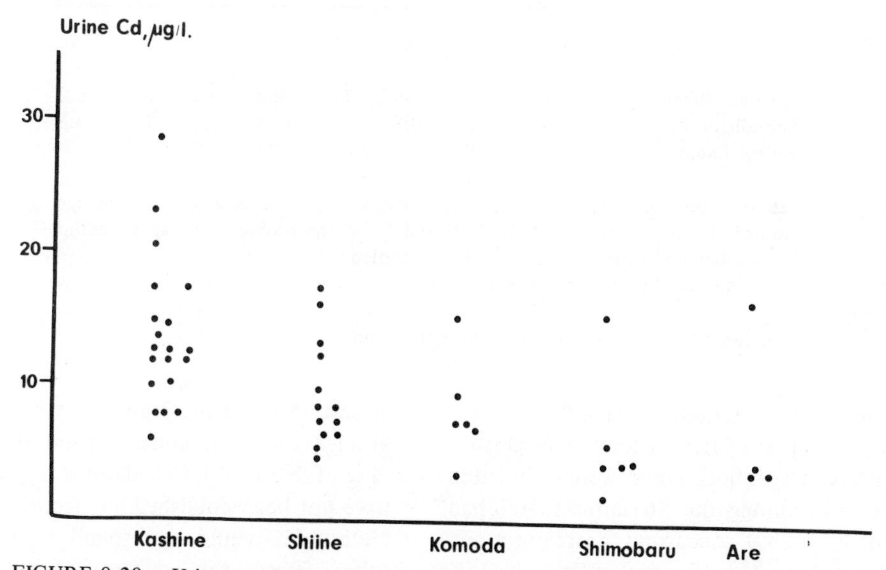

FIGURE 8:28. Urinary cadmium concentration of subjects from different areas of Tsushima Island positive for urinary protein and/or glucose. (From Nagasaki Prefecture, 1970.)

are very small, no statistically significant differences between control and polluted areas within age groups with a range of 10 years could be found.

8.3.3.3 Kakehashi Area in Ishikawa Prefecture

The source of cadmium in this area is now the Huokuriku mine, but earlier there was another, the Ogoya mine (Figure 8:29). Epidemiological studies involving determination of proteinuria, glucosuria, and cadmium in urine were performed by Ishizaki's group at Kanazawa University. The studies were discontinued in 1972. No nonpolluted area was included in the studies.

Fukushima and Nogawa, 1971, reported data from the most polluted village, Kanahira, where an

TABLE 8:29

Urinary Cadmium Concentrations of Persons in Kashine and Hikage-Kamiyama as Studied in 1970

	Kashine (polluted)			Hikage-Kamiyama (control)		
	Number of samples	Average (μg/l)	S.D.	Number of samples	Average (μg/l)	S.D.
Male farmers*	21	12.8	7.3	15	2.5	1.8
Female farmers*	32	11.9	5.9	16	2.5	1.3
Male nonfarmers	10	5.1	3.5	6	2.6	1.4
Female nonfarmers	14	6.1	3.5	7	1.4	0.3

*Difference between Kashine and Hikage-Kamiyama statistically significant.

From Takabatake, E., Keshino, M., and Matsuo, I., *J. Hyg. Chem.*, 18, 41, 1972. With permission of the publisher, the Pharmaceutical Society of Japan.

TABLE 8:30

Prevalence of Proteinuria (%) in Tsushima Area in Studies from Different Years*

	1969		1970		1971		1972	
	N†	%	N	%	N	%	N	%
Kashine, Shiine	171	13	239	7.1	268	6.7	257	6.2
Shimobaru, Komoda	223	3.6	308	3.2	614	4.7	772	3.5
Hikage-Kamiyama	79	3.8	120	0.8	177	4.5	–	–

*In the study of 1969, test tape was used and persons over 40 years of age studied. In the studies of 1970, 1971, and 1972, the trichloracetic acid method was used and persons over 30 years of age studied.
†N = number of persons examined.

Calculated from Terada, personal communication.

average cadmium concentration of 0.80 μg/g was found in 11 samples of rice. A total of 74 persons over 50 years of age, both sexes, were included in the study, and among the 56 farmers 50% had proteinuria and 29% concurrent proteinuria and glucosuria. Among the 18 nonfarmers only 17% had proteinuria and 0% concurrent proteinuria and glucosuria.

Ishizaki, 1972a, reported results from another three villages, together with some new data from Kanahira. A total of 209 persons over 50 years of age from the four villages was studied. No non-polluted area was included. Results of the study together with cadmium concentrations in rice are given in Table 8:31. Like Fukushima and Nogawa, 1971, Ishizaki confirmed that the village Kanahira, closest to the mine, is the most polluted, with an average cadmium concentration in rice of 0.80 μg/g (Table 8:31). Calculations of the daily intake have not been published but according to the Rice Method, it would correspond to 480 μg. In the other villages the concentrations were between 0.23 and 0.34 μg/g. The average urinary cadmium concentration was 23.3 μg/l in the most polluted village. Urinary cadmium concentrations were also high in the other villages. Prevalence of proteinuria (3% sulfosalicylic acid method) was 39% in the most polluted village and 22 to 30% in the other villages.

Urine protein electrophoresis on cellulose acetate membrane was performed in several cases. It was reported that proteinuria was common and

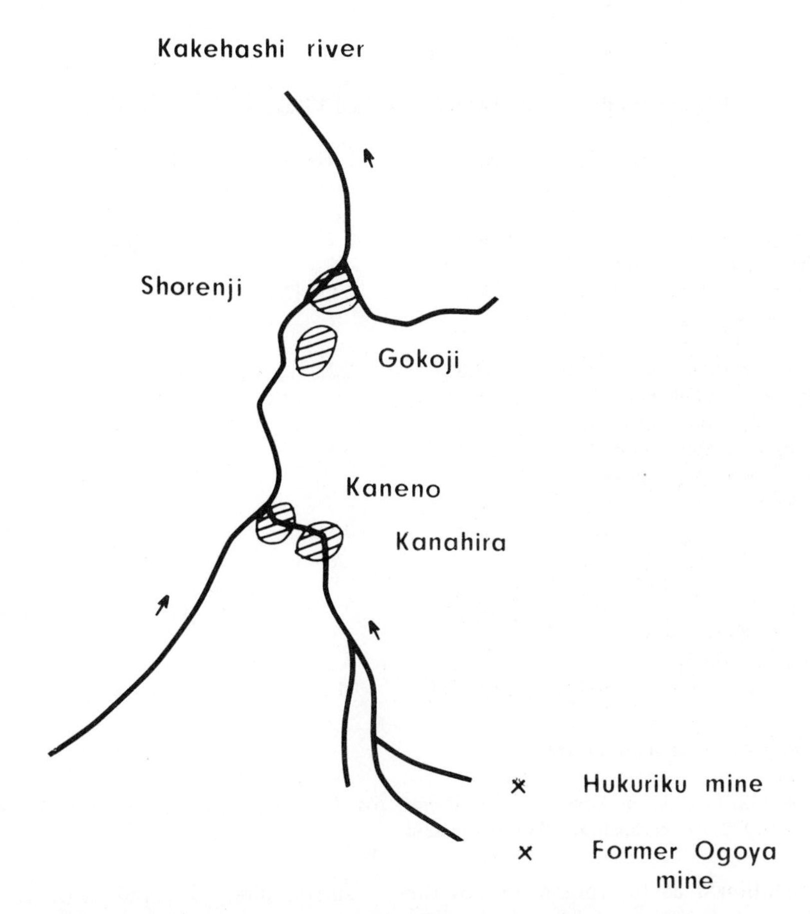

FIGURE 8:29. Location of the villages investigated in Kakehashi River basin. (From Ishizaki, A., in *Kankyo Hoken Report No. 11,* Japanese Association of Public Health, April 1972, 27. With permission.)

that electrophoretic patterns were of the tubular type in the majority of cases (Ishizaki, 1972a). Ishizaki concluded that the findings in Kanahira village are similar to those in the Itai-itai district, even though he did not find signs of actual Itai-itai disease. He considered that signs of cadmium effects had been shown in the other three villages as well. It should be remembered that the numbers of persons in this study were small and that no controls were included.

8.3.3.4 Bandai Area in Fukushima Prefecture

Bandai town in Fukushima Prefecture has been polluted by cadmium via air and sewage water outlets from a zinc refinery in operation since 1916. In 1962 zinc production was 2,200 tons per month (Nisso Smelting Co., 1972), in 1965 2,500 tons per month, and in 1970 4,000 tons per month (Fukushima Prefecture, 1972). Lead production in this town dates from 1940, and at present cadmium and other substances are produced as well.

An epidemiological survey according to the old standard method was executed in 1970, with Bandai town as the observation area and the village Kitaaizu, 10 km southwest of Bandai, as the control area. Cadmium concentration in unpolished rice from 1970 and 1971 within the observation area varied between lower than 0.2 μg/g to higher than 1.0 μg/g (Fukushima Prefecture, 1971a). Data on cadmium concentrations in rice or daily intake in the control area are not available.

The target population in the epidemiological study was 1,827 men and women over 30 years of age in the observation area and 272 women in the control area. The average prevalence of proteinuria among the 1,324 participants from the polluted area in 1970 was 12.3% and among the 215 in the control area 4.7% (Fukushima Prefecture, 1971b).

TABLE 8:31

Result of an Epidemiological Study in Four Villages in Kakehashi River Basin

	Kanahira	Gokoji	Kaneno	Shorenji
Cadmium concentrations in rice (μg/g) average ± S.D.	0.80 ± 0.30	0.23 ± 0.14	0.32 ± 0.07	0.34 ± 0.28
Persons examined and participation (in parentheses)	77 (87%)	54 (94.7%)	60 (87%)	18 (100%)
Prevalence of proteinuria (% of persons examined, sulfosalicylic acid method)	39.0	22.0	30.0	28.0
Prevalence of glucosuria (% of persons examined, orthotoluidine method)	33.0	17.0	18.0	11.0
Concurrent proteinuria and glucosuria (% of persons examined)	22.0	13.0	10.0	9.0
Number of persons examined for cadmium in urine	33*	13*	17*	18
Cadmium concentrations in urine (μg/l) average ± S.D.	23.3 ± 11.6	21.2 ± 10.5	11.5 ± 4.8	10.1 ± 6.5

*Mainly cases with proteinuria.

From Ishizaki, A., in *Kankyo Hoken Report No. 11,* Japanese Association of Public Health, April 1972, 27 (in Japanese). With permission.

Nothing is mentioned as to whether or not the studies were carried out on a blind basis. The criteria for selecting persons for the second screening were proteinuria ± in the polluted area and proteinuria + in the control area. In the polluted area persons with glucosuria were also included. This makes it impossible to compare age-specific participation in the second screening for epidemiological analysis, even though data on age distribution are available.

In the second screening, cadmium concentration in urine was analyzed for those with proteinuria. The average among the 21 men studied in the observation area was 3.7 μg/l and among the 128 women 2.3 μg/l (Fukushima Prefecture, 1971b). Urinary cadmium was not analyzed in the control area, but among 17 women in the villages with lowest cadmium exposure in the observation area (Omagari, Nagamine, and Motodera) the average was 1.5 μg/l. On those who did not undertake the screening in 1970, follow-up studies were made in 1971 and 1972, but since very small groups were examined, the results are not discussed here.

8.3.3.5 Annaka Area in Gumma Prefecture

In Annaka a large zinc refinery has polluted surrounding farmland with cadmium primarily through air. Operations started in 1937, cadmium production in 1948. The monthly production of zinc has increased rapidly during the last decade from about 2,000 tons to about 12,000 tons, according to data from Mr. Fujimori, vice-president of the refinery.

The pollution level in Annaka was studied by Kobayashi et al., 1970. Cadmium concentration in morus leaves was elevated close to the factory and decreased with increasing distance from it (Figure 3:2). The morus bushes are planted at such places that cadmium exposure can be assumed to result mainly from air pollution. Even at a distance of 2,500 m east of the factory the cadmium level in the leaves was much higher than the reported value from a remote area in Okayama Prefecture.

Kobayashi, 1970, and Kobayashi (unpublished data) studied cadmium concentration in soil at different distances and in different directions from the Annaka refinery (Figure 8:30). An enlargement of the area nearest the refinery as seen in Figure 8:30 is found in Figure 8:31. It can be observed that the cadmium concentrations (dry weight) in soil of irrigated rice fields were more than 15 μg/g downstream from the factory within

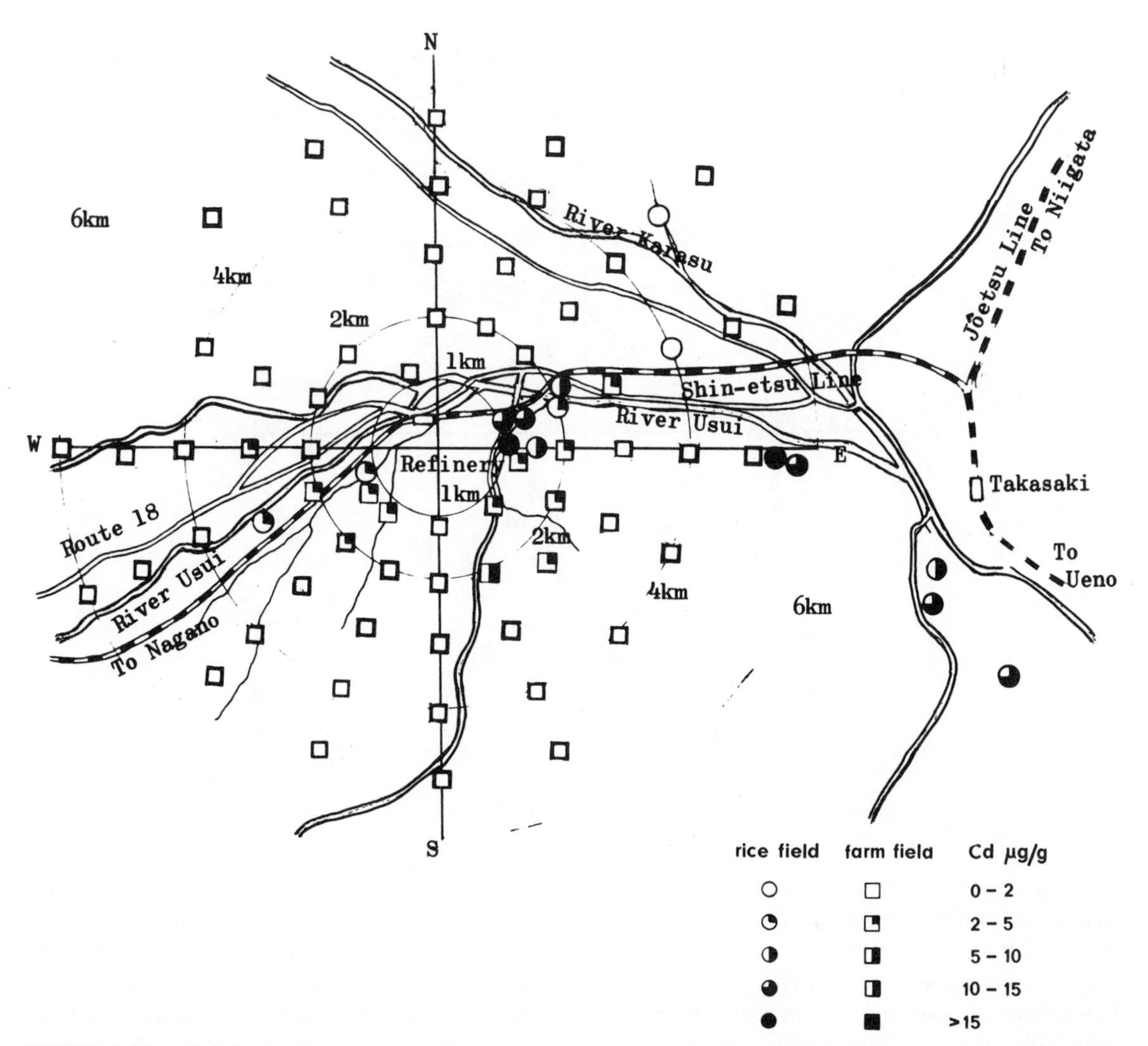

FIGURE 8:30. Cadmium concentration in soil in the vicinity of the refinery in Annaka. (From Kobayashi, unpublished data.)

a 1-km distance. The concentrations were up to between 10 to 15 µg/g as far as 10 km downstream from the factory in fields close to the polluted river. Upstream, the concentration was mostly 5 to 10 µg/g within 1 km from the factory and 2 to 5 µg/g at 1 to 3 km from it. Nonirrigated farm fields showed a cadmium concentration over 15 µg/g in most samples within 1 km to the east of the factory and 2 to 10 µg/g in other directions. At a distance over 4 km no samples from nonirrigated fields showed a concentration over 2 µg/g. The prevailing winds in the area are westerly and easterly (Gumma Prefecture, 1972). Since no control area was included and no concentration variations under 2 µg/g can be evaluated from the data, it is not known how far pollution extends in the area, but the data show that the air pollution level decreases rapidly within a radius of a few kilometers from the refinery. Water pollution extends to larger distances downstream from the refinery.

An epidemiological study was performed in 1969–1970 (Kakinuma et al., 1971) according to the old standard method. A control area about 6 km west of the Annaka refinery, where water pollution by that source can be excluded, was chosen. As reported by the Japanese Association of Public Health, 1970b, the average cadmium concentration in 20 samples of rice in 1968 was 0.38 µg/g in the observation area and 0.30 µg/g in four samples from the control area (Table 8:13). In 1971, paddy soil and unpolished rice were studied in the observation and surrounding areas (Gumma Prefecture, 1972). An average of 0.48 µg/g cadmium in 166 samples of unpolished rice in the observation area was found. The control area

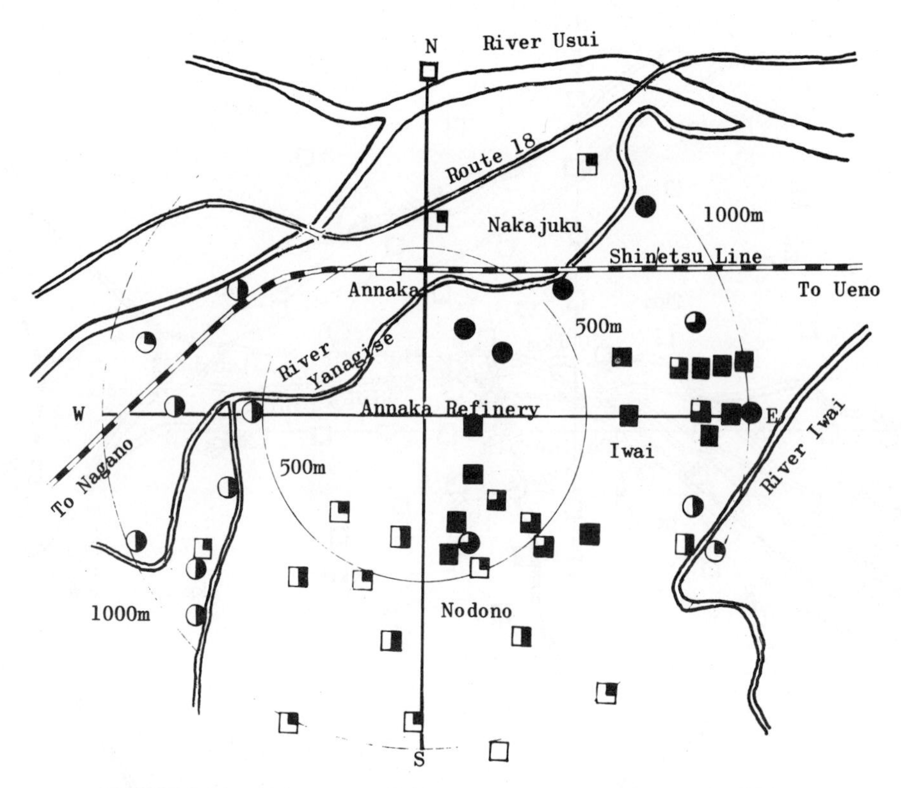

FIGURE 8:31. Enlargement of the area nearest the refinery in Figure 8:30.

was not studied, but in a village considered nonpolluted, 3 km northwest of the refinery and 3 km from the control area, the average cadmium concentration out of three samples was 0.48 μg/g. The average of all samples from parts of Annaka town other than the observation area was 0.24 μg/g as reported by I. Fukushima, 1973. The range was 0.02 to 0.83 μg/g. Thus, elevated cadmium concentrations were found in rice outside the observation area. It was calculated with the National Nutritional Census Method (Japanese Association of Public Health, 1970b) that daily cadmium intake via food was 211 μg in the polluted area where 676 persons were interviewed and 20 rice samples analyzed, and 113 μg in the control area where 138 persons were interviewed and 4 rice samples analyzed (Section 8.3.1.5). This was due to an assumed low cadmium concentration in other foodstuffs from the control area. The Total Diet Collection Method on six samples from each area gave 281 and 99 μg, respectively. This latter figure is about twice as high as the calculated Japanese average (Section 8.3.2.1).

Exposure via air can be evaluated from data on suspended particles in air and their cadmium concentration. Yamagata and Shigematsu, 1970, gave values from three sampling stations in the vicinity of the refinery: 0.104 μg/m^3 (range 0.008 to 0.235), 0.166 μg/m^3 (range: 0.044 to 0.380), and 0.055 μg/m^3 (range: 0.020 to 0.142). In Gumma Prefecture, 1972, it is reported that the average of five 24-hr samples closer than 500 m from the emission source was 0.036 μg/m^3. The average of 25 24-hr samples at a distance over 500 m (most of them between 500 and 3,000 m) was 0.015 μg/m^3.

Assuming a 25% absorption from the lungs (Section 4.1.1.3) and 6% absorption from the gut, the 0.1 μg/m^3 would correspond to a daily intake of cadmium in food of about 8 μg; 0.01 μg/m^3 corresponds to about 0.8 μg. These intakes would be insignificant compared to the calculated average daily intakes from food.

Urinary cadmium excretion has been studied (I. Fukushima, 1973) on a sample of persons over 30 years of age selected for a nutritional survey in the observation areas in Annaka town, the neighboring Takasaki town, and the control area within Annaka town. Table 8:32 shows that in the observation areas average cadmium excretion was about twice as high as in the control area. I. Fukushima, 1973, found a wide scattering when

TABLE 8:32

Distribution of Urinary Cadmium Concentration Among People from the Annaka Area

	Takasaki town – observation area	Annaka town – observation area	Annaka town – control area
Year of study	1970	1969–1970	1969–1970
Urinary cadmium (μg/l)			
0–1.9	3	30	7
2.0–3.9	21	72	21
4.0–5.9	52	48	9
6.0–7.9	40	34	4
8.0–9.9	21	23	1
10.0–11.9	18	16	–
12.0–13.9	10	18	–
14.0–15.9	8	6	–
16.0–17.9	5	3	–
18.0–19.9	3	1	–
20.0–21.9	1	–	–
22.0–23.9	–	2	–
24.0–25.9	1	2	–
26.0–27.9	1	1	–
28.0–29.9	–	–	–
30.0–31.9	–	–	–
32.0–33.9	–	–	–
Total number of persons studied	184	260	42
Average urinary cadmium in all persons studied.	8.0 μg/l	6.5 μg/l	3.8 μg/l

From Fukushima, I., in *Kankyo Hoken Report No. 24,* Japanese Association of Public Health, September 1973, 131 (in Japanese). With permission.

he compared calculated daily cadmium intake and urinary cadmium excretion among individuals. Since no distinction was made between persons with and without proteinuria, this comparison is hard to evaluate.

Summarizing available data on exposure, a certain difference exists between the polluted area and the control area. It seems obvious, however, that the control area is not ideal as judged from the average urinary excretion of cadmium, calculated daily intake, and high cadmium concentrations in rice. A factor that complicates any evaluation of exposure is the six-fold increase in the factory's production during the 1960's.

The control area used in the studies on daily intake is the same as the one used in the epidemiological study, where a total of 2,397 persons (men and women over 30 years) in the polluted area (84% participation) and 895 persons in the control area (74% participation) were examined. Age-related prevalences of proteinuria have been published by Kakinuma et al., 1971, and are shown in Table 8:33. The studies in the polluted area and in the control area were not performed at the same time nor on a blind basis. According to Kakinuma et al., 1971, proteinuria was defined as a rating of + or higher by either the trichloracetic acid or sulfosalicylic acid methods. The data in Table 8:33 do not show differences in the age-related prevalence of proteinuria between the control and the polluted area. There is an increase in prevalence of proteinuria with age.

Nomiyama, 1971b (Section 6.1.2.1.4), and Nomiyama, 1973, have reported about cadmium concentrations in organs of five deceased persons from Annaka. As the number of subjects is so small, no epidemiological conclusions can be drawn from the data.

8.3.3.6 Omuta Area in Fukuoka Prefecture

The cadmium source in this area is a refinery causing mainly airborne pollution. Epidemiological

TABLE 8:33

Prevalence of Proteinuria (%)* in Annaka Area

Age group	Females, polluted area		Females, control area	
	N[†]	%	N	%
30–39	337	9.8	85	5.9
40–49	351	8.3	156	6.4
50–59	292	11.0	125	10.4
60–69	193	17.6	87	20.7
70 +	128	22.6	54	11.1
Total	1,301	12.1	507	10.3

Age group	Males, polluted area		Males, control area	
	N	%	N	%
30–39	307	4.2	51	3.9
40–49	273	6.2	115	5.2
50–59	226	6.2	100	13.0
60–69	201	14.9	84	10.7
70 +	89	14.6	38	15.8
Total	1,096	7.9	388	9.3

*Evaluated as + or higher by either trichloracetic acid or sulfosalicylic acid method.
†N = number of persons studied.

Calculated from Kakinuma et al., 1971.

data have been collected in 1970 (Ministry of Health and Welfare, 1971c) and in 1972 (Matsushiro, personal communication).

The refinery started production in 1913 and two plants were added in 1915 and 1954. Data from the Ministry of Health and Welfare, 1971c, indicate that only a slow increase in production has occurred during the last decades. Air concentrations of cadmium in 1969, at one place reaching a maximum 24-hr value of 3 $\mu g/m^3$, were the highest measured in Japan (Yamamoto, 1972). According to Yamamoto, 1972, the average cadmium concentration in 91 samples of rice harvested in 1970 was as high as 0.72 μg/g. Intake from all food except rice was calculated to be 75 to 115 μg among farmers, depending on area. Total daily intake has been calculated according to the National Nutritional Census Method to vary between 80 and 280 μg (Ministry of Health and Welfare, 1971c).

Prevalence of proteinuria was studied in the polluted area in 1970 and 1972 (Ministry of Health and Welfare, 1971c, and Matsushiro, personal communication) and in 1970 in a control area, Fukuoka City. The exposure to cadmium in the control area is not known. Information on methods used for protein determination has not been available to us as far as the 1970 study is concerned. In 1972 the trichloracetic acid or sulfosalicylic acid method was used (Matsushiro, personal communication). The selection of target populations is not known to us, nor are the participation rates. With these reservations, the prevalences of proteinuria in polluted as well as control areas are given in Table 8:34. In the polluted area the prevalence increased with age both in 1970 and 1972, but not in the control (males). There is no evidence of increased proteinuria in the polluted area in the 1970 study. The prevalence in the polluted area is considerably higher in the 1972 study, both compared with the polluted area and with the control area from 1970. Unfortunately no data from the control area in 1972 were available. The control area studied in 1970 showed higher prevalences than the polluted area.

In the report from the Ministry of Health and Welfare, 1971c, many values on cadmium in urine are given, but no information on methods. The data are given without reference to age. It can be mentioned that in the cadmium-polluted area persons with proteinuria (130) had an average cadmium excretion of 10.3 μg/l (range 0 to 77.7 μg/l). In the control area the corresponding value for persons with proteinuria (49) was 1.8 μg/l (range: 0 to 5.1 μg/l). In the polluted area no data on cadmium excretion in persons without proteinuria were given, while such data were furnished from the control area. These values were not given as concentrations, but as total amounts excreted per day. A mean of 2.8 μg/day (range: 0 to 7.5 μg/day) was reported among 38 persons.

8.3.3.7 Uguisuzawa Area in Miyagi Prefecture

Uguisuzawa is the site of a mine and a refinery, which have polluted both air and water. A small river passes the "most polluted" area, Uguisuzawa town, close to the mine, at a rapid pace and then, converging with another river, flows more slowly through Kurigoma town, which lies about 10 km downstream from the mine. The river continues another 10 km and passes Wakayanagi, where the control area used in 1970 is situated. After another 10 km it passes through Nakata town, site of the control area used in 1969. The mine is more

TABLE 8:34

Prevalence of Proteinuria[a] (%) Among Inhabitants of the Polluted Area in Omuta and of the Control Area in Fukuoka

Female farmers

	Polluted area				Control area	
	1970		1972		1970	
Age	N[b]	%	N	%	N	%
–39	20	0	84	11.9	99	8.1
40–49	419	2.6	102	14.7	126	8.7
50–59	325	2.5	69	11.6	113	8.0
60–69	361	5.8	68	16.2	87	10.3
70 +	310	13.9	94	34.0		
Total	1,435	5.8	417	18.2	425	6.8

Male farmers

	Polluted area				Control area	
	1970		1972		1970	
Age	N	%	N	%	N	%
–39	10	0	48	4.2		
40–49	318	1.3	84	7.1		
50–59	262	1.9	76	13.2		
60–69	285	3.2	49	16.3	87	13.8
70 +	230	16.5	57	28.0		
Total	1,105	5.1	314	13.4	87	13.8

[a]1970, method not reported; in 1972 trichloracetic acid methods were used (Matsushiro, personal communication).

[b]N = number of persons examined.

Data compiled from Ministry of Health and Welfare, 1971c, and Matsushiro, personal communication.

than 100 years old, while the refinery was started in the beginning of this century.

Health screenings have been performed every year since 1969, but only data from 1969 and 1970 have been available to us. Some data on cadmium in rice have been reported in Miyagi Prefecture, 1971. In September 1970, in five villages within Uguisuzawa town, values of 0.128 to 1.352 μg/g were found in unpolished rice. In a village regarded as "control" the corresponding value was 0.522 μg/g. In order to check the cadmium content of rice in that control area, new samples were gathered in December 1970, on which occasion 0.23 to 0.26 μg/g was recorded (Miyagi Prefecture, 1971). A nutritional study from 1969 (Japanese Association of Public Health, 1970b) calculated the daily intake in the polluted area to be 245 μg cadmium and in a control area to be 85 μg (Table 8:13). This control area, however, is not the same as the ones used in the epidemiological studies. It cannot be contaminated by the river flowing from the mine.

Urinary cadmium concentrations are given in Miyagi Prefecture, 1970a, 1970b. The average among 44 persons from Uguisuzawa town was 7.11 μg/l, among 28 persons from Kurigoma town 10.59 μg/l, and among 13 persons from Nakata town (control area in 1969) 4.64 μg/l. In 1969 638 residents of the "most polluted" area, 220 residents of a neighboring area in Kurigoma town, and 239 residents of a control area in Nakata town underwent a first screening according to the old standard method (Section 8.3.1.2) (Miyagi Prefecture, 1970a). The prevalence of proteinuria (trichloracetic acid method ± or higher) in women over 30 years of age was as follows: 50% in the "most polluted" area, 60.4% in the neighboring area (Miyagi Prefecture, 1970a, 1970b), and 39.6% in the control area (Miyagi Prefecture, 1970b). No age-related data on these prevalences have been available. A diagram showing the prevalences of + or higher in the different age groups cannot be evaluated because the number of persons in each group was not reported.

The study in 1970 covered only 117 persons over 30 years of age in Uguisuzawa town, but included 206 persons in Kurigoma town (Miyagi Prefecture, 1971). It is not clear whether the persons studied had also been examined in 1969. In a new control area of Wakayanagi town 437 persons were studied. The prevalence of proteinuria, + or higher by the sulfosalicylic acid method or + or higher by the test tape procedure, was 2.6% in Uguisuzawa, 3.9% in Kurigoma, and 4.4% in Nakata.

In summary the data show that the Uguisuzawa area must be heavily polluted with cadmium, but the data on effects are extremely difficult to evaluate.

8.4 GENERAL DISCUSSION OF CADMIUM EXPOSURE AS AN ETIOLOGICAL FACTOR FOR PROTEINURIA AND ITAI-ITAI DISEASE

8.4.1 Cadmium and Proteinuria

As described in previous sections of this chapter, a large amount of epidemiological data have been collected during recent years in various parts of Japan suspected of environmental cadmium pollution. It is obvious that a considerable part of the studies was not performed in such a way as to make it possible to draw valid epidemiological conclusions concerning health effects of cadmium.

Reasons for this differ from study to study. Delineation of "polluted" and "control" areas is often not well defined or standardized. There are considerable uncertainties due to large variation in pollution levels within the observation areas and lack of detailed descriptions of selection of, for example, rice samples for analysis. Data on daily cadmium intake in nonpolluted areas of Japan as a whole are available. In some of the control areas in the epidemiological studies, reported daily intake is considerably above the Japanese average. Methods for proteinuria analysis vary among areas. Although the standard screening method sets criteria for evaluating proteinuria, very few of the studies give detailed data on qualitative ratings. In some reports, proteinuria has been based on a rating of plus-minus or higher and in other studies plus or higher. The studies on proteinuria are often not carried out on a "blind" basis, meaning possibilities for "expectation effects." Virtually no data exist from *true* low-exposure control areas involving studies of different age groups and using the same analytical methods as applied in the polluted areas.

Against this background, one has to be careful when drawing conclusions on health effects of cadmium, particularly when it comes to quantitative evaluations of dose-response relationships. To use the data in order to prove a no-effect level is impossible. This becomes all the more obvious when considering that the primary aim has not necessarily been to carry out prevalence studies of proteinuria, but instead to look for cases of Itai-itai disease.

This does not mean that the data obtained, taken as a whole, are of no use in evaluating effects of cadmium. On the contrary, there are data which make it possible to reach valid conclusions. Other data reported in this review have been referred to in a rather detailed form, particularly because they are originally reported only in Japanese and the validity of the results would be difficult for a large part of the scientific community to interpret. With these general comments in mind, the following attempt is made to evaluate the situation in Japan.

The only published Japanese evaluative review of information from various areas has been made by Hasegawa, 1972. Data from Fuchu and Ikuno were not included. His conclusion was that no effect of cadmium could be ascertained according to his epidemiological analysis, and that this lack of effect might depend upon a too-low level of exposure or the failure of the method to achieve sufficient sensitivity.

Judging from data on the cadmium concentration in rice and daily cadmium intake, the most polluted areas would be Fuchu, Ikuno, and Kakehashi, with a daily intake probably over 300 μg in their most polluted areas. Omuta, Tsushima, Annaka, Bandai, and Uguisuzawa have probably, on the average, a somewhat lower pollution level, ranging from values of 300 to less than 200 μg daily intake.

Fuchu is currently the area with the highest reported cadmium exposure. The exposure may well have been still higher in the past. There is a large difference in the average prevalence of proteinuria between polluted and control areas. Proteinuria increases considerably with age in the polluted area but also to a certain extent in the control area. Increasing proteinuria as a function of age is found in the general population of Japan, but the highest prevalences, which are found in age groups above 60 years, are two to three times higher in the polluted area of Fuchu. Concurrent proteinuria-glucosuria is also more common in the polluted area of Fuchu than in a control area. In a blind study on a large population of women over 40 years of age from Fuchu, a clear dose-response relationship between pollution level and proteinuria as well as glucosuria was reported.

In Ikuno, extensive studies have recently been made and the data are of particular interest. The exposure situation seems to be, in several aspects, similar to that in Fuchu. The average prevalence of proteinuria is extremely high in the polluted area, and the proteinuria is often of the tubular type.

Furthermore, excretion of low molecular weight retinol binding protein occurs to a considerably higher extent in the polluted area than in a control area.

In Kakehashi, the third high-pollution area, results from the limited studies performed to date are in accord with a hypothesis that cadmium-induced proteinuria exists in that district.

The data referred to above already indicate that cadmium can play an important role in the etiology of elevated prevalence of proteinuria in highly polluted areas. In the areas where somewhat lower daily intakes have been recorded, there is no unequivocal interpretation of the total data available at present. In some areas the data are indicative of an occurrence of cadmium-induced proteinuria whereas the opposite is true for other areas.

Differences between polluted and control areas in Tsushima were reported on in detail in the earlier edition of this review (Friberg, Piscator, and Nordberg, 1971). There are considerable difficulties in interpreting data from Tsushima. Nonetheless, it can be stated that in both former and recent studies the highest prevalence of proteinuria has consistently been observed in the most polluted villages. A cadmium-related effect thus seems highly likely in this area. In addition, data from Bandai may be in accord with such an effect. There the average prevalence of proteinuria is more than twice as high in the polluted area as in the control area.

There are findings, which, taken at their face value, do not support the hypothesis that cadmium causes proteinuria. In one area, Annaka, effects have not been observed despite relatively high levels of cadmium, but it may well be that the exposure time has not been long enough. In another study (in Omuta) there is no obvious explanation as to why effects have not been observed. Whether or not effects exist in those areas is a question that will have to await results of studies whose design is better suited to show a no-effect level than the studies reported thus far.

Considering all the data together, there can be no doubt that cadmium plays a role in the etiology of elevated prevalences of proteinuria in heavily exposed areas. A question of importance, of course, is whether a present daily intake of about 200 to 300 μg, an intake which has been associated with effects, could be considered sufficient to bring about chronic cadmium poisoning. If it is assumed that an exposure of about 50 μg/day during 50 years gives about 50 μg/g in kidney cortex (Section 4.3.2.1), a corresponding exposure to about 200 μg/day would give about 200 μg/g as an average. As was discussed in Section 6.1.2.4, the critical cadmium level in kidney cortex at which proteinuria might occur in a fraction of the population has been estimated to be about 200 μg/g. Low calcium or vitamin D intakes would enhance cadmium absorption, and daily intakes lower than 200 μg might then lead to effects.

Another approach in evaluating effects is to discuss cadmium excretion in the urine. It can be seen from data by Tsuchiya, Seki, and Sugita, 1972a (Section 4.3.3.1), that a mean urinary excretion of about 2 μg Cd/l corresponds to a mean cadmium level in kidney cortex of about 100 μg/g after long-term low-level exposure in a so-called nonpolluted area. As was discussed in Section 4.3.4.1, at this type of exposure there is reason to believe that urinary excretion is related to body burden. It has not been shown that at long-term, relatively low-level exposure a threshold exists at which the urinary excretion of cadmium suddenly would increase without concurrent renal tubular dysfunction. If kidney cortex concentration is assumed to be directly proportional to body burden, the urinary excretion corresponding to a kidney cortex level of 200 μg/g would be 4 μg Cd/l (200/100 × 2). For example, in the study in Ikuno, the criterion for selecting the polluted areas was a minimum average cadmium excretion in urine of 9 μg/l. In two villages in Kakehashi, mainly among persons with proteinuria, the cadmium concentration in urine was on an average about 22 μg/l. Even if the urinary excretion is not directly related to kidney cortex concentration in the same way as to body burden, human as well as animal data strongly support tubular dysfunction as the reason for such a high excretion of cadmium.

8.4.2 Etiology of Itai-itai Disease

Clinical data have shown that Itai-itai disease can be classified as a form of osteomalacia. Generally, osteomalacia can be divided into the following categories:

1. Vitamin D deficiency
2. Malabsorption of vitamin D and bone minerals
3. So-called vitamin D-resistant osteomalacia

Since hyperparathyroidism sometimes shows a clinical picture similar to osteomalacia, a differential diagnosis can be very difficult. Therefore, it also occasionally has been difficult to differentiate between Itai-itai disease and hyperparathyroidism, as discussed by Takase et al., 1967, for example. In primary hyperparathyroidism the patients have an elevated blood calcium level not seen in cases of osteomalacia. However, a more or less decreased calcium level, seen in cases of osteomalacia, can stimulate increased parathyroid activity, so-called secondary hyperparathyroidism.

The etiology of Itai-itai disease has recently been discussed by Takeuchi and Naito, 1972, and Takeuchi, 1973.

8.4.2.1 Vitamin D Deficiency

This type of osteomalacia is dependent on the combination of lack or low intake of vitamin D in the diet and deprivation of ultraviolet irradiation. This combination together with a low intake of calcium and a simultaneous high demand for calcium and vitamin D during pregnancy and lactation in some cases has given rise to osteomalacia even in countries with much sunshine, as documented by Groen et al., 1965 (see also the review by Arnstein, Frame, and Frost, 1967). In the areas where Itai-itai disease is seen, the consumption of foodstuffs rich in calcium and vitamin D such as milk and milk products is very low. The calcium intake is considerably lower than in a country such as Sweden, as seen in Table 8:8. However, the consumption is not lower than in other parts of Japan where Itai-itai disease is not seen. As we have said before, the weather in this part of Japan is very gloomy with a lot of rain and snow. The women also wear their clothes in such a manner that the main part of the sunlight is screened away. Osteomalacia and rickets also have been seen more frequently in the Toyama Prefecture than in other parts of Japan (Kajikawa et al., 1957). However, available data (Section 8.2.1) do not show agglomeration to any certain areas such as Fuchu, for example. The dietary conditions cannot be the only etiological factor behind the disease, because conditions are similar in nearby villages and towns where the disease has not been found. However, the nature of the diet may well be a subsidiary factor for the elicitation of the disease.

Takeuchi, 1973, suggested that religious practices existing only in the Fuchu area at the time of the occurrence of Itai-itai disease would have made the women more susceptible to vitamin D deficiency. According to Takeuchi, rickets and osteomalacia were regarded as diseases induced by the punishment of God among the people of Toyama Prefecture at the beginning of this century. The family confined the patient strictly within the house. Dietary deficiency and lack of sunlight caused osteomalacia. Takeuchi claims that Fuchu area was the only part of Toyama Prefecture where this religious belief persisted until recently. No data are presented to support any such close association between this practice and cadmium exposure.

8.4.2.2 Malabsorption Syndrome

Osteomalacia also can be caused by a deficient absorption of vitamin D and bone minerals. Chronic pancreatic disease, hepatobiliary disease, and resection of parts of the gastrointestinal tract have given rise to such osteomalacia, as reviewed by Arnstein, Frame, and Frost, 1967, Boström, 1967, and Muldowney, 1969. In this context the clinical observations reported by Murata et al., 1970, concerning the function of the gastrointestinal tract and pancreas are of interest. In his 1970 paper Murata mentioned that he had detected a tendency toward a declined function of the pancreas in many cases. Furthermore, examinations of the gastrointestinal tract have shown shortened ciliated epithelia as well as atrophy of the mucous membrane and submucosal cell infiltration in the small intestine. Fat absorption tests have shown decreased fat absorption in many cases. These changes in the gastrointestinal tract have been called "cadmium enteropathy" by Murata. At present it is impossible to judge the frequency of these changes. Murata did not state information about frequency. However, in cases in which changes are present, they could act as a contributory etiological factor of the Itai-itai disease.

8.4.2.3 So-called Vitamin D-resistant Osteomalacia (Renal Osteomalacia)

This form of osteomalacia is caused by a renal tubular dysfunction which gives rise to losses of bone minerals, primarily phosphate, through the kidneys. There is also an increased excretion of amino acids and glucose. Both hereditary (Dent and Harris, 1956) and acquired forms (deSeze et al., 1964) have been described. Persons with

Itai-itai disease always have proteinuria of a tubular type. There does not seem to be any clear difference between what is termed "acquired" and "hereditary," and they can be spoken of collectively as a form of osteomalacia, with which the clinical findings in Itai-itai disease are in accord. Available data on treatment (Section 8.2.6) indicate that Itai-itai disease can be classified as a vitamin D-resistant form of osteomalacia.

8.4.2.4 Cadmium as an Etiological Factor

It has been well established (Section 6.1) that exposure to cadmium can give rise to tubular damage of the kidneys with proteinuria and glucosuria. The kidney damage seen in Itai-itai disease has been very similar to that seen in classical industrial chronic cadmium poisoning, but bone changes have not been a common finding in the latter condition. There are, however, two French accounts by Nicaud, Lafitte, and Gros, 1942, and Gervais and Delpech, 1963, and two British studies by Bonnell, 1955, and Adams, Harrison, and Scott, 1969, which taken together show that both in male and female workers occupationally exposed to cadmium, bone changes similar to those seen in Itai-itai disease have occurred (Section 6.4).

Data from industrial exposure refer mostly to males while almost all of the Itai-itai patients have been females. It should be stressed, however, that in the endemic area the prevalence of proteinuria and glucosuria has been extremely high also in males and only slightly lower than in females. Proteinuria and glucosuria were already common findings among patients before vitamin D treatment was started (Section 8.2.2.2.2). When tubular damage, directly or indirectly, disturbs the metabolism of calcium and phosphorus, women will be affected more than men. This is particularly the case in the women involved who were multiparas and who lived in an area with a low intake of calcium and probably also of vitamin D. Therefore, it is not contradictory to a cadmium etiology that females almost exclusively have suffered from bone changes.

Although no data exist concerning the cadmium exposure several years ago when it was supposedly highest, available information on cadmium concentrations in rice and paddy fields from recent years has shown a close correlation between high cadmium concentrations and the occurrence of Itai-itai disease. Furthermore, available data favor a causal relationship between cadmium and proteinuria in polluted areas of Japan (Section 8.4.1).

Questions remain as to whether or not the cadmium exposure has been high enough to cause Itai-itai disease and as to why such effects have not been diagnosed in other areas as yet. First of all, whether some few cases have occurred must still be considered an open question (Sections 8.3.3.1 and 8.3.3.2). The diagnostic criteria for Itai-itai disease used in the Fuchu area have not been employed in other areas (Section 8.3.1.2). Further, all data favor a higher exposure in Fuchu than in the other polluted areas, particularly when considering the possibility of a still higher exposure in the past. Looking at the few autopsy data at hand, it seems that exposure has been as high as that of workers with cadmium intoxication. In Fuchu, liver values have been found to be high in persons having low kidney values at the same time, indicating pronounced kidney damage. The only Japanese data from a polluted area other than Fuchu that are available are a few values from Annaka. Liver values were considerably lower than in Fuchu.

To what has been said comes the fact that apparently Toyama Prefecture has long been considered an area in which rickets is common, possibly due to lack of sunlight brought about by meteorological and other conditions discussed in Section 8.2.5.9. It thus seems that there may well be reasons why Itai-itai disease has occurred to such an extent in the Fuchu area and why it has been so closely associated epidemiologically with cadmium exposure within the area.

8.5 CONCLUSIONS

Available data strongly support a conclusion that cadmium plays a causal role in renal tubular dysfunction in several areas, as manifested above all in a high prevalence of proteinuria. Concerning the Itai-itai disease, we have no doubt that it is an expression of chronic cadmium poisoning. There is reason to believe, however, that deficient consumption of certain essential food elements and vitamins has been a contributing factor in the occurrence of Itai-itai disease. A low intake of calcium and vitamin D may have been of particular importance.

Chapter 9

DOSE-RESPONSE RELATIONSHIPS BETWEEN RENAL EFFECTS AND CADMIUM CONCENTRATIONS IN THE ENVIRONMENT

9.1 GENERAL DISCUSSION OF DAILY CADMIUM RETENTION AND RENAL CONCENTRATIONS

As was discussed in Chapter 6 and in particular in Section 6.1.2.4, the kidney is considered the critical organ in chronic cadmium poisoning. The relationship between daily intake and cadmium concentrations in the kidney may therefore be the basis for setting the maximum allowable average daily intake for the general population as well as the threshold limit values for industry. Calculations based on mathematical models as described in Section 4.5.1 are useful in this respect.

Assuming that the body burden excretion rate does not vary in relation to body burden, the body burden resulting from a daily retention of 1 μg cadmium can be calculated under different excretion rate and exposure time alternatives. In Table 9:1 the mathematical method described in Section 4.5.1 has been used. Since the daily retention in this case was constant regardless of exposure time, the original differential equation was used for the calculation.

Five excretion rate alternatives have been used in Table 9:1, with the corresponding biological half-times being presented in the table. As was discussed in Section 4.6 the biological half-time in the kidney cortex, which seems most plausible, would be over 19 years.

Using Table 9:1 the expected body burden under different daily retention alternatives can be calculated as well, since it is assumed that excretion rate does not change when body burden increases. This assumption would not hold true when the critical concentration in renal cortex is exceeded and tubular proteinuria occurs (Section 4.3.3.1). As was concluded in Section 6.1.2.4, the critical concentration in renal cortex would be 200 μg/g wet weight. Cadmium concentration in renal cortex can be assumed to be 1.5 times the average concentration in the kidney (Section 4.3.2.1).

Kidney weight in a "standard" man (body weight = 70 kg) is 300 g (Section 4.5.1). Using these assumptions, it can be calculated that 200 μg/g wet weight in adult kidney cortex corresponds to a body burden of 120 mg.

Based on the figures in Table 9:1 the necessary daily retention to reach a renal cortex cadmium concentration of 200 μg/g wet weight can be calculated (Table 9:2).

The daily intake of cadmium may stem from generally four different sources: food, ambient air, tobacco smoke, and industrial air.

9.2 DAILY CADMIUM INTAKE VIA FOOD AND RENAL CADMIUM CONCENTRATIONS

Variation of cadmium concentrations in renal cortex with age depending upon food intake has been discussed in Section 4.5. Taking the difference in average calorie intake between age groups as well as differences in kidney weight into consideration, Figure 4:36 displayed the accumulation of cadmium under different body burden

TABLE 9:1

Accumulated Body Burden of Cadmium (mg) from a Daily Retention of 1 μg Under Different Excretion Rate and Exposure Time Alternatives

Exposure time in years	Excretion per day, % of body burden (corresponding biological half-time in years in parentheses)				
	0 (∞)	0.002 (95)	0.005 (38)	0.01 (19)	0.02 (9.5)
10	3.65	3.52	3.34	3.06	2.59
25	9.13	8.34	7.33	5.98	4.19
50	18.25	15.29	11.97	8.39	4.87

TABLE 9:2

Necessary Daily Cadmium Retention (μg) to Reach Renal Cortex Cadmium Concentration of 200 μg/g (Wet Weight) under Different Excretion Rate and Exposure Time Alternatives

Exposure time in years	Excretion per day, % of body burden (corresponding biological half-time in years in parentheses)				
	0 (∞)	0.002 (95)	0.005 (38)	0.01 (19)	0.02 (9.5)
10	32.9	34.1	35.9	39.2	46.3
25	13.1	14.4	16.4	20.1	28.6
50	6.6	7.8	10.0	14.3	24.6

TABLE 9:3

Necessary Daily Cadmium Intake[a] to Reach Critical Concentration (200 μg/g Wet Weight) in Kidney Cortex at Age 50. Necessary Cadmium Concentration in a Basic Foodstuff to Reach These Cadmium Intakes[b]

	Excretion per day, % of body burden (corresponding biological half-time in years in parentheses)				
	0 (∞)	0.002 (95)	0.005 (38)	0.01 (19)	0.02 (9.5)
Necessary daily cadmium intake	164	196	248	352	616
Necessary cadmium concentration in basic foodstuff (μg/g wet weight)	0.27	0.33	0.41	0.59	1.03

[a]For an adult, calorie intake = 2,500 cal.
[b]Assumptions: a. ½ of the daily cadmium intake from this foodstuff; b. 300 g of this foodstuff ingested daily.

excretion alternatives. The renal cortex cadmium concentration was set at 50 μg/g wet weight (Section 4.3.4) at age 50. Assuming 4.5% retention of daily cadmium intake and age-related data on daily calorie intake and kidney weight (Section 4.5.1) the necessary daily cadmium intake for an adult (average daily calorie intake = 2,500 cal) was calculated (Table 4:13). When comparing these figures with the estimated daily cadmium intake in the United States (Section 3.2) the three lowest excretion rates (0.005% or less) give the most plausible intake figures. It should be remembered that if a lower retention is assumed using the same excretion rates, the calculation will yield correspondingly higher figures on necessary daily intake.

Employing the same calculation method the necessary daily cadmium intakes to reach the critical concentration (200 μg/g) at age 50 have been calculated to be 196 μg for 0.002% excretion and 248 μg for 0.005% (Table 9:3).. Assuming that one half of the calorie intake comes from a basic foodstuff and that 300 g of this food is consumed daily (Section 8.3.1.5), the corresponding "critical concentrations" in the basic foodstuff are 0.33 and 0.41 μg/g, respectively (Table 9:3). The theoretical figures presented here are in accordance with Japanese data on effect levels (Section 8.4.1).

9.3 DAILY CADMIUM INTAKE VIA AMBIENT AIR AS RELATED TO RENAL CADMIUM CONCENTRATION

Assuming that a person inhales 20 m^3 of ambient air during 24 hr, the necessary cadmium concentrations in air to reach 200 μg/g in renal cortex can be calculated (Table 9:4). Pulmonary absorptions of 25 and 50%, respectively, were used in accordance with empirical data (Section 4.1.1). with an exposure time of 50 years, ambient air concentrations around 1 μg/m^3 may be sufficient for reaching the renal cortex critical concentration. Such high ambient air concentrations are seldom found over long periods even in polluted districts (Section 3.1.1).

9.4 DAILY CADMIUM INTAKE FROM TOBACCO SMOKING

Table 9:5 shows the results of calculations as to what amount of tobacco smoking is necessary to cause an accumulation of cadmium such that the renal cortex concentration is 200 μg/g. The assumptions concerning the mathematical model are the same as in Section 9.1 and the calculations start from the data in Table 9:2. It is assumed that one cigarette contains 0.1 μg Cd (Section 3.1.5). Even under the assumptions that pulmonary absorption is 50% and exposure time 50 years, more than ten cigarette packs per day are necessary for reaching the "critical" cadmium intake. If it is assumed that food intake gives 50 μg Cd/g in kidney cortex, the figures in Table 9:5 are decreased by 25%. In order to reach 50 μg Cd/g in kidney cortex (average in U.S.A., Section 4.3.2.1) from smoking only (25 years of exposure, 0.005% excretion, and 50% absorption) it is necessary to smoke 82 cigarettes per day. As the average smoking rate in the U.S.A. is about ten cigarettes per day (Beese, 1972), it can be concluded that in a randomly selected cross-sectional autopsy sample, smoking may contribute less than food to the observed cadmium accumulation with age.

9.5 DAILY CADMIUM INTAKE FROM INDUSTRIAL AIR

When calculating permissible concentrations in industrial air, it is generally assumed that a worker inhales 10 m^3 air during an 8-hr workday and that he works 225 days a year. Using these assumptions and the data in Table 9:2 as a base, it can be calculated that even if the daily body burden excretion were 0.02%, exposure time 10 years, and pulmonary absorption 25%, a kidney cortex cadmium concentration of 200 μg/g may be reached with an air concentration of 30 μg/m^3 (Table 9:6). This value is considerably below the present threshold limit value in the U.S.A. (100 μg/m^3 for cadmium fumes and 200 μg/m^3 for cadmium dust (ACGIH, 1971, 1973); an "intended change" of the TLV for cadmium fumes to 50 μg/m^3 was

TABLE 9:4

Necessary Cadmium Concentration (μg/m^3) in Ambient Air to Reach Critical Cadmium Concentration (200 μg/g Wet Weight) in Kidney Cortex under Different Absorption, Excretion, and Exposure Time Alternatives (Ventilation = 20 m^3/24 hr)*

Pulmonary absorption (%)	Exposure time in years	Excretion per day, % of body burden (corresponding biological half-time in years in parentheses)				
		0 (∞)	0.002 (95)	0.005 (38)	0.01 (19)	0.02 (9.5)
25	10	6.6	6.8	7.2	7.8	9.3
	25	2.6	2.9	3.3	4.0	5.7
	50	1.3	1.6	2.0	2.9	4.9
50	10	3.3	3.4	3.6	3.9	4.7
	25	1.3	1.5	1.7	2.0	2.9
	50	0.65	0.8	1.0	1.5	2.5

*Based on Table 9:2.

TABLE 9:5

Necessary Daily Amount of Tobacco Smoking* (Number of Cigarettes) to Reach 200 μg Cd/g Wet Weight in Renal Cortex under Different Absorption, Excretion, and Exposure Time Alternatives

Pulmonary absorption (%)	Exposure time in years	Excretion per day, % of body burden (corresponding biological half-time in years in parentheses)				
		0 (∞)	0.002 (95)	0.005 (38)	0.01 (19)	0.02 (9.5)
25	10	1,316	1,364	1,436	1,568	1,852
	25	524	576	656	804	1,144
	50	264	312	400	572	984
50	10	658	682	718	784	926
	25	262	288	328	402	572
	50	132	156	200	286	492

*Each cigarette causes an inhalation of 0.1 μg Cd.

TABLE 9:6

Cadmium Concentration in Industrial Air ($\mu g/m^3$) Necessary for Exposed Workers to Reach 200 μg Cd/g Wet Weight in Renal Cortex under Different Absorption, Excretion, and Exposure Time Alternatives. Ventilation = 10 m^3 during an 8-hr Workday; 225 Workdays Per Year

Pulmonary absorption (%)	Exposure time in years	Excretion per day, % of body burden (corresponding biological half-time in years in parentheses)				
		0 (∞)	0.002 (95)	0.005 (38)	0.01 (19)	0.02 (9.5)
25	10	21.3	22.1	23.3	25.4	30.0
	25	8.5	9.3	10.6	13.0	18.6
	50	4.3	5.1	6.5	9.3	16.0
50	10	10.7	11.1	11.7	12.7	15.0
	25	4.3	4.7	5.3	6.5	9.3
	50	2.2	2.6	3.3	4.7	8.0

reported in the 1973 issue of the U.S.A. TLV's. In the Soviet Union the maximum allowable concentration is 100 $\mu g/m^3$ (State Committee of the Council of Ministries, U.S.S.R., 1972). This value has been confirmed by the Ministry of Health of the U.S.S.R. and may not be exceeded. The recommended TLV in Japan is 100 $\mu g/m^3$ (Japanese Association of Industrial Health, 1971). Czechoslovakia has a recommended TLV of 100 $\mu g/m^3$ (Czechslovak Committee of MAC, 1969). In Finland the TLV was recently set at 10 $\mu g/m^3$. In Sweden the TLV had been the same as in the U.S.A. until September 1974, at which time it was lowered. As of now, the TLV for cadmium fumes is 20 $\mu g/m^3$. For cadmium dust the TLV is 20 $\mu g/m^3$ for the fraction with a particle size less than 5 μm and 50 $\mu g/m^3$ for total cadmium dust. In addition the new Swedish TLV will be accompanied by requirements for regular medical examinations of cadmium-exposed workers. It should be emphasized that the body burden excretion from the kidney most probably is 0.01 to 0.005% (Section 4.6). Hence, depending upon the actual pulmonary absorption, air concentrations of cadmium between 5.3 and 13.0 $\mu g/m^3$ may cause accumulation in the renal cortex up to the critical concentration of 200 μg/g after 25 years of exposure.

The calculations above did not account for the cadmium accumulation caused by food intake and

TABLE 9:7

Cadmium Concentration in Industrial Air ($\mu g/m^3$) Necessary for Exposed Workers (Smokers vs. Nonsmokers) with a Certain Intake of Cadmium in Food to Reach 200 μg Cd/g Wet Weight in Renal Cortex at 50 Years of Age under Different Excretion Rate and Exposure Time Alternatives*

	Exposure time in years	Excretion per day, % of body burden (corresponding biological half-time in years in parentheses)				
		0 (∞)	0.002 (95)	0.005 (38)	0.01 (19)	0.02 (9.5)
Nonsmoker	10	18.1	18.8	19.8	21.6	25.5
	25	7.2	7.9	9.0	11.1	15.8
Smoker	10	17.7	18.3	19.3	21.1	25.1
	25	6.7	7.5	8.6	10.6	15.3

*Conditions: pulmonary absorption = 25%; ventilation = 10 m^3 during an 8-hr workday; 225 workdays in a year; 30 μg Cd/g wet weight in renal cortex at age 50 from food intake. The smoker has smoked 30 cigarettes per day for 25 years. The inhalation of cadmium from each cigarette is 0.1 μg.

smoking. In another calculation we assumed that the food intake alone causes a renal cortex concentration of 30 μg/g wet weight in a 50-year-old person. We also assumed that the pulmonary absorption is 25% and the ventilation and number of workdays are the same as mentioned above. Table 9:7 presents the industrial air cadmium concentrations necessary to reach 200 μg Cd/g in kidney cortex under different excretion rates and exposure times as well as for different smoking statuses. The smokers were assumed to have smoked 30 cigarettes per day for 25 years, each cigarette with an inhalation of 0.1 μg Cd. The values are still lower than those in Table 9:6. It may well be concluded that industrial air concentrations of about 10 $\mu g/m^3$ can cause renal cortex cadmium concentrations exceeding the critical concentration. The daily cadmium intake from smoking adds to the risk for chronic cadmium poisoning among cadmium-exposed workers.

9.6 CONCLUSIONS

Using a one-compartment logarithmic accumulation model for cadmium in human renal cortex it was calculated that the necessary daily cadmium intake from food in order to reach the critical concentration would be 250 to 350 μg, corresponding to a cadmium concentration of 0.4 to 0.6 μg/g wet weight in basic foodstuffs.

Ambient air concentrations of 1 to 2 $\mu g/m^3$ during a 50-year exposure period may allow accumulation to reach the critical renal cortex level. Smoking alone may give such accumulation only if a person smokes more than ten packs per day. As based upon any of the assumed excretion, pulmonary absorption, and exposure time alternatives, the industrial air concentrations necessary to make renal cortex accumulation reach the critical level fall considerably below the current TLV, in most of the countries having a TLV, including the U.S.S.R. and U.S.A.

Chapter 10

GENERAL DISCUSSION AND CONCLUSIONS; NEED FOR FURTHER RESEARCH

10.1 GENERAL DISCUSSION AND CONCLUSIONS

In the preceding chapters a detailed review of the different aspects of cadmium intoxication has been given. Whenever motivated, in each chapter, there are special sections for conclusions, so these will not generally be repeated here. This chapter instead will deal with more general aspects of the problem, pointing out some of the most important conclusions and emphasizing the need for further studies.

Cadmium can undoubtedly constitute a most serious health problem. There is much evidence that exposure to this metal both in the industrial and in the general environment has given rise to serious intoxications in human beings. Such effects have been shown from inhaled as well as from ingested cadmium.

The manifestations of cadmium intoxication can take several forms. Inhalation of cadmium oxide fumes can produce acute damage in the lungs in the form of pneumonitis or pulmonary edema. Prolonged exposure to dust or fumes can cause an invalidating emphysema.

Although dose-response relationships are generally uncertain, it was concluded that an exposure to about 500 min · mg/m^3 of cadmium oxide fumes is immediately dangerous from the point of view of acute pulmonary manifestations. Chronic exposure to cadmium oxide fumes in concentrations well below 0.1 mg/m^3 is considered hazardous with reference to emphysema.

Systemic effects arise after absorption of cadmium. With the exposure that occurs from the general environment and also industrially, it is the effects due to long-term exposure to low concentrations of cadmium that are of interest. The critical organ is the kidney. Renal tubular dysfunction with proteinuria is a common manifestation. The renal tubular dysfunction may under certain circumstances give rise to severe secondary manifestations including a pronounced and invalidating osteomalacia. This form of cadmium intoxication has been seen industrially after inhalation of cadmium as well as after long-term ingestion of contaminated food. Other systemic effects include anemia and liver dysfunction.

In Japan the fully developed picture of cadmium intoxication with kidney damage and osteomalacia is known under the name "Itai-itai byo" (literally ouch-ouch disease) because of the severe pains accompanying the skeletal disorder. A large number of cases of cadmium intoxication with renal dysfunction but without known osteomalacia have also been found in Japan. Contamination of the food, particularly rice, with cadmium is considered to be the main cause of this disorder. The contamination of the food has with all probability its primary cause in contaminated river water, and in some places perhaps also in the contaminated ambient air. The reason why the manifestations of cadmium intoxication in Japan often have come to such an advanced stage is not quite clear. In one area it may be that the exposure several years ago was very excessive. There is also the possibility that poor nutritional habits, such as low intake of calcium, protein, and vitamin D, have been essential to the problem. Furthermore, the farmers and their families eat primarily rice, and this rice is only from the locally grown crop, meaning that if the rice is contaminated the exposure will be high.

In some animal experiments it has been possible to produce hypertension after prolonged exposure to cadmium. There is no conclusive evidence that cardiovascular disease in human beings is causally associated with cadmium exposure although some statistical associations found in epidemiological studies merit further investigations.

Animal experiments have shown clearly that injections of cadmium salts cause malignancies at the site of injection. There are some data from human beings that tend to show an association between cadmium exposure and cancer. However, much more evidence is needed before any conclusions can be made concerning the causality.

The nature of the effects of cadmium on the cellular level has been discussed in several sections of the report, and the intimate relationship between cadmium and zinc metabolism has been pointed out. There are mechanisms that are fairly well known, but much needs to be done before the complete picture of the different manifestations of cadmium intoxication can be understood.

What makes cadmium contamination of the environment a particularly serious hazard is its pronounced tendency to accumulate in the body. This is evident from autopsy data on human beings as well as from animal studies. In humans the biological half-time for total body is between 10 to 30 years. There are still considerable uncertainties, and a proper accumulation model for cadmium in different organs, including the kidney, has yet to be devised.

The excretion of cadmium via urine and the gastrointestinal tract is extremely low. In the feces, the excretion is only to a minor degree related to total body burden. In normal human beings the excretion of cadmium via the urine is around 2 μg per day or less, increasing with age. On a group basis, in normal populations with moderate exposure to cadmium, this increase with age parallels the increase in the total body burden. A similar relationship has not been found in exposed workers without proteinuria. Furthermore, due to the wide scatter and based on animal data, the predictive value of urinary cadmium as an index of total body burden or kidney burden on an individual basis must be considered low.

Daily, human beings ingest a substantial amount of cadmium in food, of which the average normal absorption is probably about 5%. Based on animal and human data, however, it is considered quite possible that under certain circumstances such as calcium and protein deficiency absorption in human beings may reach 10%.

The ambient air contains small amounts of cadmium, and in the vicinity of certain factories the concentration can be considerable. High exposure is not uncommon inside industries where cadmium is used. The absorption via inhalation is probably higher than that via ingestion. Depending on factors such as particle size, absorption of 10 to 50% has been estimated. It should also be remembered that estimates of absorption via inhalation are subject to considerable uncertainty.

When absorbed, cadmium will accumulate in liver and kidneys. The larger the exposure, the more will be accumulated in the liver relative to other organs. Long-term exposure to fairly small amounts of cadmium will give kidney values corresponding to about one third of the total body burden. When kidney damage occurs, the concentration of cadmium in the kidneys will decrease substantially. This probably explains why advanced cases of cadmium intoxication, as seen in the Itai-itai disease and sometimes in industrial poisoning, will have low kidney levels in spite of high liver levels of cadmium.

When renal tubular dysfunction appears, the excretion of cadmium via the urine will increase drastically. Animal experiments have shown an increase of about 100 times. The fact that urinary cadmium excretion is low before renal dysfunction appears, along with the great individual scatter, makes the analysis of cadmium in urine of limited value as an indicator of total body burden. But, if the purpose is to detect cadmium-induced renal dysfunction, it may be used since high cadmium excretion would point very strongly towards such an effect.

Cadmium levels in blood will increase with exposure, though during exposure they are not a good indicator of the body burden, as blood levels also will reflect recent exposure to cadmium.

As was stated above, at long-term exposure to fairly low concentrations of cadmium about one third of the cadmium will be found in the kidneys. Animal data as well as data from autopsies of workers with none or only a slight renal dysfunction point towards 200 μg/g wet weight of cadmium in the renal cortex as being the critical level for renal dysfunction, diagnosed through kidney function tests and occurrence of proteinuria. Based on this level and reasonable retention rates for cadmium, estimates have been made of exposures that are necessary to reach the critical concentration of cadmium in the kidneys.

It is calculated that the necessary daily cadmium intake from food in order to reach the critical concentration would be 250 to 350 μg, corresponding to a cadmium concentration of 0.4 to 0.6 μg/g wet weight in basic foodstuffs.

Ambient air concentrations of 1 to 2 $\mu g/m^3$ during a 50-year exposure period may allow accumulation to reach the critical renal cortex level. Smoking alone may give such accumulation only if a person smokes ten or more packs per day. Industrial air concentrations necessary to make renal cortex accumulation reach the critical level fall considerably below the current TLV (0.1 to 0.2 mg/m^3) as based on any of the assumed excretion, pulmonary absorption, and exposure time alternatives.

It should also be mentioned that problems with

correct analysis of cadmium in various materials are considerable. For example, the concentrations in blood and urine are in the order of some nanograms per gram or less. The method most commonly used is atomic absorption spectrophotometry, but, depending on interfering substances or type of matrix in the sample, different chemical preparation techniques are necessary for each type of material to be analyzed.

10.2 NEED FOR FURTHER RESEARCH

There is an immediate and urgent need for research on the dose-response relationships. Above all, absorption and excretion rates should be investigated further. Such studies should include exposure via the peroral route as well as inhalation. Different cadmium compounds should be studied with reference to the influence of particle size at absorption via inhalation. The information on absorption to date is very inadequate, and it goes without saying that preventive measures that need to be taken may be quite different for different absorption rates. Animal experiments can certainly provide a substantial part of the necessary information. On human beings it should be possible to carry out metabolic studies where uptake and excretion of cadmium are examined over prolonged periods.

The biological half-time of cadmium is no doubt very long, but further studies on this matter in animals and in human beings are greatly needed. Such studies should be carried out not only after single exposures but also after continuous exposures. This would make it possible to study how much of the excretion via the urine and feces is directly proportional to the total body burden and how much is proportional to the daily dose of cadmium.

More detailed studies on concentrations of cadmium relative to total body burden in kidneys and in other organs are well motivated. Such studies should be carried out at different exposure levels.

The critical level in kidneys should be studied further. It should also be borne in mind that effect levels used up until now have included only fairly gross effects. Dysfunctions on a cellular level may be found at considerably lower concentrations.

As a basis for the other necessary studies on cadmium toxicity, investigations into the methodology of cadmium analysis in various types of biological samples are called for. Comparisons between methods that differ chemically as well as comparisons between different laboratories should be carried out in order to evaluate accuracy of analysis. Such interlaboratory studies would at the same time provide an opportunity to compare epidemiological data from different areas.

Epidemiological studies should be carried out on populations exposed in different degrees to cadmium. Such populations may be found not only within industries but also in areas surrounding industries emitting cadmium. The unfortunate wide-scale contamination with cadmium in Japan will give unique possibilities for studies in that country on several aspects of dose-response relationships. There may well be other countries with large-scale mining and metal smelting operations where environmental pollution with cadmium could be a public health problem. Systematic investigations into such possibilities should be carried out.

When embarking on epidemiological studies, due attention must be paid to methodological questions in order to make results from different studies comparable. Effects to be looked for should include renal dysfunction including proteinuria, certain cardiovascular diseases, and malignancies.

The accumulation of cadmium in the body should be studied further in connection with autopsies in cases of accidental death as well as from other causes. Longitudinal studies are imperative. For the time being this seems the only way to check whether or not the long-term accumulation within the populations of different countries is on the rise. Obviously, the exposure via different routes should be studied simultaneously.

High concentrations of cadmium have not only been found in the kidneys. The significance of an accumulation of cadmium in liver, pancreas, and thyroid certainly merits further studies.

Very little is known about genetic and teratogenic effects of cadmium. There is a great need for studies in these areas.

Chelating agents may be found in the environment due to natural occurrence or as wastes from commercial products. These substances may have a deleterious effect on cadmium metabolism. It is necessary to examine further the combined effects of cadmium and such agents (e.g., EDTA and NTA) in long-term studies.

It is known that some manifestations of cadmium toxicity can be prevented by the simultaneous administration of zinc. However, concerning the possible interaction of cadmium with other metals, essential or nonessential, virtually nothing is known.

It should be emphasized that there are severe gaps in the understanding of the mechanisms behind the different manifestations of cadmium intoxication. Much of our present knowledge of cadmium intoxication is based on studies on the metabolism of the metal. By stimulating further basic research on metal toxicology, we can expect to get a better understanding of the complex effects of cadmium on human health.

Finally, it should be stressed that research needs exist concerning aspects of cadmium in the environment which are indirectly important for toxicological and epidemiological appraisals of cadmium. Although not directly within the scope of this review, it should be mentioned that there is a need for more data on the turnover of cadmium in nature, including the accumulation via air and water in food chains.

REFERENCES

Many of the references in this list are in Japanese. In some cases, the Department of Environmental Hygiene at the Karolinska Institute (S-104 01 Stockholm 60, Sweden) has a translation on file. For every case, the primary author or organization is included in a separate address list on page 229.

References with an asterisk (*) indicate data originally reported in the first edition of *Cadmium in the Environment* and at that time referred to as "unpublished data" or "personal communication."

Abdullah, M. I. and Royle, L. G., The determination of copper, lead, cadmium, nickel, zinc and cobalt in natural waters by pulse polarography, *Anal. Chim. Acta,* 58, 283, 1972.

Abdullah, M. I., Royle, L. G., and Morris, A. W., Heavy metal concentration in coastal waters, *Nature,* 235, 158, 1972.

ACGIH, *Documentation of the Threshold Limit Values for Substances in Workroom Air,* 3rd ed., American Conference of Governmental Industrial Hygienists, Cincinnati, Ohio, 1971, 35.

ACGIH, *TLV's® For Chemical Substances and Physical Agents in the Workroom Environment with Intended Changes for 1973,* American Conference of Governmental and Industrial Hygienists, Cincinnati, Ohio, 1973, 12, 36.

Adams, R. G., Harrison, J. F., and Scott, P., The development of cadmium-induced proteinuria, impaired renal function and osteomalacia in alkaline battery workers, *Q. J. Med.,* 38, 425, 1969.

Ahlmark, A., Axelsson, B., Friberg, L., and Piscator, M., Further investigations into kidney function and proteinuria in chronic cadmium poisoning, *Int. Congr. Occup. Health,* 13, 201, 1961.

Ahlmark, A., Friberg, L., and Hardy, H., The solubility in water of two cadmium dusts with regard to the risk of chronic poisoning, *Ind. Med. Surg.,* 25, 514, 1956.

Air Quality Criteria for Particulate Matter, National Air Pollution Control Administration Publication AP-49, U.S. Dept. of Health, Education, and Welfare, Public Health Service, Washington, D.C., 1969.

Albert, R. E., Lippmann, M., and Peterson, H. T., Jr. The effects of cigarette smoking on the kinetics of bronchial clearance in humans and donkeys, in *Inhaled Particles,* 3rd ed., Walton, W. H., Ed., Unwin Brothers, London, 1971, 165.

Alexander, F. W., Delves, H. T., and Clayton, B. E., The uptake and excretion by children of lead and other contaminants, in *Int. Symp. Environ. Health Aspects Lead,* Amsterdam, October 1972, Commission of the European Communities, Luxembourg, 1973, 319.

Allanson, M. and Deanesly, R., Observations on cadmium damage and repair in rat testes and the effects on the pituitary gonadotrophs, *J. Endocrinol.,* 24, 453, 1962.

Allen, H. E., Matson, W. R., and Mancy, K. H., Trace metal characterization in aquatic environments by anodic stripping voltammetry, *J. Water Pollut. Control Fed.,* 42, 573, 1970.

Alsberg, C. L. and Schwartze, E. W., Pharmacological action of cadmium, *J. Pharmacol.,* 13, 504, 1919.

Altman, P. L. and Dittmer, D. S., Eds., *Biology Data Book,* Federation of American Societies for Experimental Biology, Washington, D.C., 1964, 220.

Anbar, M. and Inbar, M., The effect of certain metallic cations on the iodide uptake in the thyroid gland of mice, *Acta Endocrinol.,* 46, 643, 1964.

Andreuzzi, P. and Odescalchi, C. P., Experimental acute intoxication from cadmium chloride in the rabbit. I. Changes in the GOT-activity in the serum, *Boll. Soc. Ital. Biol. Sper.,* 34, 1376, 1958 (in Italian).

Anke, M. and Schneider, H.-J., Der Zink-, Kadmium- und Kupferstoffwechsel des Menschen, *Arch. Veterinärmed.,* 25, 805, 1971.

Anwar, R., Langham, R., Hoppert, C. A., Alfredson, B. V., and Byerrum, R. U., Chronic toxicity studies. III. Chronic toxicity of cadmium and chromium in dogs, *Arch. Environ. Health,* 3, 456, 1961.

Arnstein, A. R., Frame, B., and Frost, H. M., Recent progress in osteomalacia and rickets, *Ann. Intern. Med.,* 67, 1296, 1967.

Association of Health Statistics (Koosei Tookei Kyokai), *Trends in National Health (Kokumin Eisei no Dookoo),* Tokyo, 1972 (in Japanese).

Athanasiu, M. and Langlois, P., L'action comparee des sels de cadmium et de zinc, *Arch. Physiol.,* 28, 251, 1896.

Athanassiadis, Y. C., *Preliminary Air Pollution Survey of Cadmium and Its Compounds,* National Air Pollution Control Administration, Raleigh, N.C., 1969.

Axelsson, B., Urinary calculus in long-term exposure to cadmium, *Int. Congr. Occup. Health,* 14, 939, 1963.

Axelsson, B., Dahlgren, S. E., and Piscator, M., Renal lesions in the rabbit after long-term exposure to cadmium, *Arch. Environ. Health,* 17, 24, 1968.

Axelsson, B. and Piscator, M., Renal damage after prolonged exposure to cadmium. An experimental study, *Arch. Environ. Health,* 12, 360, 1966a.

Axelsson, B. and Piscator, M., Serum proteins in cadmium poisoned rabbits with special reference to hemolytic anemia, *Arch. Environ. Health,* 12, 374, 1966b.

***Axelsson, B. and Piscator, M.**, in *Cadmium in the Environment,* Friberg, L., Piscator, M., and Nordberg, G., CRC Press, Cleveland, 1971, 85.

Baader, E. W., Die chronische Kadmiumvergiftung, *Dtsch. Med. Wochenschr.*, 76, 484, 1951.

Baader, E. W., Chronic cadmium poisoning, *Ind. Med. Surg.*, 21, 427, 1952.

Baker, T. D. and Hafner, W. G., Cadmium poisoning from a refrigerator shelf used as an improvised barbecue grill, *Public Health Rep.*, 76, 543, 1961.

Balkrishna, Changes of thyroid function in response to cadmium administration in rats – Studies with I^{131}, *J. Sci. Ind. Res.*, 21C, 187, 1962.

Banis, R. J., Pond, W. G., Walker, E. F., and O'Connor, J. R., Dietary cadmium, iron, and zinc interactions in the growing rat, *Proc. Soc. Exp. Biol. Med.*, 130, 802, 1969.

Barrett, H. M. and Card, B. Y., Studies on the toxicity of inhaled cadmium, II. The acute lethal dose of cadmium oxide for man, *J. Ind. Hyg. Toxicol.*, 29, 288, 1947.

Barrett, H. M., Irwin, D. A., and Semmons, E., Studies on the toxicity of inhaled cadmium, I. The acute toxicity of cadmium oxide by inhalation, *J. Ind. Hyg. Toxicol.*, 29, 286, 1947.

Barthelemy, P. and Moline, R., Intoxication chronique par l'hydrate de cadmium, son signe precoce: La Bague Jaune Dentaire, *Paris Med.*, 1, 7, 1946.

Baum, J. and Worthen, H. G., Induction of amyloidosis by cadmium, *Nature,* 213, 1040, 1967.

Baumslag, N., Keen, P., and Petering, H., Carcinoma of the maxillary antrum and its relationship to trace metal content of snuff, *Arch. Environ. Health,* 23, 1, 1971.

Beese, D. H., Tobacco Consumption in Various Countries, Research Paper No. 6, Tobacco Research Council, London, 1972.

Berggard, I. and Bearn, A. C., Isolation and properties of a low molecular weight globulin occurring in human biological fluids, *J. Biol. Chem.*, 243, 4095, 1968.

Bergner, K. G., Lang, B., and Ackermann, H., Zum Cadmiumgehalt deutscher Weine, *Mitt. Rebenwein Obstbau Fruechteverwert.*, 22, 101, 1972.

Berlin, M., Fredricsson, B., and Linge, G., Bone marrow changes in chronic cadmium poisoning in rabbits, *Arch. Environ. Health,* 3, 176, 1961.

Berlin, M. and Friberg., L., Bone-marrow activity and erythrocyte destruction in chronic cadmium poisoning, *Arch. Environ. Health,* 1, 478, 1960.

Berlin, M., Hammarström, L., and Maunsbach, A. G., Microautoradiographic localization of water-soluble cadmium in mouse kidney, *Acta Radiol.*, 2, 345, 1964.

Berlin, M. and Piscator, M., Blood volume in normal and cadmium-poisoned rabbits, *Arch. Environ. Health,* 2, 576, 1961.

Berlin, M. and Ullberg, S., The fate of ^{109}Cd in the mouse. An autoradiographic study after a single intravenous injection of $^{109}CdCl_2$, *Arch. Environ. Health,* 7, 686, 1963.

Berrow, M. L. and Webber, J., Trace elements in sewage sludges, *J. Sci. Food Agric.*, 23, 93, 1972.

Berry, J.-P. Les lésions rénales provoquées par le cadmium. Etude au microscope électronique et au micro-analyseur a sonde électronique, *Pathol. Biol.*, 20, 401, 1972.

Beton, D. C., Andrews, G. S., Davies, H. J., Howells, L., and Smith, G. F., Acute cadmium fume poisoning, five cases with one death from renal necrosis, *Br. J. Ind. Med.*, 23, 292, 1966.

Blejer, H. P., Caplan, P. E., and Alcocer, A. E., Acute cadmium fume poisoning in welders – a fatal and a nonfatal case in California, *Calif. Med.*, 105, 290, 1966.

Blix, G., Wretlind, A., Bergström, S., and Westin, S. I., The food intake of the Swedish people, *Vår Föda,* 17, 1, 1965 (in Swedish).

Bonnell, J. A., Emphysema and proteinuria in men casting copper-cadmium alloys, *Br. J. Ind. Med.*, 12, 181, 1955.

Bonnell, J. A., Kazantzis, G., and King, E., A follow-up study of men exposed to cadmium oxide fume, *Br. J. Ind. Med.*, 16, 135, 1959.

Bonnell, J. A., Ross, J. H., and King, E., Renal lesions in experimental cadmium poisoning, *Br. J. Ind. Med.*, 17, 69, 1960.

Boström, H., Osteomalacia, *Läkartidningen,* 64, 4679, 1967 (in Swedish).

Boström, H. and Wester, P. O., Full balances of trace elements in two cases of osteomalacia, *Acta Med. Scand.*, 183, 209, 1968.

Boström, H. and Wester, P. O., Pre- and postoperative excretion of trace elements in primary hyperparathyroidism, *Acta Endocrinol.*, 60, 380, 1969.

Bouissou, H. and Fabre, M. Th., Lésions provoquées par le sulfate de cadmium sur le testicule du rat, *Arch. Mal. Prof. Med. Trav. Secur.*, 26, 127, 1965.

Bowen, H. J. M., *Trace Elements in Biochemistry,* Academic Press, London, 1966.

Brune, D., Frykberg, B., Samsahl, K., and Wester, P. O., Determination of elements in normal and leukemic human whole blood by neutron activation analysis, AE-60, Aktiebolaget Atomenergi, Stockholm, 1961.

Bryan, S. E. and Hayes, E. F., Partial characterization of liver proteins following exposure to mercury, *FEBS (Fed. Eur. Biochem. Soc.) Lett.*, 21, 21, 1972.

Buchauer, M. J., Contamination of soil and vegetation near a zinc smelter by zinc, cadmium, copper and lead, *Environ. Sci. Technol.*, 7, 131, 1973.
Bühler, R., Human hepatic metallothionein, in *9th Int. Cong. Biochem.*, Stockholm, July 1973, Aktiebolaget Egnellska Boktryckeriet, Stockholm, 1973, 82 (abstr.).
Bulmer, F. M. R., Rothwell, H. E., and Frankish, E. R., Industrial cadmium poisoning, *Can. Public Health J.*, 29, 19, 1938.
Bunn, C. R. and Matrone, G., In vivo interactions of cadmium, copper, zinc and iron in the mouse and the rat, *J. Nutr.*, 90, 395, 1966.
Burch, G. E. and Walsh, J. J., The excretion and biologic decay rates of Cd^{115m} with a consideration of space, mass, and distribution in dogs, *J. Lab. Clin. Med.*, 54, 66, 1959.
Burkitt, A., Lester, P., and Nickless, G., Distribution of heavy metals in the vicinity of an industrial complex, *Nature*, 238, 327, 1972.
Butler, E. A. and Flynn, F. V., The proteinuria of renal tubular disorders, *Lancet*, 2, 978, 1958.
Butler, E. A. and Flynn, F. V., The occurrence of post-gamma protein in urine: a new protein abnormality, *J. Clin. Pathol.*, 14, 172, 1961.
Butler, E. A., Flynn, F. V., Harris, H., and Robson, E. B., A study of urine proteins by two dimensional electrophoresis with special reference to the proteinuria of renal tubular disorders, *Clin. Chim. Acta*, 7, 34, 1962.
Butt, E. M., Nusbaum, R. E., Gilmour, T. C., Didio, S. L., and Sister Mariano, Trace levels in human serum and blood, *Arch. Environ. Health*, 8, 52, 1964.
Buxton, R. St. J., Respiratory function in men casting cadmium alloys, II.: The estimation of the total lung volume, its subdivisions and the mixing coefficient, *Br. J. Ind. Med.*, 13, 36, 1956.
Cameron, E. and Foster, C. L., Observations on the histological effects of sub-lethal doses of cadmium chloride in the rabbit, *J. Anat.*, 97, 269, 1963.
Carlson, L. A. and Friberg, L., The distribution of cadmium in blood after repeated exposure, *Scand. J. Clin. Lab. Invest.*, 9, 1, 1957.
Carroll, R. E., The relationship of cadmium in the air to cardiovascular disease death rates, *J.A.M.A.*, 198, 267, 1966.
Castano, P. and Vigliani, E. C., Cadmium nephropathy: ultrastructural observations (horseradish peroxidase), *J. Occup. Med.*, 14, 125, 1972.
Caujolle, F., Oustrin, J., and Silve-Mamy, G., Fixation et circulation entérohépatique du cadmium, *Eur. J. Toxicol.*, 4, 310, 1971.
Ceresa, C., An experimental study of cadmium intoxication, *Med. Lav.*, 36, 71, 1945 (in Italian).
Chatterjee, S. N. and Kar, A. B., Chemical sterilization of stray dogs, *Indian Vet. J.*, 45, 649, 1968.
Chaube, S., Nishimura, H., and Swinyard, C. A., Zinc and cadmium in normal human embryos and fetuses, *Arch. Environ. Health*, 26, 237, 1973.
Chen, R., Wagner, P., Ganther, H. E., and Hoekstra, W. G., A low molecular weight cadmium-binding protein in testes of rats: possible role in cadmium-induced testicular damage, *Fed. Proc. Abstr.*, 31, 699, 1972.
Chernoff, N. and Courtney, K. D., Maternal and Fetal Effects of NTA, NTA and Cadmium, NTA and Mercury, NTA and Nutritional Imbalance in Mice and Rats, Nat. Inst. Environ. Health Sci. Progress Report, Dec. 1, 1970 (Jan. 18, 1971).
Chiappino, G. and Baroni, M., Morphological signs of hyperactivity of the renin-aldosterone system in cadmium induced experimental hypertension, *Med. Lav.*, 60, 297, 1969 (in Italian).
Chiappino, G., Repetto, L., and Pernis, B., A histochemical study of the changes induced by cadmium on the leucylaminopeptidase activity in the rat kidney, *Med. Lav.*, 59, 584, 1968 (in Italian).
Chiquoine, A. D., Observations on the early events of cadmium necrosis of the testis, *Anat. Rec.*, 149, 23, 1964.
Chiquoine, A. D. and Suntzeff, V., Sensitivity of mammals to cadmium necrosis of the testis, *J. Reprod. Fertil.*, 10, 455, 1965.
Church, F. W., A mixed color method for the determination of cadmium in air and biological samples by the use of dithizone, *J. Ind. Hyg. Toxicol.*, 29, 34, 1947.
Clarkson, T. W. and Kench, J. E., Urinary excretion of amino acids by men absorbing heavy metals, *Biochem. J.*, 62, 361, 1956.
Clegg, E. J. and Carr, I., Changes in the blood vessels of the rat testis and epididymis produced by cadmium chloride, *J. Pathol. Bacteriol.*, 94, 317, 1967.
Clegg, E. J., Carr, I., and Niemi, M., The effect of a second dose of cadmium salts on vascular permeability in the rat testis, *J. Endocrinol.*, 45, 265, 1969.
Cole, G. M. and Baer, L. S., "Food poisoning" from cadmium, *U.S. Nav. Med. Bull.*, 43, 398, 1944.
Cooper, W. C., Tabershaw, I. R., and Nelson, K. W., Laboratory studies of workers in lead smelting and refining, in *Int. Symp. Environ. Health Aspects Lead*, Amsterdam, October 1972, Commission of the European Communities, Luxembourg, 1973, 517.
Coppoletta, J. M. and Wolbach, S. B., Body length and organ weights of infants and children, *Am. J. Pathol.*, 9, 55, 1933.
Corneliussen, P. E., Pesticide residues in total diet samples (V), *Pestic. Monit. J.*, 3, 89, 1970.
Cotzias, G. C., Borg, D. C., and Selleck, B., Virtual absence of turnover in cadmium metabolism: ^{109}Cd studies in the mouse, *Am. J. Physiol.*, 201, 927, 1961.

Creason, J. P., McNulty, O., Heiderscheit, L. T., Swanson, D. H., and Buechley, R. W., Roadside gradients in atmospheric concentrations of cadmium, lead and zinc, in *Trace Substances in Environmental Health – V. A Symposium,* Hemphill, D. D., Ed., Univ. of Missouri Press, Columbia, 1972, 129.

Curry, A. S., and Knott, A. R., "Normal" levels of cadmium in human liver and kidney in England, *Clin. Chim. Acta,* 30, 115, 1970.

Cvetkova, R. P., Materials on the study of the influence of cadmium compounds on the generative function, *Gig. Tr. Prof. Zabol.,* 14, 31, 1970 (in Russian).

Czechoslovak Committee of MAC, *Documentation of MAC in Czechoslovakia,* Czechoslovak Committee of MAC, Prague, 1969, 29.

Dalhamn, T. and Friberg, L., The effect of cadmium on blood pressure and respiration and the use of dimercaprol (BAL) as antidote, *Acta Pharmacol.,* 10, 199, 1954.

Dalhamn, T. and Friberg, L., Dimercaprol (2,3-dimercaptopropanol) in chronic cadmium poisoning, *Acta Pharmacol.,* 11, 68, 1955.

Dalhamn, T. and Friberg, L., Morphological investigations on kidney damage in chronic cadmium poisoning, *Acta Pathol. Microbiol. Scand.,* 40, 475, 1957.

Davies, J. M., Mortality among workers at two copper works where cadmium was in use, *Br. J. Prev. Soc. Med.,* 26, 59, 1972.

Davis, J. S., Flynn, F. V., and Platt, H. S., The characterisation of urine protein by gel filtration, *Clin. Chim. Acta,* 21, 357, 1968.

Decker, C. F., Byerrum, R. U., and Hoppert, C. A., A study of the distribution and retention of cadmium-115 in the albino rat, *Arch. Biochem.,* 66, 140, 1957.

Decker, L. E., Byerrum, R. U., Decker, C. F., Hoppert, C. A., and Langham, R. F., Chronic toxicity studies. I. Cadmium administered in drinking water to rats, *A.M.A. Arch. Ind. Health,* 18, 228, 1958.

Delves, H. T., A micro-sampling method for the rapid determination of lead in blood by atomic absorption spectrophotometry, *Analyst,* 95, 431, 1970.

Delves, H., Bicknell, J., and Clayton, B., The excessive ingestion of lead and other metals by children, in *Int. Symp. Environ. Health Aspects Lead,* Amsterdam, October 1972, Commission of the European Communities, Luxembourg, 1973, 345.

Dent, C. E. and Harris, H., Hereditary forms of rickets and osteomalacia, *J. Bone Jt. Surg. Br. Vol.,* 38, 204, 1956.

Dimow, G. and Knorre, D., Fermenthistochemische und enzymelektrophoretische Untersuchungen an Rattenhoden in den ersten 48 Stunden experimenteller Cadmiumintoxikation, *Virchows Arch. Abt. A. Pathol. Anat.,* 342, 252, 1967.

Djuric, D., Kerin, Z., Graovac-Leposavic, L., Novak, L., and Kop, M., Environmental contamination by lead from a mine and smelter, *Arch. Environ. Health,* 23, 275, 1971.

Doll, R., Muir, C. S., and Waterhouse, J.A.H., Eds., *Cancer Incidence in Five Continents,* Vol. 2., Union Internationale Contre Le Cancer (UICC), Geneva, 1970 (distributed by Springer-Verlag, Berlin).

Dreizen, S., Levy, B. M., Niedermeier, W., and Griggs, J. H., Comparative concentrations of selected trace metals in human and marmoset saliva, *Arch. Oral Biol.,* 15, 179, 1970.

Duggan, R. E. and Corneliussen, P. E., Dietary intake of pesticide chemicals in the United States (III), June 1968 – April 1970, *Pestic. Monit. J.,* 5, 331, 1972.

Durbin, P. W., Scott, K. G., and Hamilton, J. G. The distribution of radioisotopes of some heavy metals in the rat, *Univ. Calif. Publ. Pharmacol.,* 3, 1, 1957.

Durum, W. H., Hem, J. H., and Heidel, S. G., Reconnaissance of Selected Minor Elements in Surface Waters of the United States, Geological Survey Circular 643, U.S. Department of the Interior, Washington, D.C., October 1970.

Ediger, R. D. and Coleman, R. L., Determination of cadmium in blood by a Delves cup technique, *Atomic. Absorp. Newsl.,* 12, 3, 1973.

Elcoate, P. V., Fischer, M. I., Mawson, C. A., and Millar, M. J., The effect of zinc deficiency on the male genital system, *J. Physiol.,* 129, 53, 1955.

Environment Agency (Kankyo cho). Pollution Hygiene Section, Countermeasures Against Environmental Pollution by Cadmium (Kadmium Kankyo Osen Taisaku) (mimeographed document including a number of earlier reports from the Ministry of Health and Welfare), March 1972 (in Japanese, translated by Seizaburo Aoki, Japanese Language Translation Service, Fujisawa, Japan).

Erickson, A. E. and Pincus, G., Insensitivity of fowl testes to cadmium, *J. Reprod. Fertil.,* 7, 379, 1964.

Eschnauer, H., Bestimmung von Cadmium im Wein, *Z. Lebensm-Unters. -Forsch.,* 127, 4, 1965.

Essing, H. G. Schaller, K. H., Szadkowski, D., and Lehnert, G., Usuelle Cadmiumbelastung durch Nahrungsmittel und Getrenke, *Arch. Hyg. Bakteriol.,* 153, 490, 1969.

Evans, G. W. and Cornatzer, W. E., Copper and zinc metalloproteins in the rat, *Fed. Proc. Abstr.,* 29, 695, 1970.

Evans, G. W., Majors, P. F., and Cornatzer, W. E., Mechanism for cadmium and zinc antagonism of copper metabolism, *Biochem. Biophys. Res. Commun.,* 40, 1142, 1970.

Evrin, P.-E., β_2-microglobulin in human biological fluids and cells, *Acta Univ. Uppsaliensis,* 150, 1973.

Evrin, P.-E., Peterson, P. A., Wide, L., and Berggård, I., Radioimmunoassay of β_2-microglobulin in human biological fluids, *Scand. J. Clin. Lab. Invest.,* 28, 439, 1971.

Evrin, P.-E. and Wibell, L., The serum levels and urinary excretion of β_2-microglobulin in apparently healthy subjects, *Scand. J. Clin. Lab. Invest.*, 29, 69, 1972.

Eybl, V., Sýkora, J., and Mertl, F., Wirkung von CaADTA und CaDTPA bei der Kadmiumvergiftung, *Acta. Biol. Med. Ger.*, 17, 178, 1966a.

Eybl, V., Sýkora, J., and Mertl, F., K prechodu Zn, Cd, Hg a chelátú techto kovu placentarni barriéou, *Cesk. Fysiol.*, 15, 36, 1966b.

Eybl, V., Sýkora, J., and Mertl, F., Einfluss von Natriumselenit, Natriumtellurit und Natriumsulfit auf Retention und Verteilung von Cadmium bei Mäusen, *Arch. Toxikol.*, 26, 169, 1970.

Favino, A., Candura, F., Chiappino, G., and Cavalleri, A., Study on the androgen function of men exposed to cadmium, *Med. Lav.*, 59, 105, 1968.

Favino, A. and Nazari, G., Renal lesions induced by a single subcutaneous $CdCl_2$ injection in rat, *Lav. Um.*, 19, 367, 1967 (in Italian).

Ferm, V. H., Hanlon, D. P., and Urban, J., The permeability of the hamster placenta to radioactive cadmium, *J. Embryol. Exp. Morphol.*, 22, 107, 1969.

Finlayson, J. S., Asofsky, R., Potter, M., and Runner, C. C., Major urinary protein complex of normal mice: origin, *Science*, 149, 981, 1965.

Fitzhugh, O. G. and Meiller, F. H., The chronic toxicity of cadmium, *J. Pharmacol. Exp. Ther.*, 72, 15, 1941.

Flynn, F. V. and Platt, H. S., The origin of the proteins excreted in tubular proteinuria, *Clin. Chim. Acta*, 21, 377, 1968.

Foster, C. L. and Cameron, E., Observations on the histological effects of sub-lethal doses of cadmium chloride in the rabbit. II: The effect on the kidney cortex, *J. Anat.*, 97, 281, 1963.

Fournier, J. E., Le depistage, et la mesure de la surdite professionnelle, *Arch. Mal. Prof. Med. Trav. Secur. Soc.*, 24, 817, 1963.

Fox, M.R.S. and Fry, B. E., Jr., Cadmium toxicity decreased by dietary ascorbic acid supplements, *Science*, 169, 989, 1970.

Fox, M.R.S., Fry, B. E., Jr., Harland, B. F., Schertal, M. E., and Weeks, C. E., Effect of ascorbic acid on cadmium toxicity in the young coturnix, *J. Nutr.*, 101, 1295, 1971.

Freedman, L. R., Seki, M., Phair, J. P., and Nefzger, M. D., Proteinuria in Hiroshima and Nagasaki, Japan, *Yale J. Biol. Med.*, 40, 109, 1967.

Friberg, L., Proteinuria and kidney injury among workmen exposed to cadmium and nickel dust, *J. Ind. Hyg. Toxicol.*, 30, 32, 1948a.

Friberg, L., Proteinuria and emphysema among workers exposed to cadmium and nickel dust in a storage battery plant, *Proc. Int. Congr. Ind. Med.*, 9, 641, 1948b.

Friberg, L., Health hazards in the manufacture of alkaline accumulators with special reference to chronic cadmium poisoning, *Acta Med. Scand.*, 138, Suppl. 240, 1950.

Friberg, L., Further investigations on chronic cadmium poisoning; a study on rabbits with radioactive cadmium, *A.M.A. Arch Ind. Hyg. Occup. Med.*, 5, 30, 1952.

Friberg, L., Iron and liver administration in chronic cadmium poisoning and studies on the distribution and excretion of cadmium. Experimental investigations in rabbits, *Acta Pharmacol.*, 11, 168, 1955.

Friberg, L., Edathamil calcium-disodium in cadmium poisoning, *A.M.A. Arch. Ind. Health*, 13, 18, 1956.

Friberg, L., Deposition and distribution of cadmium in man in chronic cadmium poisoning, *A.M.A. Arch. Ind. Hyg. Occup. Med.*, 16, 27, 1957.

Friberg, L. and Nyström, A., Aspects on the prognosis in chronic cadmium poisoning, *Läkartidningen*, 49, 2629, 1952 (in Swedish).

Friberg, L. and Odeblad, E., Localization of Cd^{115} in different organs. An autoradiographic study, *Acta Pathol. Microbiol. Scand.*, 41, 96, 1957.

Friberg, L., Piscator, M., and Nordberg, G., *Cadmium in the Environment*, CRC Press, Cleveland, 1971.

Fukushima, I., Cadmium in the environment in the basin of the rivers Usui and Watanase in Gumma Prefecture and its health effect on the inhabitants, in *Kankyo Hoken Report No. 24*, Japanese Association of Public Health, September 1973, 131 (in Japanese).

Fukushima, M., Cadmium content in various foodstuffs, in *Kankyo Hoken Report No. 11*, Japanese Association of Public Health, April 1972, 22 (in Japanese).

Fukushima, M., Kobayashi, S., and Sakamoto, R., Urinary free amino acids in Itai-itai patients and among inhabitants in a cadmium polluted area, in *Kankyo Hoken Report No. 24*, Japanese Association of Public Health, September 1973, 53 (in Japanese).

Fukushima, M. and Nogawa, K., Research about environmental pollution from heavy metals. Report 3. Relationships between ricefield pollution and urinary findings, *Jap. J. Public Health*, 18, 482, 1971 (in Japanese).

Fukushima, M. and Sugita, Y., Some urinary findings on Itai-itai disease patients (2nd report). About urinary proteins, *Jap. J. Public Health*, 17, 759, 1970 (in Japanese).

Fukushima Prefecture, Result of the Investigation of the Cadmium Content in Rice Produced in 1971, Nov. 25, 1971a (in Japanese).

Fukushima Prefecture, Result of the Health Examination on Inhabitants of the "Observation Area" for Environmental Pollution of Cadmium, March, 1971b (in Japanese).

Fukushima Prefecture, Report about an Investigation of the Nisso Refinery in Aizu, April, 1972 (in Japanese).

Fukuyama, Y., Proteinuria in Itai-itai disease and cadmium poisoning, *Med. Biol.*, 84, 41, 1972 (in Japanese).

Fukuyama, Y. and Kubota, K., Gel filtration pattern of the urine of Itai-itai disease patients, *Med Biol.*, 80, 31, 1970 (in Japanese).

Fukuyama, Y. and Kubota, K., Epidemiological survey of urinary protein from inhabitants in the Itai-itai disease district, *Med. Biol.*, 84, 249, 1972a (in Japanese).

Fukuyama, Y. and Kubota, K., Geographic and historical survey of urinary proteins in the inhabitants of the Itai-itai disease district, *Med. Biol.*, 84, 305, 1972b (in Japanese).

Fukuyama, Y. and Kubota, K., Relationship between proteinuria and heavy metal pollution in the Itai-itai disease district, *Med. Biol.*, 85, 103, 1972c (in Japanese).

Fukuyama, Y., Shiroishi, K., and Kubota, K., Evaluation of screening tests for proteinuria in Itai-itai disease, *Med. Biol.*, 83, 85, 1971 (in Japanese).

Fulkerson, W., Goeller, H. E., Gailar, J. S., and Copenhaver, E. D., Eds., *Cadmium, the Dissipated Element*, Oak Ridge National Laboratory ORNL NSF-EP–21, Oak Ridge, Tenn., 1973.

Furst, A., Cadden, J. E., and Firpo, E. J., Excretion of cadmium compounds by the rat, *Proc. West. Pharmacol. Soc.*, 15, 55, 1972.

Gabbiani, G., Action of cadmium chloride on sensory ganglia, *Experimentia*, 22, 261, 1966.

Gabbiani, G., Baic, D., and Deziel, C., Studies on tolerance and ionic antagonism for cadmium or mercury, *Can. J. Physiol. Pharmacol.*, 45, 443, 1967a.

Gabbiani, G., Baic, D., and Deziel, C., Toxicity of cadmium for the central nervous system, *Exp. Neurol.*, 18, 154, 1967b.

Gabbiani, G., Gregory, A., and Baic, D., Cadmium-induced selective lesions of sensory ganglia, *J. Neuropathol. Exp. Neurol.*, 26, 498, 1967.

Geldmacher-v. Mallinckrodt, M. and Opitz, O., Zur Diagnostik der Cadmimvergiftung der normale Cadmiumgehalt menschlicher Organe und Korperflüssigkeiten, *Arbeitsmed. Sozialmed. Arbeitshyg.*, 3, 276, 1968.

Geldmacher-v. Mallinckrodt, M. and Pooth, M., Gleichzeitige spektrographische Prüfung auf 25 Metalle und Metalloide in Biologischem Material, *Arch. Toxikol.*, 25, 5, 1969.

Gelting, G. and Pontén, A., Heavy metal pollution in the Uppsala area, Dept. of Plant Biology, Institute of Plant Biology, Uppsala, 1971 (Internal report; in Swedish).

Gerard, R. W., Studies on Cadmium, Office of Emergency Management, Committee on Medical Research, June 30, 1944.

Gervais, J. and Delpech, P., L'intoxication cadmique, *Arch. Mal. Prof. Med. Trav. Secur. Soc.*, 24, 803, 1963.

Gilman, A., Philips, F. S., Allen, R. P., and Koelle, E. S., The treatment of acute cadmium intoxication in rabbits with 2,3-dimercaptopropanol (BAL) and other mercaptans, *J. Pharmacol. Exp. Ther.*, 87, 85, 1946.

Girod, C., A propos de l'influence du chlorure de cadmium sur le testicule; recherches chez le singe macacus irus F. Cuv., *C. R. Seances Soc. Biol. Fil.*, 158, 297, 1964a.

Girod, C., Etude des cellules gonadotropes antehypophysaires du singe macacus irus F. Cuv., apres administration de chlorure de cadmium, *C. R. Seances Soc. Biol. Fil.*, 158, 948, 1964b.

Goodman, G. T. and Roberts, T. M., Plants and soil as indicators of metals in the air, *Nature*, 231, 287, 1971.

Goyer, R. A., Tsuchiya, K., Leonard, D. L., and Kahyo, H., Aminoaciduria in Japanese workers in the lead and cadmium industries, *Am. J. Clin. Pathol.*, 57, 635, 1972.

Graovac-Leposavic, L., Djuric, D., Valjarevic, V., Senicar, H., Senicar, L., Milic, S., and Delic, V., Environmental lead contamination of Meza Valley – study on lead exposure of population, in *Int. Symp. Environ. Health Aspects Lead*, Amsterdam, October, 1972, Commission of the European Communities, Luxembourg, 1973, 685.

Groen, J. J., Eshchar, J., Ben-Ishay, D., Alkan, W. J., and Ben Assa, B. I., Osteomalacia among the Bedouin of the Negev Desert, *Arch. Intern. Med.*, 116, 195, 1965.

Gumma Prefecture, Annual Report on the Status of Public Hazards, 1972 (in Japanese).

Gunn, S. A. and Gould, T. C., Selective accumulation of Cd^{115} by cortex of rat kidney, *Proc. Soc. Exp. Biol. Med.*, 96, 820, 1957.

Gunn, S. A. and Gould, T. C., Cadmium and other mineral elements, in *The Testis*, Vol. 3, Johnson, A. D., Gomes, W. R., and VanDemark, N. L., Eds., Academic Press, New York, 1970, 377.

Gunn, S. A., Gould, T. C., and Anderson, W. A. D., Zinc protection against cadmium injury to rat testis, *Arch. Pathol.*, 71, 274, 1961.

Gunn, S. A., Gould, T. C., and Anderson, W. A. D., Interference with fecal excretion of Zn65 by cadmium, *Proc. Soc. Exp. Biol. Med.*, 111, 559, 1962.

Gunn, S. A., Gould, T. C., and Anderson, W. A. D., Cadmium-induced interstitial cell tumors in rats and mice and their prevention by zinc, *J. Natl. Cancer Inst.*, 31, 745, 1963a.

Gunn, S. A., Gould, T. C., and Anderson, W. A. D., The selective injurious response of testicular and epididymal blood vessels to cadmium and its prevention by zinc, *Am. J. Pathol.*, 42, 685, 1963b.

Gunn, S. A., Gould, T. C., and Anderson, W. A. D., Effect of zinc on cancerogenesis by cadmium, *Proc. Soc. Exp. Biol. Med.*, 115, 653, 1964.

Gunn, S. A., Gould, T. C., and Anderson, W. A. D., Strain differences in susceptibility of mice and rats to cadmium-induced testicular damage, *J. Reprod. Fertil.*, 10, 273, 1965.

Gunn, S. A., Gould, T. C., and Anderson, W. A. D., Protective effect of thiol compounds against cadmium-induced vascular damage to testis, *Proc. Soc. Exp. Biol. Med.,* 122, 1036, 1966.

Gunn, S. A., Gould, T. C., and Anderson, W. A. D., Specific response of mesenchymal tissue to cancerogenesis by cadmium, *Arch. Pathol.,* 83, 493, 1967.

Gunn, S. A., Gould, T. C., and Anderson, W. A. D., Selectivity of organ response to cadmium injury and various protective measures, *J. Pathol. Bacteriol.,* 96, 89, 1968a.

Gunn, S. A., Gould, T. C., and Anderson, W. A. D., Mechanisms of zinc, cysteine and selenium protection against cadmium-induced vascular injury to mouse testis, *J. Reprod. Fertil.,* 15, 65, 1968b.

Gunn, S. A., Gould, T. C., and Anderson, W. A. D., Maintenance of the structure and function of the cauda epididymis and contained spermatazoa by testosterone following cadmium-induced testicular necrosis in the rat, *J. Reprod. Fertil.,* 21, 443, 1970.

Guthenberg, H. and Beckman, I., Investigation of leakage of lead from glazed pottery and enameled steel in household items designed for direct contact with food elements, *Vår Föda,* 5, 41, 1970 (in Swedish).

Guyton, A. C., Measurements of the respiratory volumes of laboratory animals, *Am. J. Physiol.,* 150, 70, 1947.

Haddow, A., Dukes, C. E., and Mitchley, B. C. V., Carcinogenity of iron preparations and metal-carbohydrate complexes, *Rep. Br. Emp. Cancer Campgn.,* 39, 74, 1961.

Haddow, A., Roe, F. J. C., Dukes, C. E., and Mitchley, B. C. V., Cadmium neoplasia: sarcomata at the site of injection of cadmium sulphate in rats and mice, *Br. J. Cancer,* 18, 667, 1964.

Hagino, N., About investigations on Itai-itai disease, *J. Toyama Med. Assoc.,* Suppl., Dec. 21, 1957 (in Japanese).

Hagino, N., About the so-called "Itai-itai disease," *J. Jap. Assoc. Rural Med.,* 7, 288, 1959 (in Japanese).

Hagino, N., Itai-itai disease, *Medicina,* 5, 99, 1968a (in Japanese).

Hagino, N., Itai-itai disease, *Accident Med.,* 11, 1390, 1968b (in Japanese).

Hagino, N., Cd poisoning symptoms, *Gen. Clinic,* 18, 1366, 1969 (in Japanese).

Hagino, N., Itai-itai disease and vitamin D, *Digest of Science of Labour,* 28, 32, 1973 (in Japanese).

Hagino, N. and Yoshioka, K., A study on the cause of Itai-itai disease, *J. Jap. Orthop Assoc.,* 35, 812, 1961 (in Japanese).

Hammer, D. I. and Finklea, J. F., Cadmium body burdens at autopsy in the United States, in Proceedings of the 17th International Congress on Occupational Health, to be published, 1974 (available through the Secretariat, Av. Rogue Saenz, Pena, 110 – 2 piso – Oficio 8, Buenos Aires, Argentina), 1972.

Hammer, D. I., Finklea, J. F., Creason, J. P., Sandifer, S. H., Keil, J. E., Priester, L. E., and Stara, J. F., Cadmium exposure and human health effects, in *Trace Substances in Environmental Health,* Vol. 5, Hemphill, D. D., Ed., University of Missouri Press, Columbia, 1972, 269.

Hammer, D. I., Finklea, J. F., Hendricks, R. H., Shy, C. M., and Horton, R. J. M., Hair trace metal levels and environmental exposure, *Am. J. Epidemiol.,* 93, 84, 1971.

Harada, A., Findings on urinary protein electrophoresis, in *Kankyo Hoken Report No. 11,* Japanese Association of Public Health, April 1972, 87 (in Japanese).

Harada, A., Medical examination of workers in a cadmium pigment factory, in *Kankyo Hoken Report No. 24,* September 1973, 66 (in Japanese).

Hardy, H. L. and Skinner, J. B., The possibility of chronic cadmium poisoning, *J. Ind. Hyg. Toxicol.,* 29, 321, 1947.

Harrison, H. E., Bunting, H., Ordway, N., and Albrink, W. S., The effects and treatment of inhalation of cadmium chloride in the dog, *J. Ind. Hyg. Toxicol.,* 29, 302, 1947.

Harrison, J. F., Lunt, G. S., Scott, P., Blainey, J. D., Urinary lysozyme, ribonuclease, and low-molecular weight protein in renal disease, *Lancet,* 1, 371, 1968.

Harrison, P. R. and Winchester, J. W., Area-wide distribution of lead, copper, and cadmium in air particulates from Chicago and Northwest Indiana, *Atmos. Environ.,* 5, 863, 1971.

Hasegawa, Y., Result of examinations of residents in cadmium polluted areas, in *Kankyo Hoken Report No. 11,* Japanese Association of Public Health, April 1972, 13 (in Japanese).

Hauser, T. R., Hinners, T. A., and Kent, J. L., Atomic absorption determination of cadmium and lead in whole blood by a reagent-free method, *Anal. Chem.,* 44, 1819, 1972.

Havre, G. N., Underdal, B., and Christiansen, C., The content of lead and some other heavy elements in different fish species from a fjord in western Norway, *Int. Symp. Environ. Health Aspects Lead,* Amsterdam, October 1972, Commission of the European Communities, Luxembourg, 1973, 99.

The Health Consequences of Smoking, A Report to the Surgeon General: 1971, U.S. Department of Health, Education, and Welfare, Washington, D.C., 1971.

Heath, J. C., Daniel, M. R., Dingle, J. T., and Webb, M., Cadmium as a carcinogen, *Nature,* 193, 592, 1962.

Heath, J. C. and Webb, M., Content and intracellular distribution of the inducing metal in the primary rhabdomyosarcomata induced in the rat by cobalt, nickel and cadmium, *Br. J. Cancer,* 21, 768, 1967.

Henke, G., Sachs, H. W., and Bohn, G., Cadmium-bestimmungen in Leber und Nieren von Kindern und Jugendlichen durch Neutronenaktivierungsanalyse, *Arch. Toxikol.,* 26, 8, 1970.

Hickey, R. J., Schoff, E. P., and Clelland, R. C., Relationship between air pollution and certain chronic disease death rates, *Arch. Environ. Health,* 15, 728, 1967.

Hill, C. H., Matrone, G., Payne, W. L., and Barber, C. W., In vivo interactions of cadmium with copper, zinc and iron, *J. Nutr.,* 80, 227, 1963.

Hirota, M., Influences of cadmium on the bone salt metabolism, *Acta Sci. Med. Univ. Gifu*, 19, 82, 1971 (in Japanese).

Hodgen, G. D., Butler, W. R., and Gomes, W. R., In vivo and in vitro effects of cadmium chloride on carbon anhydrase activity, *J. Reprod. Fertil.*, 18, 156, 1969.

Hodgen, G. D., Gomes, W. R., and VanDemark, N. L., Carbonic anhydrase isoenzymes in rat erythrocytes, kidney and testis, *Fed. Proc.*, 28, 773, 1969.

Hodgen, G. D., Gomes, W. R., and VanDemark, N. L., In vitro and in vivo effects of cadmium chloride on isoenzymes of carbonic anhydrase in rat testis and erythrocytes, *Biol. Reprod.*, 2, 197, 1970.

Holden, H., Cadmium toxicology, (letter), *Lancet*, 2, 57, 1969.

Holmberg, R. E. and Ferm, V. H., Interrelationships of selenium, cadmium, and arsenic in mammalian teratogenesis, *Arch. Environ. Health*, 18, 873, 1969.

Hörstebrock, Die Pathologische Anatomie der chronischen Kadmiumvergiftungen, *Bundesarbeitsblatt*, Suppl. 10, 7, 1951.

Huck, F. F., Cadmium poisoning by inhalation, report of a case, *Occup. Med.*, 3, 411, 1947.

Huey, N. A., Survey of airborne pollutants, in *Helena Valley, Montana, Area Environmental Pollution Study*, U.S. Environmental Protection Agency, Research Triangle Park, N.C., 1972, 25.

Humperdinck, K., Kadmium und Lungenkrebs, *Med. Klin.*, 63, 948, 1968.

Hundley, H. K. and Warren, E. C., Determination of cadmium in total diet samples by anodic stripping voltammetry, *J. Assoc. Off. Anal. Chem.*, 53, 705, 1970.

Hunt, W. F., Pinkerton, C., McNulty, O., and Creason, J., A study in trace element pollution of air in 77 midwestern cities, in *Trace Substances in Environmental Health*, Vol. 4, Hemphill, D. E., Ed., University of Missouri Press, Columbia, 1971, 56.

Hyogo Prefecture, Results of Studies on the Health Effects of Cadmium on the Population in the Vicinity of the Mine in Ikuno, Report No. 1, December, 1972a (in Japanese).

Hyogo Prefecture, Results of a Study on 13 Inhabitants of the Surroundings of the Ikuno Mine, Report to the Committee for Differential Diagnosis of Itai-itai Disease and Cadmium Poisoning, September 6, 1972b (in Japanese).

Hyogo Prefecture, The General Situation as Regards Countermeasures against Pollution of Cadmium etc. in the Vicinity of the Mine in Ikuno, June, 1972c (in Japanese).

Imbus, H. R., Cholak, J., Miller, L. H., and Sterling, T., Boron, cadmium, chromium, and nickel in blood and urine, *Arch. Environ. Health*, 6, 286, 1963.

Inhizawa, M., Okada, K., Yoshida, N., Yoshioka, C., Shigematsu, T., and Yoshimura, S., Cadmium content in the kidney of the Japanese adults, *Med. Biol.*, 75, 105, 1967 (in Japanese).

Insull, W., Jr., Oiso, T., and Tsuchiya, K., Diet and nutritional status of Japanese, *Am. J. Clin. Nutr.*, 21, 753, 1968.

Ishizaki, A., Observations on the influence on the body of cadmium in the food, *Clin. Nutr.*, 35, 28, 1969a (in Japanese).

Ishizaki, A., On the so-called Itai-itai disease, *J. Jap. Med. Soc.*, 62, 242, 1969b (in Japanese).

Ishizaki, A., On "Itai-itai" (ouch-ouch) disease, *Asian Med. J.*, 14, 421, 1971.

Ishizaki, A., About the result of a health examination, in *Kankyo Hoken Report No. 11*, Japanese Association of Public Health, April 1972a, 27 (in Japanese).

Ishizaki, A., About the Cd and Zn concentrations in organs of Itai-itai disease patients and in inhabitants of Hokuriku area, in *Kankyo Hoken Report No. 11*, Japanese Association of Public Health, April 1972b, 154 (in Japanese).

Ishizaki, A. and Fukushima, M., Studies on "Itai-itai" disease (review), *Jap. J. Hyg.*, 23, 271, 1968 (in Japanese).

Ishizaki, A., Fukushima, M., Kurachi, T., Sakamoto, M., and Hayashi, E., The relationship between Cd and Zn contents and the year rings of the Sugi tree in the basin of the Jintzu River, *Jap. J. Hyg.*, 25, 376, 1970 (in Japanese).

Ishizaki, A., Fukushima, M., and Sakamoto, M., Cadmium content of rice eaten in the Itai-itai disease area, in *Annual Meeting of the Japanese Association of Public Health*, Japanese Association of Public Health, Kyoto, 1968 (in Japanese).

Ishizaki, A., Fukushima, M., and Sakamoto, M., On the distribution of Cd in biological materials. I. Human hair and rice straw, *Jap. J. Hyg.*, 24, 375, 1969 (in Japanese).

Ishizaki, A., Fukushima, M., and Sakamoto, M., Distribution of Cd in biological materials. II. Cadmium and zinc contents of foodstuffs, *Jap. J. Hyg.*, 25, 207, 1970a (in Japanese).

Ishizaki, A., Fukushima, M., and Sakamoto, M., On the accumulation of cadmium in the bodies of Itai-itai patients, *Jap. J. Hyg.*, 25, 86, 1970b (in Japanese).

Ishizaki, A., Hagino, N., and Shibata, I., An advanced case of osteomalacia in a person living in Ichikawa River basin, polluted by cadmium, in *Kankyo Hoken Report No. 24*, Japanese Association of Public Health, September 1973, 43 (in Japanese).

Ishizaki, A., Nomura, K., Tanabe, S., and Sakamoto, M., Observations on urinary and fecal excretion of heavy metals (Cd, Pb and Zn) in the patients of the so-called "Itai-itai" disease, *Jap. J. Hyg.*, 20, 261, 1965 (in Japanese).

Itai-itai Research Committee, *Research about the Cause of Itai-itai Disease*, Toyama Prefecture Authorities, Toyama, Japan, 1968, 46, (in Japanese).

Ito, T. and Sawauchi, K., Inhibitory effects on cadmium-induced testicular damage by pretreatment with smaller cadmium dose, *Okajimas Folia Anat. Jap.*, 42, 107, 1966.

Itokawa, Y., Abe, T., and Tanaka, S., Bone changes in experimental cadmium poisoning, *Arch. Environ. Health*, 26, 241, 1973.

Jakubowski, M., Piotrowski, J., and Trojanowska, B., Binding of mercury in the rat: studies using $^{203}HgCl_2$ and gel filtration, *Toxicol. Appl. Pharmacol.*, 16, 743, 1970.

Japanese Association of Industrial Health, Recommendations for permissible concentrations of toxic substances in industry, *Jap. J. Ind. Health*, 13, 475, 1971.

Japanese Association of Public Health, *Research about Differential Diagnosis of Itai-itai Disease and Cadmium Poisoning*, Japanese Association of Public Health, Tokyo, 1970a (in Japanese).

Japanese Association of Public Health, *Research about Intake and Accumulation of Cadmium in Areas "Requiring Observation,"* Japanese Association of Public Health, Tokyo, March 30, 1970b (in Japanese).

Japanese Association of Public Health, *Studies of Standardization of Analytical Methods for Chronic Cadmium Poisoning*, Japanese Association of Public Health, Tokyo, March 30, 1970c (in Japanese).

Johansson, T. B., Akselsson, R., and Johansson, S. A. E., X-ray analysis: elemental trace analysis at the 10^{-12} g level, *Nucl. Instrum. Meth.*, 84, 141, 1970.

John, M. K., Uptake of soil-applied cadmium and its distribution in radishes, *Can. J. Plant Sci.*, 52, 715, 1972.

John, M. K., Van Laerhoven, C. J., and Chuah, H. H., Factors affecting plant uptake and phylotoxicity of cadmium added to soils, *Environ. Sci. Technol.*, 5, 1006, 1972.

Johnson, A. D., Gomes, W. R., and VanDemark, N. L., Early actions of cadmium in the rat and domestic fowl testis. I. Testis and body temperature changes caused by cadmium and zinc, *J. Reprod. Fertil.*, 21, 383, 1970.

Johnson, A. D. and Miller, W. J., Early actions of cadmium in the rat and domestic fowl testis. II. Distribution of injected ^{109}cadmium, *J. Reprod. Fertil.*, 21, 395, 1970.

Johnson, M. H., The effect of cadmium chloride on the blood-testis barrier of the guinea-pig, *J. Reprod. Fertil.*, 19, 551, 1969.

Just, J. and Kelus, J., Cadmium in the air atmosphere of ten selected cities in Poland, *Rocz. Panstw. Zakl. Hig.*, 22, 249, 1971 (translation obtainable from U.S. Environmental Protection Agency, Office of Air Programs, Translation Section, Research Triangle Park, N.C. 27711 U.S.A.).

Kägi, J. H. R., Hepatic metallothionein, in *8th Int. Congr. Biochem.*, Switzerland, Sept. 3 – 9, 1970, International Union of Biochemistry, 1970, 130.

***Kägi, J. H. R. and Piscator, M.,** in *Cadmium in the Environment*, Friberg, L., Piscator, M., and Nordberg, G., CRC Press, Cleveland, 1971, 67.

Kägi, J. H. R. and Vallee, B. L., Metallothionein: a cadmium and zinc-containing protein from equine renal cortex, *J. Biol. Chem.*, 235, 3460, 1960.

Kägi, J. H. R. and Vallee, B. L., Metallothionein: a cadmium and zinc-containing protein from equine renal cortex, II. Physicochemical properties, *J. Biol. Chem.*, 236, 2435, 1961.

Kajikawa, K., Pathological studies on renal lesions of Itai-itai patients, in *Kankyo Hoken Report No. 24*, Japanese Association of Public Health, September 1973, 29 (in Japanese).

Kajikawa, K., Okuno, S., Igawa, K., and Hirono, R., A bone disease which occurred in the Toyama Prefecture, so-called "Itai-itai byo" (painful disease), *Trans. Soc. Pathol. Jap.*, 46, 655, 1957 (in Japanese with English summary).

Kakinuma, K., Shimizu, K., Segareda, M., Akita, Y., Takijima, T., Fukushima, I., Sugimura, K., Fujii, A., Sakamura, K., Nagata, M., and Tsuji, T., About the prevalence of proteinuria as well as a result of a health examination of the inhabitants along the Usuigawa River, in *Meeting of the Japanese Association of Public Health*, November 1971 (in Japanese).

Kanisawa, M. and Schroeder, H. A., Life term studies on the effect of trace elements on spontaneous tumors in mice and rats, *Cancer Res.*, 29, 892, 1969a.

Kanisawa, M. and Schroeder, H. A., Renal arteriolar changes in hypertensive rats given cadmium in drinking water, *Exp. Mol. Pathol.*, 10, 81, 1969b.

Kapoor, N. K., Agarwala, S. C., and Kar, A. B., The distribution and retention of cadmium in subcellular fractions of rat liver, *Ann. Biochem. Exp. Med.* (Calcutta), 21, 51, 1961.

Kar, A. B., Chemical sterilization of male rhesus monkeys, *Endocrinology*, 69, 1116, 1961.

Kar, A. B., Chemical sterilization of male goats, *Indian J. Vet. Sci. Anim. Husb.*, 32, 70, 1962.

Kar, A. B. and Das, R. P., Testicular changes in rats after treatment with cadmium chloride, *Acta Biol. Med. Ger.*, 5, 153, 1960.

Kar, A. B. and Das, R. P., Effect of cadmium chloride on fertility of rats, *Indian J. Vet. Sci. Anim. Husb.*, 32, 210, 1962a.

Kar, A. B. and Das, R. P., Sterilization of males by intratesticular administration of cadmium chloride, *Acta Endocrinol.*, 40, 321, 1962b.

Kar, A. B. and Das, R. P., The nature of protective action of selenium on cadmium-induced degeneration of the rat testis, *Proc. Natl. Inst. Sci. India Part B Biol. Sci.*, 29, 297, 1963.

Kar, A. B., Das, R. P., and Mukerji, B., Prevention of cadmium induced changes in the gonads of rat by zinc and selenium A study in antagonism between metals in the biological system, *Proc. Natl. Inst. Sci. India Part B Biol. Sci.*, 26, Suppl. 40, 1960.

Kar, A. B., Dasgupta, P. R., and Das, R. P., Effect of cadmium chloride on gonadotrophin content of the pituitary of male and female rats, *J. Sci. Ind. Res. (India) Sect. C Biol. Sci.*, 19, 225, 1960.

Kar, A. B. and Kamboj, V. P., Sterilization of rats and rhesus monkeys by scrotal inunction of cadmium chloride, *Excerpta Med. Int. Congr. Ser.*, 72, 465, 1963.

Kar, A. B. and Kamboj, V. P., Cadmium damage to the rat testis and its prevention, *Indian J. Exp. Biol.*, 3, 45, 1965.

Katagiri, Y., Tati, M., Iwata, H., and Kawai, M., Concentration of cadmium in urine by age, *Med. Biol.*, 82, 239, 1971 (in Japanese).

Kato, T. and Kawano, S., Review of past and present of Itai-itai disease. On the process of research development, *Curr. Med.,* 16, 29, 1968 (in Japanese).

Kazantzis, G., Respiratory function in men casting cadmium alloys, I: Assessment of ventilatory function, *Br. J. Ind. Med.,* 13, 30, 1956.

Kazantzis, G., Induction of sarcoma in the rat by cadmium sulphide pigment, *Nature,* 198, 1213, 1963.

Kazantzis, G., Industrial hazards to the kidney and urinary tract, in *Sixth Symposium on Advanced Medicine,* Slater, J. D. H., Ed., Pitman and Sons, London, 1970, 263.

Kazantzis, G., Osteomalacia following exposure to cadmium, to be published, 1974.

Kazantzis, G., Flynn, F. V., Spowage, J. S., and Trott, D. G., Renal tubular malfunction and pulmonary emphysema in cadmium pigment workers, *Q. J. Med.,* 32, 165, 1963.

Kazantzis, G. and Hanbury, W. J., The induction of sarcoma in the rat by cadmium sulphide and by cadmium oxide, *Br. J. Cancer,* 21, 190, 1966.

Kench, J. E., Gain, A. C., and Sutherland, E. M., A biochemical study of the minialbumin to be found in the urine of men and animals poisoned by cadmium, *S. Afr. Med. J.,* 39, 1191, 1965.

Kench, J. E. and Sutherland, E. M., The nature and origin of the minialbumin found in cadmium-poisoned animals, *S. Afr. Med. J.,* 40, 1109, 1966.

Kench, J. E., Wells, A. R., and Smith, J. C., Some observations on the proteinuria of rabbits poisoned with cadmium, *S. Afr. Med. J.,* 36, 390, 1962.

Kennedy, A., The effect of L-cysteine on the toxicity of cadmium, *Br. J. Exp. Pathol.,* 49, 360, 1968.

Kimura, H., Experimental investigation on glycosuria induced by cadmium, *Acta Sch. Med. Univ. Gifu,* 19, 342, 1971 (in Japanese with English summary).

King, E., An environmental study of casting copper-cadmium alloys, *Br. J. Ind. Med.,* 12, 198, 1955.

Kipling, M. D. and Waterhouse, J. A. H., Cadmium and prostatic carcinoma (letter), *Lancet,* 1, 730, 1967.

Kitamura, S., Sumino, K., and Kamatani, N., Cadmium concentrations in livers, kidneys and bones of human bodies, *Jap. J. Public Health,* 17, Abstract 507, page 177(761), 1970 (in Japanese).

Kitamura, S., Cadmium absorption and accumulation (mainly about humans), in *Kankyo Hoken Report No. 11,* Japanese Association of Public Health, April 1972, 42 (in Japanese).

Kjellström, T., A mathematical model for the accumulation of cadmium in human kidney cortex, *Nord. Hyg. Tidskr.,* 53, 111, 1971.

Kjellström, T., An Epidemiological Exposure and Effect Study on Cadmium. An Investigation of the General and Industrial Environment around a Swedish Copper and Lead Refinery, Report to the Swedish Environment Protection Board, August 30, 1973 (in Swedish).

Kjellström, T. and Friberg, L., Interpretation of Empirically Documented Body Burdens by Age of Metals with Long Biological Half-times with Special Reference to Past Changes in Exposure, in Proceedings of the 17th International Congress on Occupational Health, to be published, 1974 (available through the Secretariat, Av. Rogue Saenz, Pẽna, 110 – 2 piso – Oficio 8, Buenos Aires, Argentina), 1972.

Kjellström, T., Friberg, L., Nordberg, G., and Piscator, M., Further considerations on uptake and retention of cadmium in human kidney cortex, in *Cadmium in the Environment,* Friberg, L., Piscator, M., and Nordberg, G., CRC Press, Cleveland, 1971, 140.

Kjellström, T., Lind, B., Linnman, L., and Nordberg, G., A comparative study of methods for cadmium analysis of grain with an application to pollution evaluation, to be published in *Environ. Res.,* 1974.

Klein, D. H., Mercury and other metals in urban soils, *Environ. Sci. Technol.,* 6, 560, 1972.

Kleinfeld, M., Acute pulmonary edema of chemical origin, *Arch. Environ. Health,* 10, 942, 1965.

Kneip, T. J., Eisenbud, M., Strehlow, C. D., and Freudenthal, P. C., Airborne particulates in New York City, *J. Air Pollut. Control Assoc.,* 20, 144, 1970.

Kobayashi, J., Agricultural Damage in the Jintsu River Basin Caused by the Kamioka Mine, a Report to the Ministry of Agriculture, July, 1943 (in Japanese).

Kobayashi, J., Relation between the "Itai-itai" disease and the pollution of river water by cadmium from a mine, *Proc. Int. Water Pollution Research Conf.,* I-25, 1970.

Kobayashi, J., Air and water pollution by cadmium, lead and zinc attributed to the largest zinc refinery in Japan, in *Trace Substances in Environmental Health,* Vol. 5, Hemphill, D. E., Ed., University of Missouri Press, Columbia, 1972, 117.

Kobayashi, J., Morii, F., Muramoto, S., and Nakashima, S., Effects of air and water pollution on agricultural products by Cd, Pd, and Zn attributed to a mine and refinery in Annaka City, Gumma Prefecture, *Jap. J. Hyg.,* 25, 364, 1970 (in Japanese).

Kobayashi, J., Nakahara, H., and Hasegawa, Y., Accumulation of cadmium in organs of mice fed on cadmium-polluted rice, *Jap. J. Hyg.,* 26, 401, 1971 (in Japanese, with English summary).

Kono, M., Fukuzawa, K., Yamagiri, K., and Fujii, A., On progress in the treatment of the so-called Itai-itai disease, *J. Orthop. Soc. Jap.,* 32, 727, 1958 (in Japanese; translation by Seizaburo Aoki, Japanese Language Translation Service, Fujisawa, Japan).

Kormano, M., Microvascular supply of the regenerated rat testis following cadmium injury, *Virchows Arch.,* 349, 229, 1970.

Kory, R. C., Callahan, R., Boren, H. G., and Syner, J. C., The Veterans Administration-Army cooperative study of pulmonary function. I. Clinical spirometry in normal men, *Am. J. Med.,* 30, 243, 1961.

Kropf, R. and Geldmacher-v. Mallinckrodt, M., Der Cadmiumgehalt von Nahrungsmitteln und die tägliche Cadmiumaufnahme, *Arch. Hyg. Bakteriol.,* 152, 218, 1968.

Kubota, J., Lazar, V. A., and Losee, F., Cooper, zinc, cadmium, and lead in human blood from 19 locations in the United States, *Arch. Environ. Health,* 16, 788, 1968.

Kubota, J., Tsuchiya, K., Koizumi, A., and Inoue, T., Environmental Health in Japan (Exhibition, Tokyo, The Organizing Committee of the 16th International Congress on Occupational Health, obtainable from the "Occupational Health Service Center," Japanese Industrial Safety Association, Tokyo, September 22–27, 1969.

Laamanen, A., Functions, progress and prospects for an environmental subarctic base level station, *Work Environ. Health,* 9, 17, 1972.

Lamy, P., Heully, F., Pernot, C., Anthoine, D., Couillaut, S., and Thomas, G., Pneumopathies aigues par vapeurs de cadmium, *J. Fr. Med. Chir. Thorac.,* 17, 275, 1963.

Lane, R. E. and Campbell, A. C. P., Fatal emphysema in two men making a copper cadmium alloy, *Br. J. Ind. Med.,* 11, 118, 1954.

Larsson, S.-E. and Piscator, M., Effect of cadmium on skeletal tissue in normal and calcium-deficient rats, *Isr. J. Med. Sci.,* 7, 495, 1971.

Lease, J. G., Effect of graded levels of cadmium on tissue uptake of ^{65}Zn by the chick over time, *J. Nutr.,* 96, 294, 1968.

Lee, R. E., Goranson, S. S., Enrione, R. E., and Morgan, G. B., National air surveillance cascade impactor network. II. Size distribution measurements of trace metal components, *Environ. Sci. Technol.,* 6, 1025, 1972.

Lee, R. E., Patterson, R. K., and Wagman, J., Particle-size distribution of metal components in urban air, *Environ. Sci. Technol.,* 2, 288, 1968.

Lehnert, G., Klavis, G., Schaller, K. H., and Haas, T., Cadmium determination in urine by atomic absorption spectrometry as a screening test in industrial medicine, *Br. J. Ind. Med.,* 26, 156, 1969.

Lehnert, G., Schaller, K. H., and Haas, T., Atomabsorptionsspektrometrische Cadmiumbestimmung in Serum und Harn, *Z. Klin. Chem.,* 6, 174, 1968.

Lener, J., Determination of Trace Amounts of Cadmium in Biological Materials from the Viewpoint of Their Biological Consequences and Their Significance in Hygiene, Ph.D. thesis, Charles University, Prague, 1968 (in Czech).

Lener, J. and Bibr, B., Spurenmengen von Kadmium in Nahrungsmitteln tierischer Herkunft, *Vitalst. Zivilisationskr.,* 14, 125, 1969.

Lener, J. and Bibr, B., Cadmium content in some food stuffs in respect of its biological effects, *Vitalst. Zivilisationskr.,* 15, 139, 1970.

Lener, J. and Bibr, B., Determination of traces of cadmium in biological materials by atomic absorption spectrophotometry, *J. Agric. Food Chem.,* 19, 1011, 1971a.

Lener, J. and Bibr, B., Cadmium and hypertension, *Lancet,* 1(7706), 970, 1971b.

Lener, J. and Bibr, B., A contribution to the study of cadmium effects in the sphere of experimental hypertension, *Cesk. Hyg.,* 18, 282, 1973 (in Czech with English and Russian summaries).

Lener, J. and Musil, J., Cadmium influence on the excretion of sodium by kidneys, *Experientia* (Basel), 26, 902, 1970.

L'Epée, P., Lazarini, H., Franchome, J., N'Doky, Th., and Larrivet, C., Contribution a l'etude de l'intoxication cadmique, *Arch. Mal. Prof. Med. Trav. Secur. Soc.,* 29, 485, 1968.

Lewis, G. P., Coughlin, L., Jusko, W., and Hartz, S., Contribution of cigarette smoking to cadmium accumulation in man, *Lancet,* 1, 291, 1972a.

Lewis, G. P., Jusko, W. J., Coughlin, L. L., and Hartz, S., Cadmium accumulation in man: influence of smoking, occupation, alcoholic habit and disease, *J. Chronic Dis.,* 25, 717, 1972b.

Lewis, G. P., Lyle, H., and Miller, S., Association between elevated hepatic water-soluble protein-bound cadmium levels and chronic bronchitis and/or emphysema, *Lancet,* 2, 1330, 1969.

Lieberman, K. W. and Kramer, H. H., Cadmium determination in biological tissue by neutron activation analysis, *Anal. Chem.,* 42, 266, 1970.

Linnman, L., Andersson, A., Nilsson, K. O., Lind, B., Kjellström, T., and Friberg, L., Cadmium uptake by wheat from sewage sludge used as a plant nutrient source, *Arch. Environ. Health,* 27, 45, 1973.

Livingston, H. D., Measurement and distribution of zinc, cadmium, and mercury in human kidney tissue, *Clin. Chem.,* 18, 67, 1972.

Ljunggren, B., Sjöstrand, B., Johnels, A. G., Olsson, M., Otterlind, G., and Westermark, T., Activation analysis of mercury and other environmental pollutants in water and aquatic ecosystems, in *Nuclear Techniques in Environmental Pollution,* IAEA-SM-142a/22, International Atomic Energy Agency, Vienna, 1971, 373.

Lovett, R. J., Gutenmann, W. H., Pakkala, I. S., Youngs, W. D., Lisk, D. J., Burdick, G. E., and Harris, E. J., A survey of the total cadmium content of 406 fish from 49 New York State fresh waters, *J. Fish Res. Board Can.,* 29, 1283, 1972.

Lucas, H. F., Jr., Edgington, D. N., and Colby, P. J., Concentrations of trace elements in Great Lakes fishes, *J. Fish Res. Board Can.,* 27, 677, 1970.

Lucis, O. J. and Lucis, R., Distribution of cadmium 109 and zinc 65 in mice of inbred strains, *Arch. Environ. Health,* 19, 334, 1969.

Lucis, O. J., Lucis, R., and Shaikh, Z. A., Cadmium and zinc in pregnancy and lactation, *Arch. Environ. Health,* 25, 14, 1972.

Lucis, O. J., Lynk, M. E., and Lucis, R., Turnover of cadmium 109 in rats, *Arch. Environ. Health,* 18, 307, 1969.
Lucis, O. J., Shaikh, Z. A., and Embil, J. A., Cadmium as a trace element and cadmium binding components in human cells, *Experientia* (Basel), 26, 1109, 1970.
Lufkin, N. H. and Hodges, F. T., Cadmium poisoning, *U.S. Nav. Med. Bull.,* 43, 1273, 1944.
MacFarland, H. N., The use of dimercaprol (BAL) in the treatment of cadmium oxide fume poisoning, *Arch. Environ. Health,* 1, 487, 1960.
McCabe, L. J., Symons, J. M., Lee, R. D., and Robeck, G. G., Survey of community water supply systems, *J. Am. Water Works Assoc.,* 62, 670, 1970.
McKenzie, J. M., Urinary excretion of zinc, cadmium, sodium, potassium, and creatinine in ninety-six students, *Proc. Univ. Otago Med. Sch.,* 50, 16, 1972a.
McKenzie, J. M., Variation in urinary excretion of zinc and cadmium, *Proc. Univ. Otago Med. Sch.,* 50, 15, 1972b.
Maehara, T., Studies on the calcium metabolism of rat bone after long term administration of heavy metal (cadmium), *J. Jap. Orthop. Assoc.,* 42, 287, 1968 (in Japanese).
Malcolm, D., Potential carcinogenic effect of cadmium in animals and man, *Ann. Occup. Hyg.,* 15, 33, 1972.
Mappes, R., Normaler Cadmiumgehalt des Urins, *Arbeitsmed. Sozialmed. Arbeitshyg.,* 5, 142, 1969.
Margoshes, M. and Vallee, B. L., A cadmium protein from equine kidney cortex, *J. Am. Chem. Soc.,* 79, 4813, 1957.
Mason, K. E. and Young, J. O., Effectiveness of selenium and zinc in protecting against cadmium-induced injury of rat testis, in *Symposium: Selenium in Biomedicine,* Muth, O. H., Ed., Avi, Westport, Conn., 1967, 383.
Meek, E. S., Cellular changes induced by cadmium in mouse testis and liver, *Br. J. Exp. Pathol.,* 40, 503, 1959.
Menden, E. E., Elia, V. J., Michael, L. W., and Petering, H. G., Distribution of cadmium and nickel of tobacco during cigarette smoking, *Environ. Sci. Technol.,* 6, 830, 1972.
Mertz, D. P., Koschnick, R., Wilk, G., and Pfeilsticker, K., Untersuchungen über den Stoffwechsel von Spurenelementen beim Menschen. I. Serumverte von Kobalt, Nickel, Silber, Cadmium, Chrom, Molybdän, Mangan, *Z. Klin. Chem.,* 6, 171, 1968.
Mertz, D. P., Koschnick, R., and Wilk, G., Renale Ausscheidungsbedingunge von Cadmium beim normotensiven und hypertensiven Menschen, *Z. Klin. Chem.,* 10, 21, 1972.
Miller, W. J., Blackmon, D. M., Gentry, R. P., and Pate, F. M., Effect of dietary cadmium on tissue distribution of cadmium following a single oral dose in young goats, *J. Dairy Sci.,* 52, 2029, 1969.
Miller, W. J., Blackmon, D. M., and Martin, W. G., Cadmium absorption, excretion and tissue distribution following single tracer oral and intravenous doses in young goats, *J. Dairy Sci.,* 51, 1836, 1968.
Miller, W. J., Lampp, B., Powell, G. W., Salotti, C. A., and Blackmon, D. M., Influence of a high level of dietary cadmium on cadmium content in milk, excretion and cow performance, *J. Dairy Sci.,* 50, 1404, 1967.
Ministry of Agriculture, Declaration of Methodology for Measuring the Amount of Cadmium in Polluted Agricultural Soil, Declaration No. 47, Tokyo, June 24, 1971 (in Japanese).
Ministry of Health and Welfare, National Nutrition Survey, 1963 (in Japanese).
Ministry of Health and Welfare (Statistics Division), Report on Studies during 1961–62 on Adult Diseases, March 20, 1964 (in Japanese).
Ministry of Health and Welfare, Opinion of the Welfare Ministry with regard to "Itai-itai" disease in Toyama Prefecture, May 8, 1968a, in *Environment Agency,* 1972, 115 (in Japanese).
Ministry of Health and Welfare, Opinion of the Welfare Ministry with regard to "ouch-ouch" disease and its causes, May 8, 1968b, in *Environment Agency,* 1972, 125 (in Japanese).
Ministry of Health and Welfare, The essentials of tentative countermeasures against environmental pollution by cadmium, 1969a, in *Environment Agency,* 1972, 3 (in Japanese).
Ministry of Health and Welfare, The opinion of the Ministry of Health and Welfare as regards environmental pollution by cadmium and countermeasures in the future, March 27, 1969b, in *Environment Agency,* 1972, 149 (in Japanese).
Ministry of Health and Welfare, The opinon of the Welfare Ministry with regard to environmental pollution by cadmium in the Okudake River area of Oita Prefecture and countermeasures in the future, May 28, 1969c, in *Environment Agency,* 1972, 175 (in Japanese).
Ministry of Health and Welfare, Results of Studies on Cadmium Pollution around Mines and Refineries, April 5, 1971a (in Japanese; also see *Environment Agency,* 1972, 329).
Ministry of Health and Welfare, Method of Health Examination. A part of Provisional Countermeasures against Environmental Pollution by Cadmium, May 19, 1971b (in Japanese; see also *Environment Agency,* 1972, 81).
Ministry of Health and Welfare, The Opinion of the Ministry and Countermeasures in the Future Concerning Environmental Pollution by Cadmium in the Omuta Area, January 30, 1971c (in Japanese; see also *Environmental Agency,* 1972, 283).
Miyagi Prefecture, Report on the Health Examination in 1969 of the Inhabitants of the Area under Observation for Cadmium Pollution, February, 1970a (in Japanese).
Miyagi Prefecture, Studies on cadmium pollution, in *Report on Studies of Environmental Pollution,* Vol. 7, September 1970b, 280 (in Japanese).
Miyagi Prefecture, Investigations in the cadmium-polluted area, in *Report on Studies of Environmental Pollution,* Vol. 8, December 1971, 397 (in Japanese).
Molokhia, M. M. and Smith, H., Trace elements in the lung, *Arch. Environ. Health,* 15, 745, 1967.

Moore, W., Jr., Stara, J. F., and Crocker, W. C., Gastrointestinal absorption of different compounds of ^{115m}cadmium and the effect of different concentrations in the rat, *Environ. Res.*, 6, 159, 1973.
Morgan, J. M., Tissue cadmium concentrations in man, *Arch. Intern. Med.*, 123, 405, 1969.
Morgan, J. M., Cadmium and zinc abnormalities in bronchogenic carcinoma, *Cancer,* 25, 1394, 1970.
Morgan, J., Tissue cadmium and zinc content in emphysema and bronchogenic carcinoma, *J. Chronic Dis.*, 24, 107, 1971.
Moritsugi, M. and Kobayashi, J., Study on trace metals in biomaterials. II. Cadmium content in polished rice, *Ber. Ohara Inst. Landwirtsch. Biol. Okayama Univ.*, 12, 145, 1964 (in Japanese: in *Nogaku Kenkyu,* 50, 37, 1963).
Muldowney, F. P., Metabolic bone disease secondary to renal and intestinal disorders, *Calif. Med.*, 110, 397, 1969.
Murata, I., Chronic entero-osteo-nephropathy cadmium, *J. Jap. Med. Assoc.*, 65, 15, 1971 (in Japanese).
Murata, I., Hirono, T., Saeki, Y., and Nakagawa, S., Cadmium enteropathy, renal osteomalacia ("Itai-itai" disease in Japan), *Proc. Int. Cong. Radiol.*, 12th, (Tokyo), October, 1969.
Murata, I., Hirono, T., Saeki, Y., and Nakagawa, S., Cadmium enteropathy, renal osteomalacia ("Itai-itai" disease in Japan), *Bull. Soc. Int. Chir,* 1, 34, 1970.
Murata, I., Nakagawa, S., and Hirono, T., Clinical progress of Itai-itai disease, in *Kankyo Hoken Report No. 11,* Japanese Association of Public Health, April 1972, 132 (in Japanese).
Murata, I., Nakagawa, S., and Yoshimoto, A. (Furumoto, S.), A study of 50 patients with osteomalacia and tubular absorption damage, *J. Jap. Orthop. Surg. Soc.*, 32, 726, 1958 (in Japanese).
Murthy, G. K., Rhea, U., and Peeler, J. T., Levels of antimony, cadmium, chromium, cobalt, manganese, and zinc in institutional total diets, *Environ. Sci. Technol.*, 5, 436, 1971.
Nagasaki Prefecture, Dept. of Health, Report of Medical Examinations of the Population in the River Sasu and Shiine Basin, March 30, 1970.
Nagasaki Prefecture, Countermeasures against Cadmium Pollution in Tsushima, July 1971 (in Japanese).
Nagasaki Prefecture, Pollution Bureau, Report from the First Screening in 1972: Age-related Participation and Proteinuria, July 17, 1972a (in Japanese).
Nagasaki Prefecture, Countermeasures against Heavy Metal Pollution in Tsushima Performed 1971, May 25, 1972b (in Japanese).
Nagasawa, T., Nagasawa, S., Kawada, Y., Horiguchi, Y., Nomura, T., and Otsuka, R., The extensive occurrence of the rheumatic disease in some villages along the Jintsu River in the Toyama Prefecture, *J. Juzen Med. Soc.*, 50, 232, 1947 (in Japanese, cited in Kato and Kawano, 1968).
Nagata, R., Hirono, T., Yamazaki, H., Asakuno, K., Nakano, K., and Odaira, T., Air pollution by heavy metals contained in particulate matter in Tokyo, in *Annual Report of the Tokyo Metropolitan Research Institute for Environmental Protection,* Tokyo Metropolitan Research Institute for Environmental Protection, Tokyo, 1972, 5 (in Japanese).
Nakagawa, S., A study of osteomalacia in Toyama Prefecture (so-called Itai-itai disease), *J. Radiol. Phys. Therap. Univ. Kanizawa,* Vol. 56, 1960 (in Japanese with English summary).
Nakagawa, S. and Furumoto, S., About osteomalacia poisoning occurring in Toyama Prefecture, *J. Toyama Med. Assoc.*, Dec. 21, 1957, 7 (in Japanese).
Nandi, M., Slone, D., Jick, H.,Shapiro, S., and Lewis, G. P., Cadmium content of cigarettes, *Lancet,* 2, 1329, 1969.
National Institute of Industrial Health, Change of acute toxicity of metals in mice treated with their low dose, in *Recent Researches,* Ministry of Labor, Tokyo, September 1969, 14.
National Institute of Public Health, Sweden, Recommended Daily Calorie Consumption, 1969, based upon *The Recommended Dietary Allowances,* 7th ed., U.S.A. National Academy of Sciences, 1968, table following p. 102.
Nelson, W. E., Ed., *Textbook of Pediatrics,* 8th ed., W. B. Saunders, Philadelphia, 1964, 48.
Nicaud, P., Lafitte, A., and Gros, A., Les troubles de l'intoxication chronique par le cadmium, *Arch. Mal. Prof. Med. Trav. Secur. Soc.*, 4, 192, 1942.
Niemeier, B., Der Einfluss von Chelatbildnern auf Verteilung und Toxicitat von Cadmium, *Int. Arch. Gewerbepathol. Gewerbehyg.*, 24, 160, 1967.
Niemi, M. and Kormano, M., An angiographic study of cadmium induced vascular lesions in the testis and epididymis of the rat, *Acta Pathol. Microbiol. Scand.*, 63, 513, 1965.
***Nishiyama, K.,** in *Cadmium in the Environment,* Friberg, L., Piscator, M., and Nordberg, G., CRC Press, Cleveland, 1971, 59.
Nishiyama, K. and Nordberg, G., Adsorption and elution of cadmium on hair, *Arch. Environ. Health,* 25, 92, 1972.
Nishizumi, M., Electron microscopic study of cadmium nephrotoxicity in the rat, *Arch. Environ. Health,* 24, 215, 1972.
Nisso Smelting Company, Outline of Nisso Smelting Company Ltd., Tokyo, 1972 (in Japanese).
Nitta, T., Geochemical study on heavy metals in the Jintsu River area; especially cadmium, *J. Soc. Mining Geol. Jap.*, 22, 191, 1972 (in Japanese).
Nogawa, K. and Kawano, S., A survey of the blood pressure of women suspected of Itai-itai disease, *J. Juzen Med. Soc.*, 77, 357, 1969 (in Japanese).
Nolen, G. A., Bohne, R. L., and Buehler, E. V., Effects of trisodium nitrilotriacetate, trisodium citrate and a trisodium nitrilotriacetate-ferric chloride mixture on cadmium and methyl mercury toxicity and teratogenesis in rats, *Toxicol. Appl. Pharmacol.*, 23, 238, 1972.

Nolen, G. A., Buehler, E. V., Geil, R. G., and Goldenthal, E. I., Effects of trisodium nitrilotriacetate on cadmium and methyl mercury toxicity and teratogenicity in rats, *Toxicol. Appl. Pharmacol.,* 23, 222, 1972.
Nomiyama, K., Cadmium neuropathy, in Symposium on Cadmium Intoxication Tokyo, 1971a (later published under the title: About renal damage in cadmium poisoning, in *Kankyo Hoken Report No. 11,* Japanese Association of Public Health, April 1972, 78 [in Japanese]).
Nomiyama, K., On Exposure to Cadmium and the Amount of Cadmium in Human Organs (Autopsies of the Inhabitants in the Cadmium Polluted Areas of Annaka City), Report II, 1971b.
Nomiyama, K., Development mechanism and diagnosis of cadmium poisoning. *Kankyo Hoken Report No. 24,* Japanese Association of Public Health, September 1973, 11 (in Japanese).
Nomiyama, K., Sato, C., and Yamamoto, A., Early signs of cadmium intoxication in rabbits, *Toxicol. Appl. Pharmacol.,* 24, 625, 1973.
Nordberg, G. F., Effects of acute and chronic cadmium exposure on the testicles of mice, *Environ. Physiol.,* 1, 171, 1971a.
Nordberg, G. F., Effects of NTA on the Toxicity and Turnover of Cadmium, Report to the Research Council, Swedish National Environment Protection Board, 1971b (in Swedish).
Nordberg, G. F., Cadmium metabolism and toxicity, *Environ. Physiol. Biochem.,* 2, 7, 1972a.
Nordberg, G. F., Effects of NTA on the Toxicity and Turnover of Cadmium, the Second Report on this Project to the Research Council, Swedish National Environment Protection Board, 1972c (in Swedish).
Nordberg, G. F., Urinary blood and fecal cadmium concentrations as indices of exposure and accumulation, in Proceedings of the 17th International Congress on Occupational Health, to be published, 1974 (available through the Secretariat, Av. Rogue Saenz, Pẽna, 110 – 2 piso – Ofico 8, Buenos Aires, Argentina) 1972b.
Nordberg, G. F., Models used for calculation of accumulation of toxic metals, in Proceedings of the 17th International Congress on Occupational Health, to be published, 1974 (available through the Secretariat, Av. Rogue Saenz, Pẽna, 110 – 2 piso – Oficio 8, Buenos Aires, Argentina) 1972d.
***Nordberg, G., Friberg, L., and Piscator, M.,** in *Cadmium in the Environment,* Friberg, L., Piscator, M., and Nordberg, G., CRC Press, Cleveland, 1971, 30, 44.
Nordberg, G. F., Goyer, R. A., and Nordberg, M., Renal tubular effects of cadmium-metallothionein in the mouse, to be published, 1974.
Nordberg, G. F. and Nishiyama, K., Whole-body and hair retention of cadmium in mice, *Arch. Environ. Health,* 24, 209, 1972.
Nordberg, M. and Nordberg, G., Isolation and identification of metallothionein from mice, in *9th Int. Cong. Biochem.,* Stockholm, July 1973, Aktiebolaget Egnellska Boktryckeriet, Stockholm, 1973, 82 (abstr.).
Nordberg, G. F., Nordberg, M., Piscator, M., and Vesterberg, O., Separation of components of rabbit metallothionein by isoelectric focusing, *Biochem. J.,* 126, 491, 1972.
Nordberg, G. F. and Piscator, M., Influence of long-term cadmium exposure on urinary excretion of protein and cadmium in mice, *Environ. Physiol. Biochem.,* 2, 37, 1972.
Nordberg, G. F., Piscator, M., and Lind, B., Distribution of cadmium among protein fractions of mouse liver, *Acta Pharmacol. Toxicol.,* 29, 456, 1971.
Nordberg, G. F., Piscator, M., and Nordberg, M., On the distribution of cadmium in blood, *Acta Pharmacol. Toxicol.,* 30, 289, 1971.
Nordberg, G., Slorach, S., and Stenström, T., Cadmium poisoning caused by a cooled soft-drink machine, *Läkartidningen,* 70, 601, 1973.
Odén, S., Berggren, B., and Engwall, A. -G., Heavy metals and chlorinated hydrocarbons in sludge, *Grundförbättring,* 23, 55, 1970 (in Swedish).
Ogawa, E., Suzuki, S., Tsuzuki, H., and Kawajiri, M., Experimental studies on the absorption of cadmium chloride, *Jap. J. Pharmacol.,* 22, Suppl. 63, 1972a.
Ogawa, E., Suzuki, A., Tsuzuki, H., and Kawajiri, M., Experimental studies on cadmium chloride intoxication, *Jap. J. Pharmacol.,* 22, Suppl. 116, 1972b.
Ohsawa, M. and Kimura, M., Isolation of β_2-microglobulin from the urine of patients with Itai-itai disease, *Experientia,* 29, 556, 1973.
Olhagen, B., Special investigations of urine protein, in Friberg, L., Health hazards in the manufacture of alkaline accumulators with special reference to chronic cadmium poisoning, *Acta Med. Scand.,* 138, Suppl. 240, 36, 1950.
Olofsson, A. Report on investigations of atmospheric-emissions from the Finspong plant, *Sven. Metallverken,* No. HL 757, 12, 1970.
Olson, K. S., Heggen, G., Edwards, C. F., and Gorham, L. W., Trace element content of cancerous and noncancerous human liver tissue, *Science,* 119, 772, 1954.
Pařizek, J., Effect of cadmium salts on testicular tissue, *Nature,* 177, 1036, 1956.
Pařizek, J., The destructive effect of cadmium ion on testicular tissue and its prevention by zinc, *J. Endocrinol.,* 15, 56, 1957.
Pařizek, J., Sterilization of the male by cadmium salts, *J. Reprod. Fertil.,* 1, 294, 1960.
Pařizek, J., Vascular changes at sites of oestrogen biosynthesis produced by parenteral injection of cadmium salts, the destruction of placenta by cadmium salts, *J. Reprod. Fertil.,* 7, 263, 1964.

Pařízek, J., The peculiar toxicity of cadmium during pregnancy – an experimental 'toxaemia of pregnancy' induced by cadmium salts, *J. Reprod. Fertil.*, 9, 111, 1965.

Parizek, J., Benes, I., Ostadalova, I., Babicky, A., Benes, J., and Lener, J., Metabolic interrelations of trace elements, the effect of some inorganic and organic compounds of selenium on the metabolism of cadmium and mercury in the rat, *Physiol. Bohemoslov.*, 18, 95, 1969.

Pařízek, J., Ostadalova, I., Benes, I., and Babicky, A., Pregnancy and trace elements: the protective effect of compounds of an essential trace element–selenium–against the peculiar toxic effects of cadmium during pregnancy, *J. Reprod. Fertil.*, 16, 507, 1968.

Pařízek, J. and Zahor, Z., Effect of cadmium salt on testicular tissue, *Nature*, 177, 1036, 1956.

Pate, F. M., Johnson, A. D., and Miller, W. J., Testicular changes in calves following injection with cadmium chloride, *J. Anim. Sci.*, 31, 559, 1970.

Paterson, J. C., Studies on the toxicity of inhaled cadmium. III. The pathology of cadmium smoke poisoning in man and in experimental animals, *J. Ind. Hyg. Toxicol.*, 29, 294, 1947.

Peacock, P., The Association of Serum Cadmium Values with Indices of Cardiovascular and Other Disease and Certain Environmental Factors, report submitted to the Environmental Health Service of the Department of Health, Education, and Welfare, Public Health Service, contract PH 86-67-39, 1970.

Perry, H. M., Jr. and Erlanger, M., Hypertension and tissue metal levels after intraperitoneal cadmium, mercury and zinc, *Am. J. Physiol.*, 220, 808, 1971.

Perry, H. M., Jr. and Erlanger, M. W., Elevated circulating renin activity in rats following doses of cadmium known to induce hypertension, *J. Lab. Clin. Med.*, 82, 399, 1973.

Perry, H. M., Jr., Erlanger, M., Yunice, A., Schoepfle, E., and Perry, E. F., Hypertension and tissue metal levels following intravenous cadmium, mercury and zinc, *Am. J. Physiol.*, 219, 755, 1970.

Perry, H. M., Jr., Perry, E. F., and Purifoy, J. E., Antinatriuretic effect of intramuscular cadmium in rats, *Proc. Soc. Exp. Biol. Med.*, 136, 1240, 1971.

Perry, H. M., Jr. and Schroeder, H. A., Concentration of trace metals in urine of treated and untreated hypertensive patients compared with normal subjects, *J. Lab. Clin. Med.*, 46, 936, 1955.

Perry, H. M., Jr., Tipton, I. H., Schroeder, H. A., Steiner, R. L., and Cook, M. J., Variation in the concentration of cadmium in human kidney as a function of age and geographic origin, *J. Chronic Dis.*, 14, 259, 1961.

Perry, H. M., Jr. and Yunice, A., Acute pressor effects of intra-arterial cadmium and mercuric ions in anesthetized rats, *Proc. Soc. Exp. Biol. Med.*, 120, 805, 1965.

Petering, H. G., Johnson, M., and Stemmer, K., Effect of cadmium on dose response relationships of zinc in rats, *Fed. Proc.*, 28, 691, 1969.

Petering, H., Johnson, M. A., and Stemmer, K. L., Studies of zinc metabolism in the rat. I. Dose-response effects of cadmium, *Arch. Environ. Health*, 23, 93, 1971.

Peterson, P. A., Evrin, P. E., and Berggard, I., Differentiation of glomerular, tubular, and normal proteinuria: determinations of urinary excretion of β_2-microglobulin, albumin, and total protein, *J. Clin. Invest.*, 48, 1189, 1969.

Pindborg, J. J., The effect of chronic poisoning with fluorine and cadmium upon the incisors of the white rat with special reference to the enamel organ, *Tandlaegebladet*, 53, Suppl. 1, 136, 1950 (in Danish).

Pinkerton, C., Creason, J. P., Shy, C. M., Hammer, D. I., Buechley, R. W., and Murthy, G. K., Cadmium content of milk and cardiovascular disease mortality, in *Trace Substances in Environmental Health*, Vol. 5, Hemphill, D. D., Ed., University of Missouri Press, Columbia, 1972b, 285.

Pinkerton, C., Hammer, D. I., Hinners, T., Hasselblad, V., Kent, J., Lagerwerff, J. V., Ferrand, E., and Creason, J., Trace Metals in Urban Soils and Housedust, paper presented to the Environment Section, 100th Annual Meeting of the American Public Health Association, November 12 – 16, 1972a, Atlantic City, N.J.

Piotrowski, J. K., Trodanowska, B., Wisniewska-Knypl, J. M., and Bolanowska, W., Further investigations on binding and release of mercury in the rat, in *Mercury, Mercurials and Mercaptans*, Miller, M. W. and Clarkson, T. W., Eds., Charles C Thomas, Springfield, Ill., 1972, 247.

Piscator, M., Proteinuria in chronic cadmium poisoning. II. The applicability of quantitative and qualitative methods of protein determination for the demonstration of cadmium proteinuria, *Arch. Environ. Health*, 5, 325, 1962a.

Piscator, M., Proteinuria in chronic cadmium poisoning. I. An electrophoretic and chemical study of urinary and serum proteins from workers with chronic cadmium poisoning, *Arch. Environ. Health*, 4, 607, 1962b.

Piscator, M., Hemolytic anemia in cadmium-poisoned rabbits, *Excerpta Med. Int. Congr. Ser.*, 62, 925, 1963.

Piscator, M., Cadmium in the kidneys of normal human beings and the isolation of metallothionein from liver of rabbits exposed to cadmium, *Nord. Hyg. Tidskr.*, 45, 76, 1964 (in Swedish).

Piscator, M., Proteinuria in chronic cadmium poisoning. III. Electrophoretic and immunoelectrophoretic studies on urinary proteins from cadmium workers, with special reference to the excretion of low molecular weight proteins, *Arch. Environ. Health*, 12, 335, 1966a.

Piscator, M., Proteinuria in chronic cadmium poisoning. IV. Gel filtration and ion-exchange chromatography of urinary proteins from cadmium workers, *Arch. Environ. Health*, 12, 345, 1966b.

Piscator, M., *Proteinuria in Chronic Cadmium Poisoning*, Beckman's, Stockholm, 1966c.

***Piscator, M.**, in *Cadmium in the Environment*, Friberg, L., Piscator, M., and Nordberg, G., CRC Press, Cleveland, 1971, 17 – 19, 22, 31, 51, 52, 53, 60 – 64, 81, 83, 85, 91.

Piscator, M., Cadmium toxicity – industrial and environmental experience, in Proceedings of the 17th International Congress on Occupational Health, to be published, 1974 (available through the Secretariat, Av. Rogue Saenz, Pēna, 110 – 2 piso – Oficio 8, Buenos Aires, Argentina), 1972.

Piscator, M. and Axelsson, B., Serum proteins and kidney function after exposure to cadmium, *Arch. Environ. Health*, 21, 604, 1970.

Piscator, M. and Larsson, S. E., Retention and toxicity of cadmium in calcium-deficient rats, in Proceedings of the 17th International Congress on Occupational Health, to be published. 1974 (available through the Secretariat, Av. Rogue Saenz, Peña , 110 – 2 piso – Oficio 8, Buenos Aires, Argentina), 1972.

***Piscator, M. and Lind, B.**, in *Cadmium in the Environment,* Friberg, L., Piscator, M., and Nordberg, G., CRC Press, Cleveland, 1971, 51, 54.

***Piscator, M. and Rylander, R.**, in *Cadmium in the Environment,* Friberg, L., Piscator, M., and Nordberg, G., CRC Press, Cleveland, 1971, 24.

***Piscator, M. and Tsuchiya, K.** in *Cadmium in the Environment,* Friberg, L., Piscator, M., and Nordberg, G., CRC Press, Cleveland, 1971, 112.

Plantin, L. -O., Assessment of the trace elements of blood and tissues by neutron radioactivation and gamma spectroscopy, in *l'Analyse par radioactivation et ses Applications aux Sciences Biologiques,* Presses Universitaires de France, Paris, 1964, 211.

Pond, W. G. and Walker, E. F., Jr., Cadmium-induced anemia in growing rats: prevention by oral or parenteral iron, *Nutr. Rep. Int.*, 5, 365, 1972.

Potts, A. M., Simon, F. P., Tobias, J. M., Postel, S., Swift, M. N., Patt, H. M., and Gerard, R. W., Distribution and fate of cadmium in the animal body, *Ind. Hyg. Occup. Med.*, 2, 175, 1950.

Potts, C. L., Cadmium proteinuria – the health of battery workers exposed to cadmium oxide dust, *Ann. Occup. Hyg.*, 8, 55, 1965.

Powell, G. W., Miller, W. J., and Blackmon, D. M., Effects of dietary EDTA and cadmium on absorption, excretion and retention of orally administered ^{65}Zn in various tissues of zinc deficient and normal goats and calves, *J. Nutr.*, 93, 203, 1967.

Powell, G. W., Miller, W. J., Morton, J. D., and Clifton, C. M., Influence of dietary cadmium level and supplemental zinc on cadmium toxicity in the bovine, *J. Nutr.*, 84, 205, 1964.

Preston, A., Jefferies, D. F., Dutton, J. W. R., Harvey, B. R., and Steele, A. K., British Isles coastal waters: the concentrations of selected heavy metals in sea water, suspended matter and biological indicators – a pilot survey. *Environ. Pollut.*, 3, 69, 1972.

Princi, F., A study of industrial exposures to cadmium, *J. Ind. Hyg. Toxicol.*, 29, 315, 1947.

Princi, F. and Geever, E. F., Prolonged inhalation of cadmium, *Arch. Ind. Hyg. Occup. Med.*, 1, 651, 1950.

Pringle, B. H., Hissong, D. E., Katz, E. L., and Mulawka, S. T., Trace metal accumulation by estuarine mollusks, *J. Sanit. Eng. Div. Proc. Am. Soc. Civ. Eng.*, 94, 455, No. SA3, Proc. Paper 5970, 1968.

Prodan, L., Cadmium poisoning: II. Experimental cadmium poisoning, *J. Ind. Hyg. Toxicol.*, 14, 174, 1932.

Public Health Reports, *Cadmium Poisoning,* Div. Ind. Hyg., Nat. Inst. Health, United States Public Health Service, 57, 601, 1942.

Pujol, M., Arlet, J., Bollinelli, R., and Carles, P., Tubulopathie des intoxications chroniques par le cadmium, *Arch. Mal. Prof. Med. Trav. Secur. Soc.*, 31, 637, 1970.

Pulido, P., Fuwa, K., and Vallee, B. L., Determination of cadmium in biological materials by atomic absorption spectrophotometry, *Anal. Biochem.*, 14, 393, 1966.

Pulido, P., Kägi, J. H. R., and Vallee, B. L., Isolation and some properties of human metallothionein, *Biochemistry,* 5, 1768, 1966.

Rahola, T., Aaran, R. -K., and Miettinen, J. K., Half-time Studies of Mercury and Cadmium by Whole Body Counting, I.A.E.A. Symposium on the Assessment of Radioactive Organ and Body Burdens, Stockholm, November 22 – 26, 1971 (in *Assessment of Radioactive Contamination in Man,* I.A.E.A., Vienna, 1972, 553).

Rautu, R. and Sporn, A., Bëitrage zur Bestimmung der Cadmiumzufuhr durch Lebensmittel, *Nahrung,* 14, 25, 1970.

Reinl, W., Ueber eine Massenvergiftung durch Cadmiumoxydnebel, *Arch. Toxikol.*, 19, 152, 1961.

Réme and Peres (sic), A propos d'une intoxication collective par le cadmium, *Arch. Mal. Prof. Med. Trav. Secur. Soc.*, 20, 783, 1959.

Rennie, I. D., Proteinuria, *Med. Clin. North Am.*, 55, 213, 1971.

Report of committee II on permissible dose for internal radiation (1959), *Health Phys.*, 3, 1, 1960.

Reynolds, C. V. and Reynolds, E. B., Cadmium in crabs and crabmeat, *J. Assoc. Public Anal.*, 9, 112, 1971.

Ribelin, W. E., Atrophy of rat testis as index of chemical toxicity, *Arch. Pathol.*, 75, 229, 1963.

Richardson, M. E. and Fox, M. R. S., Dietary ascorbic acid protection of cadmium-inhibited spermatogenesis, *Fed. Proc.*, 29, 300, 1970.

Richmond, C. R., Findlay, J. S., and London, J. E., Whole-body Retention of Cadmium-109 by Mice Following Oral, Intraperitoneal, and Intravenous Administration, in Biological and Medical Research Group (H-4) of the Health Division – Annual Report, July 1965 through June 1966, Los Alamos Scientific Laboratory report LA-3610-MS, 1966, 195.

Roe, F. J. C., Dukes, C. E., Cameron, K. M., Pugh, R. C. B., and Mitchley, B. C. V., Cadmium neoplasia: testicular atrophy and leydig cell hyperplasia and neoplasia in rats and mice following the subcutaneous injection of cadmium salts, *Br. J. Cancer*, 18, 674, 1964.

Rojahn, T., Determination of copper, lead, cadmium, and zinc in estuarine water by anodic-stripping alternating-current voltammetry on the hanging mercury drop electrode, *Anal. Chim. Acta,* 62, 438, 1972.

Roosemont, J. L., Properties of metallothionein-like fractions from equine renal cortex, *Arch. Int. Physiol. Biochim.*, 80, 407, 1972.

Roy, A. K. and Neuhaus, O. W., Proof of the hepatic synthesis of a sex-dependent protein in the rat, *Biochim. Biophys. Acta,* 127, 82, 1966.

Roy, A. K., Neuhaus, O. W., and Harmison, C. R., Preparation and characterization of a sex-dependent rat urinary protein, *Biochim. Biophys. Acta,* 127, 72, 1966.

Rühling, Å., Contamination by heavy metals in the Oskarshamn Area, Lund University Dept. of Ecological Botany, 1969.

Rühling, Å. and Tyler, G., Regional Differences in Deposit of Heavy Metals in Scandinavia, Report 10, Dept. of Ecological Botany, Lund University, 1970.

Rühling, Å. and Tyler, G., Deposition of Heavy Metals over Scandinavia, Dept. of Plant Biology, Lund University, 1972 (internal report; in Swedish).

Rükimäki, V., Cadmium occurrence and effects, *World Environ. Health,* 9, 91, 1972.

Rylander, R., Alterations of lung defense mechanisms against airborne bacteria, *Arch. Environ. Health,* 18, 551, 1969.

Saltzman, B. E., Colorimetric micro-determination of cadmium with dithizone, *Anal. Chem.*, 25, 493, 1953.

Sano, S., Iiguchi, H., and Kawanishi, S., Urinary protein of Itai-itai disease patients, in *Kankyo Hoken Report No. 24,* Japanese Association of Public Health, September 1973, 91 (in Japanese).

Scharpf, L. G., Hill, I. D., Wright, P., Plank, J. B., Keplinger, M. L., and Calandra, J. C., Effect of sodium nitrilotriacetate on toxicity, teratogenicity, and tissue distribution of cadmium, *Nature,* 239, 231, 1972.

Scharpf, L. G., Ramos, F. J., and Hill, I. D., Influence of nitrilotriacetate (NTA) on the toxicity, excretion and distribution of cadmium in female rats, *Toxicol. Appl. Pharmacol.*, 22, 186, 1972.

Schmidt, P. and Gohlke, R., Tierexperimentelle Untersuchungen zur Toxizität des Kadmiumstearats, *Z. Gesamte Hyg.*, 17, 827, 1971.

Schroeder, H. A., Cadmium hypertension in rats, *Am. J. Physiol.*, 207, 62, 1964.

Schroeder, H. A., Cadmium as a factor in hypertension, *J. Chronic Dis.*, 18, 647, 1965.

Schroeder, H. A., Cadmium, chromium, and cardiovascular disease, *Circulation,* 35, 570, 1967.

Schroeder, H. A. and Balassa, J. J., Abnormal trace metals in man: cadmium, *J. Chronic Dis.*, 14, 236, 1961.

Schroeder, H. A. and Balassa, J. J., Cadmium: uptake by vegetables from superphosphate in soil, *Science,* 140, 819, 1963.

Schroeder, H. A., Balassa, J. J., and Vinton, W. H. Jr., Chromium, lead, cadmium, nickel and titanium in mice: effect on mortality, tumors and tissue levels, *J. Nutr.*, 83, 239, 1964.

Schroeder, H. A., Balassa, J. J., and Vinton, W. H. Jr., Chromium, cadmium and lead in rats: effects on life span, tumors and tissue levels, *J. Nutr.*, 86, 51, 1965.

Schroeder, H. A. and Buckman, J., Cadmium hypertension, *Arch. Environ. Health,* 14, 693, 1967.

Schroeder, H. A., Kroll, S. S., Little, J. W., Livingston, P. O., and Myers, M. A. G., Hypertension in rats from injection of cadmium, *Arch. Environ. Health,* 13, 788, 1966.

Schroeder, H. A. and Mitchener, M., Toxic effects of trace elements on the reproduction of mice and rats, *Arch. Environ. Health,* 23, 102, 1971.

Schroeder, H. A. and Nason, A. P., Trace metals in human hair, *J. Invest. Dermatol.*, 53, 71, 1969.

Schroeder, H. A., Nason, A. P., and Balassa, J. J., Trace metals in rat tissues as influenced by calcium in water, *J. Nutr.*, 93, 331, 1967.

Schroeder, H. A., Nason, A. P., and Mitchener, M., Action of a chelate of zinc on trace metals in hypertensive rats, *Am. J. Physiol.*, 214, 796, 1968.

Schroeder, H. A., Nason, A. P., Tipton, I. H., and Balassa, J. J., Essential trace metals in man: zinc. Relation to environmental cadmium, *J. Chronic Dis.*, 20, 179, 1967.

Schroeder, H. A. and Vinton, W. H., Hypertension induced in rats by small doses of cadmium, *Am. J. Physiol.*, 202, 515, 1962.

Schwartze, E. W. and Alsberg, C. L., Studies on the pharmacology of cadmium and zinc with particular reference to emesis, *J. Pharmacol. Exp. Ther.*, 21, 1, 1923.

Seiffert, Die Erkrankungen der Zinkhuttenarbeiter und hygienische Maasregeln dagegen, *Dtsch. Vierteljahrschrift für öffentliche Gesundheitspflege,* 29, 419, 1897.

Setchell, B. P. and Waites, G. M. H., Changes in the permeability of the testicular capillaries and of the "blood-testis" barrier after injection of cadmium chloride in the rat, *J. Endocrinol.*, 47, 81, 1970.

Setty, B. S. and Kar, A. B., Chemical sterilization of male frogs, *Gen. Comp. Endocrinol.*, 4, 353, 1964.

Severs, R. K. and Chambers, L. A., Differences in metal areal distribution displayed by trend-surface analysis, *Arch. Environ. Health,* 25, 139, 1972.

deSèze, S., Lichtwitz, A., Hioco, D., Bordier, P., and Miravet, L., Hypophosphataemic osteomalacia in the adult with defective renal tubular function, *Ann. Rheum. Dis.*, 23, 33, 1964.

Shabalina, L. P., Problems of industrial hygiene in the production and use of cadmium stearate, *Gig. Sanit.*, 33, 29, 1968.

Shaikh, Z. A. and Lucis, O. J., Distribution and Binding of ^{65}Zn and ^{109}Cd in Experimental Animals, presented at the 12th Annual Meeting of the Canadian Federation of Biological Societies, 1969.

Shaikh, Z. A. and Lucis, O. J., Induction of cadmium binding protein, *Fed. Proc. Abstr.*, 29, 298, 1970.

Shaikh, Z. A. and Lucis, O. J., The nature and biosynthesis of cadmium binding proteins, *Fed. Proc. Abstr.*, 30, 238, 1971.
Shaikh, Z. A. and Lucis, O. J., Biological differences in cadmium and zinc turnover, *Arch. Environ. Health*, 24, 410, 1972a.
Shaikh, Z. A. and Lucis, O. J., Cadmium and zinc binding in mammalian liver and kidneys, *Arch. Environ. Health*, 24, 419, 1972b.
Shiraishi, Y., Kurahashi, H., and Yosida, T. H., Chromosomal aberrations in cultured human leucocytes induced by cadmium sulfide, *Proc. Jap. Acad.*, 48, 133, 1972.
Shiraishi, Y. and Yosida, T. H., Chromosomal abnormalities in cultured leucocyte cells from Itai-itai disease patients, *Proc. Jap. Acad.*, 48, 248, 1972.
Shiroishi, K., Tanii, M., and Kubota, K., The urinary protein and sugar in the aged, *Med. Biol.*, 85, 197, 1972 (in Japanese).
Singh, K. and Nath, R., Studies on the identification of the cadmium-binding protein in rat testis, *Biochem. J.*, 128, 48p, 1972.
Sinko, I. and Gomiscek, S., Die Bestimmung von Blei, Cadmium, Kupfer, Thallium, Wismut und Zink im Blutserum im Wege der Anodic-Stripping-Polarographie, *Microchim. Acta*, 2, 163, 1972.
Skog, E. and Wahlberg, J. E., A comparative investigation of the percutaneous absorption of metal compounds in the guinea pig by means of the radioactive isotopes: ^{51}Cr, ^{58}Co, ^{65}Zn, ^{110m}Ag, ^{115m}Cd, ^{203}Hg, *J. Invest. Dermatol.*, 43, 187, 1964.
Smith, J. C. and Kench, J. E., Observations on urinary cadmium and protein excretion in men exposed to cadmium oxide dust and fume, *Br. J. Ind. Med.*, 14, 240, 1957.
Smith, J. C., Kench, J. E., and Lane, R. E., Determination of cadmium in urine and observations on urinary cadmium and protein excretion in men exposed to cadmium oxide dust, *Biochem. J.*, 61, 698, 1955.
Smith, J. C., Kench, J. E., and Smith, J. P., Chemical and histological post-mortem studies on a workman exposed for many years to cadmium oxide fume, *Br. J. Ind. Med.*, 14, 246, 1957.
Smith, J. C., Wells, A. R., and Kench, J. E., Observations on the urinary protein of men exposed to cadmium dust and fume, *Br. J. Ind. Med.*, 18, 70, 1961.
Smith, J. P., Smith, J. C., and McCall, A. J., Chronic poisoning from cadmium fume, *J. Pathol. Bacteriol.*, 80, 287, 1960.
Society of Actuaries, *Build and Blood Pressure Study*, Vol. 1, Chicago, 1959, 16.
Sporn, A., Dinu, I., and Stoenescu, L., Influence of cadmium administration on carbohydrate and cellular energetic metabolism in the rat liver, *Rev. Roum. Biochim.*, 7, 299, 1970.
Sporn, A., Dinu, I., Stoenescu, L., and Cirstea, A., Beitrage zur Ermittlung der Wechselwirkungen zwichen Cadmium und Zink, *Nahrung*, 13, 461, 1969.
Starcher, B. C., Studies on the mechanism of copper absorption in the chick, *J. Nutr.*, 97, 321, 1969.
State Committee of the Council of Ministries, U.S.S.R., *Sanitary Norms for Industrial Enterprise Design*, Publishing House of Literature on Construction, Moscow, 1972 (in Russian).
Statistical Yearbook (Statistical Abstract of Sweden), Vol. 58, National Central Bureau of Statistics, Stockholm, 1971, 77.
Stephens, G. A., Cadmium poisoning, *J. Ind. Hyg.*, 2, 129, 1920.
Stowe, H. D., Wilson, M., and Goyer, R. A., Clinical and morphological effects of oral cadmium toxicity in rabbits, *Arch. Pathol.*, 94, 389, 1972.
Sudo, Y. and Nomiyama, K., Long-term observations on urinary cadmium excretion of a former cadmium worker, *Ind. Med.*, 14, 117, 1972 (in Japanese with English summary).
Suzuki, S., Suzuki, T., and Ashizawa, M., Proteinuria due to inhalation of cadmium stearate dust, *Ind. Health*, 3, 73, 1965.
Suzuki, S. and Taguchi, T., Sex difference of cadmium content in spot urines, *Ind. Health*, 8, 150, 1970.
Suzuki, S., Taguchi, T., and Yokohashi, G., Dietary factors influencing upon the retention rate of orally administered $^{115m}CdCl_2$ in mice with special reference to calcium and protein concentrations in diet, *Ind. Health*, 7, 155, 1969.
Suzuki, T., Iwanaga, R., Togo, C., Katsunuma, H., and Suzuki, S., Distribution of $^{109}CdCl_2$ in pregnant mice fed with the diet deficient in vitamin A and vitamin D, *Ind. Health*, 9, 46, 1971.
Swedish National Board of Health and Welfare, *Advice and Instructions for the Use of Sewage Sludge in Soil Improvement*, Allmänna Förlagets Distribution, Stockholm, 1973, 8 (in Swedish). (To order, give Swedish title, *Använding av rötslam som jordforbättringsmedel*, No. 113-73-125; publisher: Allmänna Förlagets Distribution, Fack, 162 10 Vällingby 1, Sweden.)
Swedish National Food Administration, *General Swedish Regulations Concerning Food*, Allmänna Förlagets Distribution, No. 513-72-036, Stockholm, 1972, 23 (available in English from Allmänna Förlagets Distribution, Fack, 162 10 Vällingby 1, Sweden).
Swensson, A., Changes in blood, bone-marrow and spleen in chronic cadmium poisoning, *Proc. Int. Congr. Occup. Health*, 12, (3), 183, 1957.
Szadkowski, D., Cadmium – eine ökologische Noxe am Arbeitsplatz, *Med. Monatsschr.*, 26, 553, 1972.
Szadkowski, D., Schaller, K.-H., and Lehnert, G., Renale Cadmiumausscheidung, Lebensalter und Arterieller Blutdruck, *Z. Klin. Chem. Klin. Biochem.*, 7, 551, 1969.
Szadkowski, D., Schultze, H., Schaller, K. H., and Lehnert, G., Zur ökologischen Bedeutung des Schwermetallgehaltes von Zigaretten, *Arch. Hyg. Bakteriol.*, 153, 1, 1969.
Taga, I., Murata, I., Nakagawa, S., Furumoto, S., and Hagino, N., About a common disease in Kumano village of Toyama Prefecture, *J. Jap. Orthop. Assoc.*, 30, 381, 1956 (in Japanese).

Takabatake, E., Keshino, M., and Matsuo, I., Urinary cadmium concentration of population in Sasu area, Tsushima, Nagasaki, Japan, *J. Hyg. Chem.,* 18, 41, 1972 (in Japanese with English summary).
Takase, B., et al., Facts on the Itai-itai disease originating in the Toyama Prefecture, *Jap. J. Clin. Med.,* 25, 378, 1967 (in Japanese).
Takeuchi, J., The etiology of Itai-itai disease, *Jap. J. Clin. Med.,* 31, 132, 1973 (in Japanese).
Takeuchi, J. and Naito, P., About the etiology of Itai-itai disease, *Strides Med.,* 80, 609, 1972 (in Japanese).
Takeuchi, J., Shinoda, A., Kobayashi, K., Nakamoto, Y., Takazawa, I., and Kurosaki, M., Renal involvement in Itai-itai disease, *Internal Med.,* 21, 876, 1968 (in Japanese).
Tanaka, M., Matsusaka, N., Yuyama, A., Kobayashi, H., Transfer of cadmium through placenta and milk in the mouse, *Radioisotopes,* 21, 50, 1972.
Task Group on Lung Dynamics, Deposition and retention models for internal dosimetry of the human respiratory tract, *Health Phys.,* 12, 173, 1966.
Task Group on Metal Accumulation, Accumulation of toxic metals with special reference to their absorption, excretion and biological half-times, *Environ. Physiol. Biochem.,* 3, 65, 1973.
Taylor, D. J., Report on a survey of fish products for metallic contamination undertaken in south west England and South Wales during early 1971, *J. Assoc. Public Anal.,* 9, 76, 1971.
Teitz, H. W., Hirsch, E. F., and Neyman, B., Spectrographic study of trace elements in cancerous and noncancerous tissues, *J.A.M.A.,* 165, 2187, 1957.
Tepperman, H. M., The effect of BAL and BAL-glucoside therapy on the excretion and tissue distribution of injected cadmium, *J. Pharmacol.,* 89, 343, 1947.
Terhaar, C. J., Vis, E., Roudabush, R. L., and Fassett, D. W., Protective effects of low doses of cadmium chloride against subsequent high oral doses in the rat, *Toxicol. Appl. Pharmacol.,* 7, 500, 1965.
Thind, G. S., Role of cadmium in human and experimental hypertension, *J. Air. Pollut. Control Assoc.,* 22, 267, 1972.
Thind, G. S., Karreman, G., Stephan, K. F., and Blakemore, W. S., Vascular reactivity and mechanical properties of normal and cadmium-hypertensive rabbits, *J. Lab. Clin. Med.,* 76, 560, 1970.
Thomas, B., Roughan, J. A., and Watters, E. D., Lead and cadmium content of some vegetable foodstuffs, *J. Sci. Food Agric.,* 23, 1493, 1972.
Thurlbeck, W. M. and Foley, F. D., Experimental pulmonary emphysema. The effect of intratracheal injection of cadmium chloride solution in the guinea pig, *Am. J. Pathol.,* 42, 431, 1963.
Tipton, I. H., The distribution of trace metals in the human body, in *Metal-binding in Medicine,* Seven, M. J. and Johnson, L. A., Eds., Lippincott, Philadelphia, 1960, 27.
Tipton, I. H. and Cook, M. J., Trace elements in human tissue, II. Adult subjects from the United States, *Health Phys.,* 9, 103, 1963.
Tipton, I. H., Cook, M. J., Steiner, R. L., Boyes, C. A., Perry, H. M., and Schroeder, H. A., Trace elements in human tissue, I. Methods, *Health Phys.,* 9, 89, 1963.
Tipton, I. H., Schroeder, H. A., Perry, H. M., and Cook, M. J., Trace elements in human tissue, III. Subjects from Africa, the Near East and Far East and Europe, *Health Phys.,* 11, 403, 1965.
Tipton, I. H. and Shafer, J. J., Statistical analysis of lung trace element levels, *Arch. Environ. Health,* 8, 58, 1964.
Tipton, I. H. and Stewart, P. L., Patterns of elemental excretion in long-term balance studies, II, in Synder, W. S. Internal dosimetry, *Health Phys. Div. Ann. Progr. Rep.,* for period ending July 31, 1969, ORNL-4446, 1970, 303.
Tipton, I. H., Stewart, P. L., and Dickson, J., Patterns of elemental excretion in long-term balance studies, *Health Phys.,* 16, 455, 1969.
Tobias, J. M., Lushbaugh, C. C., Patt, H. M., Postel, S., Swift, M. N., and Gerard, R. W., The pathology and therapy with 2, 3-dimercaptopropanol (BAL) of experimental Cd poisoning, *J. Pharmacol.,* 87, Suppl. 102, 1946.
Tomita, K., Sex differences of the time of biological half-life of cadmium injected subcutaneously to mice, *Jap. J. Ind. Health,* 13, 46, 1971.
Tomita, K., Cadmium in Japanese cigarettes, in *Kankyo Hoken Report No. 11,* Japanese Association of Public Health, April, 1972, 25 (in Japanese).
Townshend, R. H., A case of acute cadmium pneumonitis: lung function tests during a four-year follow-up, *Br. J. Ind. Med.,* 25, 68, 1968.
Toyoda, B., Wada, K., Takahashi, M., Igawa, K., and Saino, S., On the so-called "Itai-itai byo," a kind of osteomalacia, *J. Jap. Assoc. Rural Med.,* 6, 55, 1957.
Toyoshima, I., Seino, A., and Tsuchiya, K., Urinary amino acids in cadmium workers, in inhabitants of a cadmium-polluted area, and in Itai-itai disease patients, in *Kankyo Hoken Report No. 24,* Japanese Association of Public Health, September 1973, 65 (in Japanese).
Truhaut, R. and Boudene, C., Recherches sur le Sort du Cadmium dans l'Organisme au Cours des Intoxications Interet en Medecine du Travail, *Arh. Hig. Rad. Toksikol.,* 5, 19, 1954.
Tsuchiya, K., Proteinuria of workers exposed to cadmium fume, the relationship to concentration in the working environment, *Arch. Environ. Health,* 14, 875, 1967.
Tsuchiya, K., Causation of ouch-ouch disease, an introductory review, *Keio J. Med.,* 18, 181, 1969.
Tsuchiya, K., Distribution of cadmium in humans, in *Kankyo Hoken Report No. 3,* Japanese Association of Public Health, 1970 (1971), 43 (in Japanese).

Tsuchiya, K., Environmental pollution by cadmium and its health effect in Japan, in *Working Papers of Japanese Participants for a Planning Conference – United States – Japan,* Direct Cooperative Program, East-West Center, Honolulu, Hawaii, 1971, 65.

Tsuchiya, K., Results of long-time observation of cadmium workers, in *Kankyo Hoken Report No. 11,* Japanese Association of Public Health, April, 1972, 72 (in Japanese).

Tsuchiya, K., Seki, Y., and Sugita, M., Organ and Tissue Cadmium Concentration of Cadavers from Accidental Deaths, in Proceedings of the 17th International Congress on Occupational Health, to be published, 1974 (available through the Secretariat, Av. Rogue Saenz, Pẽna, 110 – 2 piso – Oficio 8, Buenos Aires, Argentina), 1972a.

Tsuchiya, K., Seki, Y., and Sugita, M., Biologic Threshold Limits of Lead and Cadmium, in Proceedings of the 17th International Congress on Occupational Health, to be published, 1974, (available through the Secretariat, Av. Rogue Saenz, Pẽna, 110 – 2 piso – Oficio 8, Buenos Aires, Argentina), 1972b.

Tsuchiya, K. and Sugita, M., A mathematical model for deriving the biological half-life of a chemical, *Nord. Hyg. Tidskr.*, 53, 105, 1971.

Tsuchiya, K., Sugita, M., and Seki, Y., A Mathematical Approach to Deriving Biological Half-time of Cadmium in Some Organs, Calculation from Observed Accumulation of Metal in Organs, in Proceedings of the 17th International Congress on Occupational Health, to be published, 1974 (available through the Secretariat, Av. Rogue Saenz, Pẽna, 110 – 2 piso – Oficio 8, Buenos Aires, Argentina), 1972.

Underwood, E. J., *Trace Elements in Human and Animal Nutrition,* Academic Press, New York, 1962, 429.

Vallee, B. L. and Ulmer, D. D., Biochemical effects of mercury, cadmium, and lead, *Annu. Rev. Biochem.*, 41, 91, 1972.

Vander, A. J., Cadmium enhancement of proximal tubular sodium reabsorption, *Am. J. Physiol.*, 203, 1005, 1962.

Vens, M. D. and Lauwerys, R., Détermination simultanée du plomb et du cadmium dans le sang et l'urine par le couplage des techniques de chromatographie sur résine échangeuse d'ions et de spectrophotométrie d' absorption atomique, *Arch. Mal. Prof. Med. Trav. Secur. Soc.*, 33, 97, 1972.

Vesterberg, O., Isoelectric focusing of proteins in polyacrylamide gels, *Biochim. Biophys. Acta,* 257, 11, 1972.

Vesterberg, O. and Nise, G., Urinary proteins studied by use of isoelectric focusing. I. Tubular malfunction in association with exposure to cadmium, *Clin. Chem.*, 19, 1179, 1973.

Vigliani, E. C., The biopathology of cadmium, *Am. Ind. Hyg. Assoc.*, 30, 329, 1969.

Vigliani, E. C., Pernis, B., and Luisa, A., Études Biochimiques et Immunologiques sur la Nature de la Proteinurie Cadmique, *Med. Lav.*, 57, 322, 1966.

Vorobjeva, R. S., On occupational lung disease in prolonged action of aerosol of cadmium oxide, *Arch. Patologii,* 8, 25, 1957a (in Russian).

Vorobjeva, R. S., Investigation of the nervous system function in workers exposed to cadmium oxide, *In. Nevropat. Psikhiat,* 57, 385, 1957b (in Russian).

Waites, G. M. H. and Setchell, B. P., Changes in blood flow and vascular permeability of the testis, epididymis and accessory reproductive organs of the rat after the administration of cadmium chloride, *J. Endocrinol.*, 34, 329, 1966.

Walsh, J. J. and Burch, G. E., The rate of disappearance from plasma and subsequent distribution of radiocadmium (Cd^{115m}) in normal dogs, *J. Clin. Med.*, 54, 59, 1959.

Watanabe, H., A Study of Health Effect Indices in Populations in Cadmium-polluted Areas, presented at Meeting on Research on Cadmium Poisoning, Tokyo, March 25, 1973 (in Japanese).

Watanabe, H., Murayama, H., Matsushita, J., Ohno, I., et al., Findings on urine protein electrophoresis in cadmium polluted regions, in *Kankyo Hoken Report No. 11,* Japanese Association of Public Health, April 1972, 120 (in Japanese).

Webb, J., Heath, J. C., and Hopkins, T., Intranuclear distribution of the inducing metal in primary rhabdomyosarcomata induced in the rat by nickel, cobalt and cadmium, *Br. J. Cancer,* 26, 274, 1972.

Webb, M., Biochemical effects of Cd^{2+}-injury in the rat and mouse testis, *J. Reprod. Fertil.*, 30, 83, 1972a.

Webb, M., Binding of cadmium ions by rat liver and kidney, *Biochem. Pharmacol.*, 21, 2751, 1972b.

Webb, M., Protection by zinc against cadmium toxicity, *Biochem. Pharmacol.*, 21, 2767, 1972c.

Wester, P. O., Trace elements in heart tissue, *Acta Med. Scand.*, Suppl. 439, 1965.

Wester, P. O., Trace elements in the coronary arteries in the presence and absence of atherosclerosis, *Atherosclerosis,* 13, 395, 1971.

Westermark, T., *Metals and Ecology,* Symposium, Stockholm, March 24, 1969, Ecological Research Committee, Bulletin No. 5, Swedish Natural Science Research Council, Box 23 136, S-104 32 Stockholm, 1969, 43.

Westermark, T. and Sjöstrand, B., Activation analysis of cadmium in small biopsy samples, *Int. J. Appl. Radiat. Isot.*, 9, 78, 1960.

White, I. G., The toxicity of heavy metals to mammalian spermatozoa, *Aust. J. Exp. Biol. Med. Sci.*, 33, 359, 1955.

Whitnack, G. C. and Sasselli, R., Application of anodic-stripping volammetry to the determination of some trace elements in sea water, *Anal. Chim. Acta,* 47, 267, 1969.

Williams, C. H. and David, D. J., The effect of superphosphates on the cadmium content of soils and plants, *Aust. J. Soil Res.*, 11, 43, 1973.

Wilson, R. H., DeEds, F., and Cox, A. J., Effects of continued cadmium feeding, *J. Pharmacol. Exp. Ther.*, 71, 222, 1941.

Winge, D. R. and Rajagopalan, K. V., Purification and some properties of Cd-binding protein from rat liver, *Arch. Biochem. Biophys.*, 153, 755, 1972.

Winkelstein, W. Jr. and Kantor, S., Prostatic cancer: relationship to suspended particulate air pollution, *Am. J. Public Health,* 59, 1134, 1969.

Wiśniewska, J. M., Trojanowska, B., Piotrowski, J., and Jakubowski, M., Binding of mercury in the rat kidney by metallothionein, *Toxicol. Appl. Pharmacol.,* 16, 754, 1970.

Wisniewska-Knypl, J. M. and Jabłońska, J., Selective binding of cadmium in vivo on metallothionein in rat's liver, *Bull. Acad. Pol. Sci. Ser. Sci. Biol.,* 18, 321, 1970.

Wisniewska-Knypl, J. M., Jablonska, J., and Myslak, Z., Binding of cadmium on metallothionein in man: an analysis of a fatal poisoning by cadmium iodide, *Arch. Toxikol.,* 28, 46, 1971.

Wisniewska-Knypl, J. M., Trojanowska, B., Piotrowski, J. K., and Jablonska, J. K., Binding of mercury in rat liver by metallothionein, *Acta Biochim. Pol.,* 19, 11, 1972.

Worker, N. A. and Migicovsky, B. B., Effect of vitamin D on the utilization of zinc, cadmium and mercury in the chick, *J. Nutr.,* 75, 222, 1961.

World Health Organization, Evaluation of certain food additives and the contaminants mercury, lead and cadmium, *W.H.O. Tech. Rep. Ser.,* 505, 1972.

World Health Organization, Long-term Programme in Environmental Pollution Control in Europe. The Hazards to Health of Persistent Substances in Water, EURO 3109W(1), Regional Office for Europe, W.H.O., Copenhagen, 1973, 23.

Wretlind, A., Nutritional status of the Swedish people, in *Clinical Nutrition,* Reizenstein, P., Ed., Norstedts, Stockholm, 1968, 7.

Yamagata, N. and Iwashima, S., Average cadmium intake of Japanese people, in *Kankyo Hoken Report No. 24,* Japanese Association of Public Health, September 1973, 27 (in Japanese).

Yamagata, N., Iwashima, K., Kuzuhara, Y., and Yamagata, T., A model surveillance for cadmium pollution, *Bull. Inst. Public Health* (Tokyo), 20, 170, 1971.

Yamagata, N. and Shigematsu, I., Cadmium pollution in perspective, *Bull. Inst. Public Health* (Tokyo), 19, 1, 1970.

Yamamoto, Y., Present status of cadmium environmental pollution in *Kankyo Hoken Report No. 11,* Japanese Association of Public Health, April 1972, 7 (in Japanese).

Zook, E. G., Greene, F. E., and Morris, E. R., Nutrient composition of selected wheats and wheat products. VI. Distribution of manganese, copper, nickel, zinc, magnesium, lead, tin, cadmium, chromium, and selenium as determined by atomic absorption spectroscopy and colorimetry, *Cereal Chem.,* 47, 720, 1970.

ADDRESS LIST FOR JAPANESE AUTHORS AND ORGANIZATIONS

Association of Health Statistics
Roppongi 5-13-14
Minato-ku
Tokyo
Japan

Environment Agency
Kasumigaseki 3-1-1
Chiyoda-ku
Tokyo
Japan

Mr. Mithuaki Fujimori
Vice Manager
Annaka Refinery
Tohozinc Co., Ltd.
No. 1443, Nakajuku, Annaka
Gumma-ken
Japan

Dr. Ichiro Fukushima
Director .
Gumma Prefectural Institute of Public Health
3-21-19, Iwagami-cho
Maebaeshi-shi
Gumma-ken
Japan

Dr. Masaaki Fukushima
Department of Public Health
Kanazawa Medical University
1-1 Daigaku-machi
Uchinada-machi
Kahoku-gun
Ishikawa-ken
920-20
Japan

Fukushima Prefecture:
The Section for Public Health of the Fukushima Prefecture Authorities
Sugitsuma-cho 2-16
Fukushima City
Fukushima Prefecture
Japan

Dr. Yuzo Fukuyama
Institute of Hygiene and Medical Microbiology
930 Otemachi, 1-15
Toyama City
Toyama Prefecture
Japan

Gumma Prefecture:
The Section for Public Health at the Gumma Prefecture Authorities
Maebashi City
Gumma Prefecture
Japan

Dr. N. Hagino
Fuchu-cho Neigun
Toyama Prefecture
Japan

Dr. Yutaka Hasegawa
Hyogo Prefectural Institute of Public Health
2-1, Arata-cho, Hyogo-ku
Kobe City
Japan

Dr. M. Hirota
School of Medicine
Gifu University
Tsukasamachi 40
Gifu Prefecture
Japan

Hyogo Prefecture:
The Section for Hygiene of the Hyogo Prefecture Authorities
Kobe City
Hyogo Prefecture
Japan

Dr. M. Inhizawa (= Ishizawa)
Department of Public Health
School of Medicine
Tottori University
1-1 Koyama-cho
Tottori City
Japan

Dr. Arinobu Ishizaki
Professor
Department of Public Health
Kanazawa Medical University
1-1 Daigaku-machi
Uchinada-machi
Kahoku-gun
Ishikawa-ken
920-20
Japan

Itai-itai Research Committee *see* Toyama Prefecture Authorities

Japanese Association of Public Health
Shinjuku 1-29-8
Shinjuku-ku
Tokyo
Japan

Japanese Industrial Health Association
c/o Nihon Daigaku
Igakubu
Koshu Eiseigaki Kyoshitsu, 30
Oyaguchi Kami-cho
Itabashi-ku
Tokyo
Japan

Japanese Language Translation Service
Attention: Mr. S. Aoki
2027-15, Chogo
Fujisawa-city
Kanagawa Prefecture
Japan

Prof. Kinichiro Kajikawa
Department of Pathology
Department of Medicine
Kanazawa University
Kanazawa
Japan

Dr. K. Kakinuma
Annaka Health Center
2390-2, Annaka
Annaka-city
Gumma Prefecture
Japan

Dr. Y. Katagiri
Department of Public Health
School of Medicine
Gifu University
Tsukasamachi 40
Gifu Prefecture
Japan

Dr. Takashi Kato
Associate Professor
Department of Public Health
Kanazawa University
School of Medicine
1-1 Daigaku-machi
Uchinada-machi
Kahoku-gun
Ishikawa-ken
920-20
Japan

Dr. Hidemichi Kimura
School of Medicine
Gifu University
Tsukasamachi 40
Gifu Prefecture
Japan

Dr. Shoji Kitamura
Professor
Department of Public Health
Kobe University School of Medicine
12 Kusunoki-cho, 7-chome
Ikuta-ku
Kobe
Japan

Dr. Jun Kobayashi
Institute of Agricultural and Biological Sciences
Okayama University
Kurashiki, Okayama-ken
Japan

Dr. Michimasa Kono
Kono Clinical Research Institute
1-2319, Kitashinagawa
Shinagawa-ku
Tokyo
Japan

Dr. Juko Kubota
Director
Japan Industrial Safety Assoc.
Occupational Health Service Center
35-4, Shiba 5-chome
Minato-ku
Tokyo
Japan

Dr. Kentaro Kubota
National Institute for Environmental Pollution Research
Ooaza Tateno
Yatabe-cho
Tsukuba-gun
Ibaraki Prefecture
Japan

Dr. T. Maehara
Department of Orthopedic Surgery
School of Medicine
Okayama University
Okayama City
Okayama
Japan

Dr. Matsushiro
Department of Pollution Control
Omuta City Authorities
Omuta City
Fukuoka-ken
Japan

Ministry of Agriculture
2, Kasumigaseki 1-chome
Chiyoda-ku
Tokyo
Japan

Ministry of Health and Welfare
2, Kasumigaseki 1-chome
Chiyoda-ku
Tokyo
Japan

Miyagi Prefecture
Department of Public Health and Environmental Protection
Prefecture Authorities (Ken-cho)
Sendai City
Miyagi Prefecture
Japan

Dr. Isamu Murata
Department of Surgery
Toyama Chuo Hospital
220 Nishinagae
Toyama
Japan

Nagasaki Prefecture
Department of Public Health and Environmental Protection
Prefecture Authorities (Ken-cho)
Nagasaki City
Nagasaki Prefecture
Japan

Dr. Taro Nagasawa
7-18, Nishisanno-machi
Toyama City
Japan

Dr. Shochu Nakagawa
Toyama Chuo Hospital
Department of Surgery
220 Nishinagae
Toyama
Japan

National Institute of Industrial Health of Japan
2051 Kizukisumiyoshi-cho
Nakahara-ku
Kawasaki 211
Japan

Dr. Keitaro Nishiyama
Department of Hygiene
School of Medicine
Tokushima University
Tokushima
Japan

Nisso Smelting Company
Attention: Toshio Kumagai, Director
3rd Floor
New Ohtemachi Building
No. 4, 2-chome, Ohtemachi
Chiyoda-ku
Tokyo
Japan

Mr. T. Nitta
Kamioka Mining Branch
Mitsui Mining and Smelting Co., Ltd.
Kamioka-cho
Yoshishiro-gun
Gifu Prefecture
Japan

Dr. Koji Nogawa
Department of Public Health
Kanazawa Medical University
1-1 Daigaku-machi
Uchinada-machi
Kahoku-gun
Ishikawa-ken
920-20
Japan

Dr. Kazuo Nomiyama
Department of Environmental Health
Jichi Medical College
3311-1 Yakushiji, Minamikawachi-machi
Kawachi-gun, Tochigi-ken 329-04
Japan

Dr. Motoyasu Ohsawa
Department of Occupational Diseases
National Institute of Industrial Health
Kizuki-Sumiyoshi
Kawasaki
Japan

Prof. Seiyo Sano
Department of Public Health
Faculty of Medicine
Kyoto University
Kyoto, 606
Japan

Dr. Kazuko Shiroishi
Toyama Institute of Hygiene and Medical Microbiology
1-15 Ohtemachi
Toyama City
Toyama Prefecture
Japan

Dr. Yoshio Sudo
President, Surgical Hospital
3458 Annaka
Annaka City
Gumma Prefecture
Japan

Dr. Shosuke Suzuki
Department of Health Administration
School of Health Sciences
Faculty of Medicine
University of Tokyo
Hongo
Bunkyo-ku
Tokyo
Japan

Dr. Tsuguyoshi Suzuki
Professor
Department of Public Health
School of Medicine
Tohoku University
1-2 Seiryo-cho
Sendai
Japan

Taga, I. *see* Murata, I.

Dr. Eigo Takabatake
Department of Public Health, Pharmaceutics
Institute of Public Health
6-1, Shirokanedai 4-chome
Minatoku, Tokyo 108
Japan

Prof. Buhei Takase
Department of Medicine
Kanazawa University
Kanazawa
Japan

Dr. Jugoro Takeuchi
Professor of Medicine
Tokyo Medical and Dental University
Ochanomizu
Tokyo
Japan

Dr. Seisuke Terada
Director
Environmental Pollution Control Div.
Environment Conservation Bureau
Nagasaki Prefecture
Edo-machi
Nagasaki
Japan

Dr. Kunio Tomita
Mitsui Mining and Smelting Co., Ltd.
2-Chome, Nihonbashi-Muromachi, Chuo-ku
Tokyo
Japan

Toyama Prefecture Authorities (Itai-itai Research Committee)
Section for Public Health
Toyama P.A.
Toyama City
Toyama Prefecture
Japan

Dr. Bunichi Toyoda
President, Kanazawa University
Kanazawa City
Japan

Dr. I. Toyoshima
Department of Preventive Medicine and Public Health
School of Medicine
Keio University
35 Shinanomachi, Shinjuku-ku
Tokyo
Japan

Prof. Kenzaburo Tsuchiya
Department of Preventive Medicine and Public Health
School of Medicine
Keio University
Shinanomachi, Shinjuku-ku
Tokyo
Japan

Dr. Hiromu Watanabe
Director
Hyogo Institute of Public Health
2-1 Arata-cho
Hyogo-ku
Kobe
Japan

Dr. Noburo Yamagata
Head, Radio Hygiene Section
National Institute of Public Health
6-1 Shiroganedai 4-chome
Minato-ku
Tokyo
Japan

Dr. Yoshimasa Yamamoto
Kogaihoken-ka
Kikakuchosei Kyoku
Kankyo-cho
Chiyoda-ku
Tokyo
Japan

AUTHOR INDEX

Note: Names within parentheses refer to associate authors in a multiple-author group.

C

D

E

*Erroneously referred to as Yoshimoto, A., in the reference Murata, Nakagawa, and Yoshimoto, 1958; see also Yoshimoto, A.

I

J

K

L

M

N

O

P

R

S

T

U

V

W

Y

Z

SUBJECT INDEX

F

G

H

I

J

K

L

M

N

O

P

R

S

T

U

V

W

Z